The Hypothalamic-Pituitary-Adrenal Axis
in Health and Disease

Eliza B. Geer

Editor

The Hypothalamic-Pituitary-Adrenal Axis in Health and Disease

Cushing's Syndrome and Beyond

 Springer

Editor
Eliza B. Geer
Division of Endocrinology
Multidisciplinary Pituitary and Skull Base Tumor Center
Memorial Sloan Kettering Cancer Center
New York, NY, USA

ISBN 978-3-319-45948-6   ISBN 978-3-319-45950-9   (eBook)
DOI 10.1007/978-3-319-45950-9

Library of Congress Control Number: 2016955864

Printed on acid-free paper

This Springer imprint is published by Springer Nature
The registered company is Springer International Publishing AG
The registered company address is: Gewerbestrasse 11, 6330 Cham, Switzerland

Contents

Part I
Hypothalamic-Pituitary-Adrenal Axis Regulation of the Body, Brain, and Inflammation

Glucocorticoid Regulation of Body Composition and Metabolism

Alexandria Atuahene Opata, Khadeen C. Cheesman, and Eliza B. Geer

Abstract Glucocorticoids (GCs) are critical in maintaining energy homeostasis. Chronic excessive GC exposure, as seen in Cushing's syndrome (CS), profoundly impacts body composition and metabolism by causing whole-body insulin resistance and abdominal adiposity. Peripheral insulin resistance occurs due to impaired insulin signaling and glucose uptake. Excess GCs lead to muscle atrophy which is associated with elevated plasma fatty acids and triglycerides, altered hepatic carbohydrate and lipid metabolism, and impaired pancreatic β-cell function. GCs also reduce bone density by increasing bone resorption while inhibiting bone formation, in part by decreasing osteoblast number and function. Lastly, a variety of skin manifestations result from GC excess. The current review explores GC regulation of body composition and metabolism. While physiological exposure to GCs and a dynamic HPA axis that is responsive to metabolic and environmental cues are essential for the survival of any organism, chronic exposure to even subtle GC excess causes the development of excess abdominal and ectopic adipose tissue, dyslipidemia, cardiovascular disease, and ultimately decreased survival.

Keywords Glucocorticoids • Cushing's syndrome • Adipose tissue • Lipolysis • Insulin resistance • Glucocorticoid-induced myopathy • Glucocorticoid-induced osteoporosis • β-cell • Bone remodeling

A.A. Opata, M.D. • K.C. Cheesman, M.D.
Division of Endocrinology, Metabolism and Diabetes, Icahn School of Medicine at Mount Sinai, One Gustave Levy Place, New York, NY 10029, USA

E.B. Geer, M.D. (✉)
Division of Endocrinology, Multidisciplinary Pituitary and Skull Base Tumor Center, Memorial Sloan Kettering Cancer Center, Box 419, 1275 York Ave, New York, NY 10065, USA
e-mail: geere@mskcc.org

© Springer International Publishing Switzerland 2017
E.B. Geer (ed.), *The Hypothalamic-Pituitary-Adrenal Axis in Health and Disease*, DOI 10.1007/978-3-319-45950-9_1

3

Introduction

Since the discovery of glucocorticoids (GCs) for the treatment of adrenal insufficiency over 80 years ago, the phenotypic and metabolic effects of GCs have been studied extensively [1]. Excess GC exposure can have a profound impact on body composition; this has been demonstrated most dramatically in patients with Cushing's syndrome (CS), an endocrine disorder characterized by chronic endogenous or exogenous GC exposure [2]. Although endogenous CS is rare, more subtle forms of GC excess are seen in the setting of chronic stress and depression due to activation of the hypothalamic-pituitary-adrenal (HPA) axis. "Common" or diet-induced obesity has also been suggested to be associated with excess endogenous GC exposure due to increased local production of GCs in adipose tissue, alterations of cortisol circadian rhythm, and heightened susceptibility of the HPA axis to activation [3]. Furthermore, the prevalence of oral GC use in the U.S. has been reported to be as high as 3.5 % based on data obtained from the National Health and Nutrition Examination Survey (NHANES) from 1999 to 2008 [4]. GC overexposure, whether endogenous or exogenous, results in increased visceral and trunk subcutaneous fat which in turn is implicated in insulin resistance and development of diabetes mellitus [5, 6]. The aim of the present review is to describe the mechanisms by which GCs regulate body composition, insulin action, and insulin sensitivity (Fig. 1).

GC Regulation of Adipose Tissue

Globally, the prevalence of obesity has reached epidemic proportions with over one billion adults being overweight, and of these, roughly 300 million being obese [7]. The rapid rise in obesity and its associated comorbidities pose a major public health concern and have made the study of obesity and its adverse metabolic profile increasingly important. Research in the past 20 years has led to an understanding that adipose tissue is a complex and highly active endocrine organ which contributes to the regulation of insulin action and with functions that are altered by obesity [8]. In addition, individuals who are obese have a higher all-cause mortality [9]. Those with GC overexposure have a mortality rate four times higher than the general population, primarily due to cardiovascular disease which is in part due to GC-induced obesity and insulin resistance [10]. In an effort to better understand the effects of GCs on adipose tissue, we will first discuss adipose tissue types with a focus on distribution and mass. This will be followed by a review of the common phenotypical changes seen in adipose tissue as a result of GC excess and the subsequent effects on glucose metabolism.

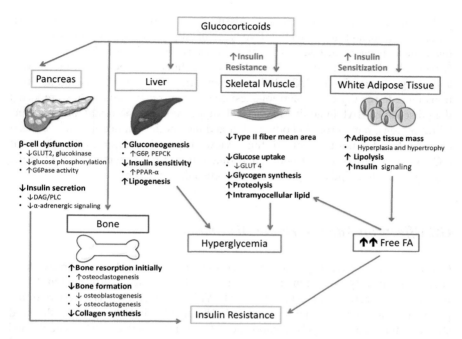

Fig. 1 Effects of GCs on body composition and metabolism. GCs promote whole-body insulin resistance via visceral adipogenesis, mobilization, and release of free fatty acids into the circulation and development of hepatic steatosis. In addition, hyperglycemia results from β-cell dysfunction, decreased insulin secretion, and increased gluconeogenesis. In skeletal muscle, GCs cause type II fiber atrophy and decreased glucose uptake. Bone loss occurs due to increased bone resorption followed by decreased formation from reduced osteoblast function and number. DAG, Diacylglycerol; PLC, phospholipase C; G6P, glucose-6 phosphatase; PPAR, peroxisome proliferator-activated receptor

Adipose Tissue, Mass and Distribution

Adipose tissue is a complex, multicellular organ that influences the functions of other organ systems and includes numerous discrete anatomical depots with variable masses, ranging from 5 to 60 % of total body weight [11]. Subcutaneous adipose tissue is responsible for storing over 80 % of total body fat, with the most described depots being abdominal, gluteal, and femoral [12]. Visceral adipose tissue refers to adipose tissue surrounding the digestive organs and can be further divided into intra-abdominal or retroperitoneal depots. In men, visceral adipose tissue typically accounts for 10–20 % of total body fat, whereas in women it is about 5–10 % [12]. Smaller depots include epicardial and inter-muscular, which may have specialized functions related to their neighboring cells. Adipose tissue is composed of adipocytes as well as stromal vascular cells, which include preadipocytes, endothelial cells, pericytes, and immune cells (macrophages, T-cells, neutrophils, and lymphocytes) [12–14].

Further distinction within adipose tissue depends on cell structure, location, vascularization, and function [15]. Two types of adipose tissue, present in all mammals, have been identified, white and brown adipose tissue (WAT and BAT). Broadly, WAT and BAT are involved in opposing functions: energy storage in WAT and energy dissipation in BAT. BAT is present primarily in newborns and its functions include regulation of thermogenesis. However, recent studies have reported the presence of BAT in adults, in cervical-supraclavicular, perirenal, and paravertebral regions, but its role in body weight and metabolic function has not yet been elucidated and is not the focus of this review [12]. A discussion of the effects of GCs on adipose tissue distribution and function will be presented here. Of note, any reference to adipose tissue refers to white adipose tissue.

GC Effects on Adipose Tissue Distribution

Chronic, excessive GC exposure has been shown to increase body fat mass; these changes are clearly evident in patients with CS who experience profound increases in total and visceral adipose tissue [5, 16]. While some studies have reported that peripheral fat stores may be reduced [17, 18], using whole-body MRI, the gold standard for assessing body composition, Cushing's disease (CD) patients had more trunk subcutaneous and visceral adipose tissue, but similar masses of total and limb subcutaneous adipose tissue, compared to healthy controls [5]. CD patients also had an increased ratio of visceral to total fat compared to healthy controls [3]. Normalization of cortisol concentrations in patients with CD resulted in a significant reduction in trunk, subcutaneous, and visceral adipose tissue [6]. Furthermore, the distribution of adipose tissue changed: visceral/total fat and visceral fat/skeletal muscle ratios decreased, further demonstrating the effects of GCs on adipose tissue distribution [6].

Mechanisms Underlying GC-Mediated Adiposity

The lipolytic effects of GCs have been well-established, yet excess GCs are associated with increased adiposity [19]. While most obese individuals do not show evidence of elevated morning serum GC levels, considerable evidence suggests that tissue GC levels may not adequately reflect plasma levels [20]. Although it is difficult to measure tissue-specific GCs, adipose tissue is thought to have levels 10–15 times that of circulating levels [21], possibly due to 11 beta hydroxysteroid deyhdrogenase type 1 (11BHSD1) activity, which converts inactive cortisone to active cortisol, and thus enhances GC action [22]. Not only has visceral fat accumulation been associated with upregulation of 11BHSD1 [22] and a higher density of glucocorticoid receptors (GR) [23, 24], but both 11BHSD1 and GR levels are higher in the visceral compared to subcutaneous adipose tissue depots, suggesting a greater susceptibility to GCs [23] and providing a plausible explanation for site-specific adiposity [25]. One implication of enhanced GC action may include increased lipoprotein lipase (LPL) activity in adipose tissue [26]. GCs are thought to increase LPL activity via transcriptional or posttranslational modifications [26–28]. Fried et al. demonstrated increased LPL activity in omental adipose tissue of obese men and women cultured in dexamethasone [28]. The increase in activity was largely explained by the ability of dexamethasone to increase LPL expression and allows for more fatty acids (FA) being available for storage in this depot [19].

GCs increase adipose tissue mass via hypertrophy and hyperplasia [29]. Hypertrophy results from fatty acid synthesis and storage within adipocytes, whereas hyperplasia results from differentiation of preadipocytes to mature adipocytes [29]. The latter has been shown to occur in the setting of cortisol and dexamethasone exposure [30]. Also, the presence of 11BHSD1 and the resulting increase in tissue GCs promotes the differentiation of human adipose stromal cells to mature adipocytes, further confirming the adipogenic effects of GCs [29]. Interestingly, if adipogenesis were exclusively responsible for increased adiposity, individuals with GC excess would have numerous small adipocytes, which is not the case; assessment of adipose morphology in patients with CS reveals enlarged, hypertrophic adipocytes [16]. In addition, expansion of the extracellular matrix and stromal vascular cells may be involved in the accumulation of adipose tissue in response to GCs [3]. Further study of the effects of chronic GC exposure on adipose tissue morphology in humans is needed.

GC Regulation Of Glucose Metabolism and Insulin Resistance in Adipose Tissue

Adipose tissue is a major site for metabolism of GCs. The functions of adipose tissue are crucial determinants of whole-body glucose and lipid homeostasis. The importance of this is emphasized by the adverse metabolic consequences of both adipose tissue excess and deficiency [31]. For example, obesity, particularly in the visceral compartment, is associated with insulin resistance, hyperglycemia, and dyslipidemia [32]. The role of GCs in regulating adipose tissue function is complex and depends on the species, concentration, specific adipose depot [33], and chronicity of GC exposure.

Human and animal studies have shown that GCs induce pre-adipocyte differentiation and whole-body lipolysis [34–37]. Corticosterone increased pre-adipocyte differentiation in 3T3-L1 cells with increased expression of adipose triglyceride lipase and hormone-sensitive lipase (HSL) [34]. HSL contributes to the hydrolysis of triglycerides (TG) in adipocytes. Similarly, lipolytic hormones increased when dexamethasone was added acutely to rat adipocytes [35]. After rats were treated for 10 days with corticosterone, free FA and glycerol levels were elevated in both fed and fasted states [34]. Thus, acute and subacute exposure to GCs increases lipolysis in vivo. Increased lipolysis results in elevated circulating free FA levels, which in turn is associated with insulin resistance [35]. Therefore, the diabetogenic effects of GCs are not only secondary to enhanced hepatic gluconeogenesis, impaired glucose uptake in muscle, and inhibition of insulin secretion, but also to elevated circulating free FA which originate from adipose tissue lipolysis [3].

Of note, it has been shown that GCs and insulin work synergistically to activate LPL, another lipolytic hormone [26]. Elevated LPL activity and intravascular lipolysis stimulate uptake of FA and glycerol into adipose tissue, leading to expansion of adipose tissue mass, as mentioned earlier [17]. This GC-dependent increase in LPL activity is thought to be due to increased transcription of LPL mRNA or post-translational modifications such as inhibiting the degradation of newly synthesized LPL [26–28].

Chronic GC exposure leads to adipose tissue expansion which suggests enhanced total body lipogenesis despite a possible increase in lipolysis [3]. To our knowledge, only two small studies have examined lipolysis in the setting of chronic endogenous GC exposure caused by CS [16, 38]. When examined ex vivo, glycerol release was reduced in femoral and abdominal adipose tissue from women with active CS, suggesting decreased lipolysis [16]. Conversely, glycerol concentrations were elevated in in vivo subcutaneous adipose tissue from patients with CS consistent with increased lipolysis [38]. Therefore, GCs possibly regulate factors such as hormone or neuronal signals in tissues other than adipose, which indirectly control adipose tissue functionality and may override the direct effects of GCs on adipose tissue [3].

Exposure to GCs leads to whole-body insulin resistance; however, the individual action in each tissue may vary. In fact, studies have shown that dexamethasone enhances insulin signaling and activity in human adipose tissue [39–42]. Forty-eight-hour dexamethasone pre-treatment led to a dose- and time-dependent increase in insulin-stimulated protein kinase B/akt phosphorylation and insulin receptor substrate (IRS)-1 phosphorylation in human adipocytes, but the reverse effect in skeletal muscle. These effects were mediated through induction of insulin receptor (IR), IRS-1, IRS2, and the p85 regulatory subunit of phosphoinositide-3-3-kinase, which led to augmented insulin-mediated activation of akt [40, 41]. Subsequent investigation showed that both short-term (24 h) and longer-term (7 day) exposure of differentiated human adipocytes ex vivo to dexamethasone increased insulin signaling, consistent with increased sensitization, whereas chronic high-dose GC exposure led to insulin resistance [42]. This was consistent with an in vivo study which showed that overnight administration of hydrocortisone induced systemic insulin resistance, but enhanced insulin signaling and uptake in subcutaneous adipose tissue [43]. These studies imply that the effect of GCs on insulin action may be tissue-dependent—increasing insulin sensitization in subcutaneous adipose tissue while inducing insulin resistance in muscle.

Lastly, GC treatment was shown to inhibit 5′ adenosine monophosphate-activated protein kinase (AMPK) activity in rat visceral, but not subcutaneous adipose tissue [44]. AMPK is a key regulatory enzyme of lipid and carbohydrate metabolism as well as appetite. This observation is supported by data showing that, compared to control patients, patients with CS demonstrated a 70 % lower AMPK activity in visceral adipose tissue [45].

In conclusion, the long-term exposure to elevated GC levels results in adipose tissue accumulation and altered distribution. These body composition changes are associated with insulin resistance in part via increased lipolysis, enhanced systemic elevations in FA and TG, and impaired insulin signaling.

GC Regulation of Skeletal Muscle

GC Regulation of Skeletal Muscle Composition

After Dr. Harvey Cushing's first description of muscle weakness in his case report of Minnie G in 1910 [46], it was not until Drs. Muller and Kugelberg's 1959 case series of six patients with CS when GC-induced myopathy was further described [2, 47]. Since then, a few clinical studies examining muscle function, histology, and metabolism in patients with CS have provided some framework for understanding the effects of GCs on muscle [48, 49]. GC-induced myopathy typically presents as proximal weakness, with predominant involvement of the lower extremities, and is seen in 56–90 % of patients with CS [2, 50]. Patients with CD also have reduced skeletal muscle mass compared to weight-matched controls [5], and surprisingly, skeletal muscle mass may continue to decrease over time after surgical remission [6]. Effects of GCs on muscle may be related to dose, type of GC (when given exogenously), duration of exposure, and specific muscle fiber type [48, 49]. In order to better understand the role of GCs in muscle mass and function, a brief review of muscle fibers followed by mechanisms underlying GC-mediated myopathy will be discussed.

Muscle fibers are categorized into slow twitch oxidative (Type I), fast twitch oxidative (Type IIa), and fast twitch glycolytic fibers (Type IIb); additional fiber types (Ic, IIc, IIab, IIac) are based on myosin ATPase histochemical staining [51]. Type 1 fibers are characterized by high levels of slow isoform contractile proteins, mitochondria, myoglobin and capillary densities, and oxidative capacity. Type IIa fibers are defined as having a high oxidative capacity with fast contraction, whereas type IIb fibers are described by low volumes of mitochondria, high glycolytic enzyme activity, increased rate of contraction, and low fatigue resistance [2].

More than 30 years ago, investigators used myometers and strain gauge techniques to quantitatively assess proximal weakness in patients with GC-induced myopathy, in addition to needle biopsy of muscle and 24 h urinary 3-methylhistidine excretion [50]. Fiber atrophy, specifically of type II fibers, is the classic histological abnormality associated with GC-mediated myopathy; interestingly, this finding is also present in other endocrinopathies including thyrotoxicosis, myxedema, and osteomalacia [2, 52]. Patients with CS have reduced type II fiber mean area, myopathic changes (including increased polyphasic muscle potentials on EMG), and an elevated 24 h urinary 3-methlyhistidine/creatinine ratio (an assessment of myofibrillar protein breakdown) [50]. Other ultrastructural changes associated with CS myopathy include pronounced mitochondrial damage, thickening and deep invaginations of the sarcolemmal basement membrane, and thickening of the basement membrane capillaries [53]. Muscle fibers from CS patients also demonstrate marked disarray and wide interfibrillar spaces containing large vacuoles which represent degenerated mitochondria [53]. Interestingly, Khaleeli et al. noted that histological abnormalities were more pronounced in the group of patients exposed to exogenous GCs for the treatment of inflammatory conditions compared to patients with endogenous CS. This was thought to be secondary to the high cumulative exposure of GCs, but alternatively it was also suggested that induction of 11BHSD1, which is present in human skeletal muscle, might also be increased in inflammatory conditions, as has been demonstrated in adipose tissue and bone [54].

Mechanisms Underlying GC-Mediated Myopathy

GC excess is associated with a decreased rate of protein synthesis and an increased rate of whole body proteolysis, even in patients who receive GC treatment for a short duration [2]. Skeletal muscle atrophy is a well-described adverse consequence of excess GC exposure [5]. Age is thought to impact the severity and mechanism of these catabolic effects; studies in rats have shown that GCs caused more severe atrophy in older compared to younger rats [55].

The inhibitory effects of GCs on protein synthesis are multifactorial. First, GCs inhibit the transportation of amino acids into the muscle [56]. Second, GCs inhibit the stimulatory action of insulin, insulin like growth factor-I (IGF-1), and amino acids (specifically leucine) on the phosphorylation of eIF4E binding protein (4E-BP1) and the ribosomal protein S6 kinase 1 (S6K1), two factors that are instrumental in the initiation of translation of mRNA responsible for the protein synthesis machinery [57]. Finally, GCs may inhibit myogenesis by down-regulating myogenin, a transcription factor mandatory for the differentiation of satellite cells to muscle fibers [58].

The stimulatory effect of GCs on muscle proteolysis is a result of the activation of major cellular proteolytic systems, specifically the ubiquitin-proteasome system (UPS), the lysosomal system (cathespins), and the calcium-dependent system (calpains) [59]. Thus, there is enhanced degradation of myofibrilliary fibers, which is evident by increased 3-methlyhistidine excretion. GCs activate protein degradation by stimulating the expression of several components of the UPS, which are either directly involved in protein degradation by a proteasome or by conjugation of protein to ubiquitin marking it for degradation [55].

Other factors that have been implicated in the development of GC-mediated myopathy include altered production of growth factors that locally control muscle development, specifically IGF-1. GCs inhibit IGF-1 production in muscle [60]. IGF-1 stimulates muscle mass by increasing protein synthesis and myogenesis while decreasing proteolysis and apoptosis [61, 62], linking decreased IGF-1 expression to GC-induced muscle atrophy [55]. Recent studies have shown that IGF-1 down-regulates the lysosomal, proteosomal, and calpain-dependent proteolytic systems [63–65], suppresses muscle cell atrophy caused by GCs [66], and interestingly prevents GC-induced muscle atrophy as evidenced by systemic administration or local overexpression of IGF-1 in rat skeletal muscle [67].

GCs also stimulate the production of myostatin (Mstn), a member of the transforming growth factor-beta family, and a potent inhibitor of muscle growth which down-regulates the proliferation and differentiation of satellite cells and protein synthesis [68]. In vitro data show that Mstn contributes to muscle cell atrophy by reversing the IGF-1/PI3K/Akt hypertrophy pathway; this finding was further solidified by a murine model that revealed that targeted disruption of Mstn gene expression in mice led to significant increases in skeletal muscle mass due to fiber hyperplasia and/or hypertrophy [55, 69, 70]. Interestingly, in humans, loss of function mutations of Mstn cause muscle hypertrophy, a rare condition characterized by reduced body fat and increased muscle size [71]. Furthermore, transgenic mice over-expressing Mstn in skeletal muscle have muscle atrophy [72, 73], and rats that were treated with dexamethasone in an effort to induce muscle atrophy were found to have significantly increased levels of Mstn mRNA expression and protein concentrations [74]. Further, in contrast to wild-type mice, Mstn knockout mice did not develop reduced muscle mass or fiber cross-sectional area after treatment with GCs [75]. Thus, increased muscle Mstn has been implicated as a key player in GC-induced muscle atrophy.

GC Regulation of Glucose Metabolism and Insulin Resistance in Skeletal Muscle

Skeletal muscle is the largest source of glycogen storage in the human body and accounts for 80 % of insulin-mediated, postprandial glucose uptake [76, 77]. GCs inhibit glucose uptake and utilization largely through antagonizing the actions of insulin in skeletal muscle. GCs also alter lipid and protein metabolism within skeletal muscle which leads to reduced insulin sensitivity [78–80].

One of the mechanisms by which GCs impede glucose uptake is by inhibiting insulin-stimulated translocation of the glucose transporter GLUT 4 to the plasma membrane, as demonstrated in mice treated with dexamethasone [81, 82]. GCs have also been shown to interfere with the insulin signaling cascade in skeletal muscle both in vitro and in vivo [83–86]. Insulin binds to the cell-surface IR, a tyrosine kinase that autophosphorylates and phosphorylates the IRS. Tyrosine-phosphorylated IRS associates with IR and activates downstream signaling [87]. Dexamethasone-treated mice have decreased expression and activity of tyrosine phosphorylated IR and IRS-1 [83]. The activity of phosphatidylinositol 3-kinase (PI3-K) and protein kinase B (PKB)/Akt, key signaling molecules that act downstream, is also reduced after GC exposure [83, 84, 86, 88]. Furthermore, GCs decrease glycogen synthesis and promote insulin resistance by suppressing glycogen synthase-3 phosphorylation [88]. A randomized cross-over study to determine the effect of 6 days of prednisone in 7 young healthy volunteers showed that although insulin infusion increased glucose uptake in both groups, uptake was 65 % lower in the prednisone-treated group versus the placebo group [89].

GCs also reduce insulin sensitivity, and subsequently glucose uptake, in skeletal muscle through effects on lipid metabolism. Elevated GCs stimulate adipose tissue lipolysis, which results in increased circulating levels of FA and TG [90, 91]. This enhances the accumulation of intramyocellular lipids (droplets of TG in skeletal muscle fibers) such as fatty acyl CoA and diacylglycerol (DAG), which are strongly correlated with reduced glucose uptake and insulin resistance [92, 93]. It has been shown, using magnetic resonance spectroscopy, that intramyocellular lipids decrease insulin signaling by activation of a serine/threonine kinase cascade involving protein kinase C, IKK-B and c-Jun amino-terminal kinases (JNKs). Phosphorylation of these serine sites leads to formation of proteins that are unable to activate PI3-K, which results in decreased glucose transport as discussed earlier [80].

Inter-muscular adipose tissue is another recently recognized ectopic adipose depot that is located beneath the muscle fascia but between the muscle groups (i.e. fat "marbling" within the muscle). It has been associated with development of insulin resistance [94], but was not found to be different in patients with CD vs. weight matched controls as measured by whole-body MRI [5].

As discussed above, enhanced protein breakdown and consequently elevated circulating amino acids (AA) have been reported after short-term high-dose GC treatment [95]. Elevated AA can impede insulin signaling by inhibiting insulin-stimulated IRS phosphorylation and activation of P13-K in cultured hepatoma cells and myocytes [96], leading to reduced glucose uptake and glycogen synthesis [78, 79]. Hence, the combined effects of reduced total muscle area and increased circulating AA lead to decreased insulin-mediated glucose uptake after prolonged GC exposure.

GC Effects on Liver

Excess GC exposure can increase glucose production and promote insulin resistance through regulation of carbohydrate and lipid metabolism in the liver. Several mechanisms have implicated GCs in the stimulation of hepatic lipogenesis and insulin resistance both directly and indirectly. Similar to skeletal muscle, excess GCs disrupt the insulin signaling cascade in hepatic tissue [84, 97, 98]. Dexamethasone-treated rats have reduced IR binding in hepatocytes [97] and down-regulation of the IR [98]. Additionally, livers of rats treated with dexamethasone exhibit decreased tyrosine phosphorylation of the IR and IRS-1 [84].

GCs also increase endogenous glucose production (EGP) by the liver [99, 100]. In the basal state, this is driven by increased gluconeogenesis via various mechanisms. First, GCs induce rate-limiting enzymes for gluconeogenesis such as phosphoenolpyruvate carboxykinase (PEPCK) and glucose-6-phosphatase (G6P) [99, 100]. PEPCK is required to generate glucose-6-phosphate, whereas G6P cleaves the phosphate allowing for glucose release into the circulation. The PEPCK gene contains GC response elements in its promoter region and plays a crucial role in GC-induced hyperglycemia. Of note, GC-mediated expression of gluconeogenic enzymes, such as PEPCK, is dependent on the cholesterol-sensing liver X receptors (LXRa and LXRb), which influence the recruitment of GR to gluconeogenic promoters. Mice lacking LXRb, but not LXRa, were resistant to dexamethasone-induced hyperglycemia, hyperinsulinemia, and hepatic steatosis [101]. Second, since GCs promote muscle wasting, lipolysis, and breakdown of protein and fat stores, the availability of substrates, such as alanine and glycerol, is increased, for gluconeogenesis in the liver [102–104]. Third, hepatic activation of the nuclear receptor peroxisome proliferator-activated receptor (PPAR-α) is associated with GC-induced hepatic insulin resistance and hyperglycemia. One study showed that PPAR-α knockout mice treated with dexamethasone did not develop hyperglycemia or hyperinsulinemia, concluding that PPAR-α expression is necessary for GC-induced increases in EGP [105]. Other mechanisms for enhanced EGP include increased metabolite transport across the mitochondrial membrane and potentiation of the effects of other gluco-regulatory hormones such as glucagon and epinephrine [103]. GCs also affect hepatic glucose metabolism by directly antagonizing the actions of insulin. For example, ceramides which are lipid-derived signaling molecules mediate GC-induced hepatic insulin resistance by blocking Akt phosphorylation and activation [106].

In addition to altering hepatic carbohydrate metabolism, GCs play an important role in hepatic lipid metabolism. Intrahepatic lipids are associated with insulin resistance and obesity and represent an important marker of cardiovascular risk, potentially even more so than visceral fat [107]. GC treatment leads to accumulation of intrahepatic lipids through various mechanisms including lipolysis of visceral adipose tissue, which leads to increased TG synthesis and delivery of free FA to the liver [108]. This leads to systemic hyperinsulinemia and hyperglycemia which drives de novo hepatic lipogenesis [109]. The critical role of GCs in hepatic lipid metabolism is demonstrated by improvement in hepatic steatosis and normalization of hepatic TG concentration in a fatty liver mouse model after liver-specific disruption of GR action [110].

GCs also enhance insulin-stimulated hepatic lipogenesis through upregulation of acetyl-CoA carboxylase and fatty acid synthase and increased very low-density lipoprotein (VLDL) production, resulting in increased TG levels, via inhibition of hepatic lipolysis [3, 111]. One small study reported enhanced VLDL secretion by the liver in patients with CD, which normalized after reduction of cortisol levels [91], and increased VLDL in healthy patients treated with prednisone [112], although these results have not been replicated.

Clinical data implicating GCs in the pathogenesis of hepatic steatosis are limited. Obese patients with nonalcoholic hepatic steatosis, measured via ultrasonography, had higher post-dexamethasone-suppressed cortisol values and insulin resistance compared to patients without steatosis [113]. Additionally, altered cortisol metabolism has been reported in patients with hepatic steatosis [32, 114], which suggests a relationship between hepatic fat and altered cortisol sensitivity and regulation in the general population [3]. Although hepatic steatosis is a known sequelae of prolonged GC exposure, only one study has investigated this in CD patients and reported a prevalence of 20 % [115]. The prevalence of hepatic steatosis in the asymptomatic general population varies widely, with results ranging from 2.8 to 24 % [116–119], and as high as 33.6 % in one study [120]. A few case reports have additionally linked the effect of excess GCs to fatty liver [121, 122]. As previously noted, H-magnetic resonance spectroscopy, the gold standard for determining hepatic lipid content, has never been investigated in humans exposed to excess GCs [3]. Therefore, although data suggest a link between chronic GC exposure and development and progression of hepatic steatosis, this topic warrants further investigation in clinical studies.

GC Regulation of the Pancreas/β-Cell

The pancreas plays a vital role in glucose metabolism and is the major sensor of circulating glucose. β-cells respond to increasing plasma glucose by secreting insulin in order to maintain euglycemia. The effects of GCs on β-cell function and insulin secretion are complex and depend on the duration, dosage, and type of GC exposure.

Glucose uptake and its oxidation in β-cell mitochondria lead to a cascade of events including elevated adenosine triphosphate (ATP)/adenosine monosphosphate (ADP) ratio, influx of calcium, and activation of signaling pathways including protein kinase A (PKA) and protein kinase C (PKC) which stimulate insulin secretion [123]. GCs impair β-cell glucose metabolism by reducing the levels of GLUT2 and glucokinase (GK), therefore reducing glucose uptake and phosphorylation and downstream events [124, 125]. GCs also amplify glucose cycling by enhancing G6P activity [103, 126].

In vitro studies have shown that corticosterone inhibits the release of insulin in rodent islets following acute (within minutes) exposure [127, 128]. On the other hand, this rapid inhibitory effect is not seen in vitro with dexamethasone, a synthetic GC [129]. Only after a three-hour incubation period, isolated rat islet cells demonstrated up to 75 % reduced glucose-induced insulin secretion. These events were mediated through impaired activation of the DAG-phospholipase C (PLC)/protein kinase C signaling system. Additionally, dexamethasone reduced cyclic adenosine monophosphate (cAMP) levels, leading to reduced PKA activity and hence reduced insulin secretion [129]. GCs have also been shown to inhibit insulin secretion via upregulation of voltage gated K+ channel activity, thereby leading to decreased calcium transport [130, 131].

In humans, GCs may also inhibit insulin secretion after acute exposure. A single dose of prednisolone administered to healthy subjects resulted in reduced insulin secretion during a meal with reduced insulinogenic index (ratio between change in insulinemia and change in glycemia) [132]. Another study showed that one dose of dexamethasone administered during an oral glucose tolerance test caused impaired glucose clearance, but had no effect on insulin sensitivity [133]. It should be noted, however, that this acute inhibitory effect has not always been noted, with another study showing a rise in circulating insulin after acute administration of intravenous hydrocortisone [134].

Both in vitro and in vivo experiments suggest that this rapid inhibitory effect on insulin secretion may be due to increased sympathetic drive via activation of α-adrenergic signaling [135, 136]. For example, when hydrocortisone was administered to Swiss-Webster mice, the glucose-stimulated insulin levels were suppressed in both fed and fasted mice compared to mice not given hydrocorti-cone [135]. However, if the mice were given chlorisondamine or phentolamine (non-selective α-adrenergic antagonists) prior, this resulted in higher insulin lev-els in response to the hydrocortisone-induced hyperglycemia [135].

Longer exposure (2–15 days) to dexamethasone or prednisolone in healthy subjects can lead to hyperinsulinemia with increased C-peptide and decreased insulin sensitiv-ity [132, 133, 137]. In these studies, healthy subjects were able to compensate for the GC-induced insulin resistance, resulting in euglycemia or only modest increases in fasting hyperglycemia. Other studies have shown that this hyperinsulinemic state after prolonged GC treatment is mediated by augmented β-cell function and mass, which counteracts the insulin resistance caused by GCs [138, 139]. However, normoglyce-mic subjects with reduced insulin sensitivity, first degree relatives of patient with Type 2 diabetes mellitus, obese, and other susceptible subjects may not be able to compen-sate [140–142]. In these settings, β-cell function does not correspond to the insulin demand and the imbalance of glucose homeostasis is more pronounced, resulting in hyperglycemia. These studies reinforce the concept that individual background is a critical factor when predicting the effects of GC exposure.

GC Regulation of Bone

GCs have a significant effect on bone physiology, and long-term exposure of the skeleton to GCs can result in osteoporosis and increased risk for fractures [2, 143]. Thus, the prevalence of osteoporosis in patients with CS is very high: approximately 55 % of women with CS have osteoporosis [144], and 19–50 % develop fractures, most commonly vertebral and rib fractures [2, 145–147]. Although bone mineral density may be decreased throughout the skeleton in CS patients, bone loss is most significant in areas rich in trabecular bone [146, 147]. First, the bone remodeling process and key cells will be briefly discussed, followed by a review of the mecha-nisms of bone loss secondary to GCs.

Bone Remodeling

Bone is dynamic tissue that is constantly undergoing catabolism (bone resorption) and anabolism (bone formation). Bone remodeling is the coupled process of bone break-down followed by new bone formation; it occurs in bone multicellular units (BMU) consisting of osteoclasts, osteoblasts, and surrounding tissue, and is regulated by bio-chemical and mechanical factors [148, 149]. Bone remodeling involves three consecu-tive phases: resorption, reversal, and formation. Resorption begins with the migration of mononuclear preosteoclasts to the surface of bone. Then under the influence of cytokines, hormones, physical, and chemical stimuli, preosteoclasts mature into osteo-clasts, which are multinucleated cells which are able to decalcify bone by creating resorption pits [148, 149]. After osteoclastic resorption is complete, mononuclear cells appear on the bone surface in preparation for bone formation and to provide the necessary signals for osteoblast differentiation and migration. The formation phase consists of osteoblasts which cover the resorbed bone with osteoid, a compound that becomes bone once calcified. These phases vary in length of time, with resorption lasting about two weeks, reversal continuing up to 5 weeks, and a timeframe up to 4 months for the completion of formation [148]. Typically, bone formation and resorption occur in concert, but in conditions where bone resorption predominates or bone formation is compromised osteoporosis occurs [143].

Osteoblasts are specialized bone-forming cells with several important roles in bone remodeling which include expression of osteoclastogenic factors, production of bone matrix proteins, and bone mineralization [150]. Osteoclast maturation or osteoclastogenesis is regulated by various stimuli; one in particular is receptor activator of nuclear factor KB ligand (RANKL). RANKL is a transmembrane glycoprotein expressed on the surface of osteoblasts/stromal cells in the bone, and its expression leads to increased osteoclast maturation and activity, as well as suppressed apoptosis [149]. Interestingly, knockout mice lacking RANKL completely lack osteoclasts and the ability to resorb bone [151]. Additional factors stimulating osteoclastogenesis include sustained hyperparathyroidism, decreased sex steroids, and an increase in inflammatory cytokines [149]. Balancing the effects of RANKL, osteoblasts secrete a decoy receptor, osteoprotegerin (OPG), or osteoclast inhibitory factor (OCIF), which binds to RANKL preventing its interaction with RANK, subsequently leading to decreased osteoclast maturation and survival.

Mechanisms Underlying GC-Induced Osteoporosis

It is well-established that excess GCs reduce bone formation [152–164], which is the predominant mechanism of GC-induced osteoporosis, whereas studies on the effects on bone resorption have been conflicting [153–155, 157–160, 164]. One reason for these contradictory results is that many studies included patients with GC excess secondary to exogenous GC treatment for various disorders that impact bone turnover and mass independently [152, 159, 165–171]. Also, previous studies have included both eugonadal and hypogonadal patients, thus introducing another confounding factor in GC regulation of bone turnover and mass [172, 173]. Lastly, bone resorption has been studied with nonspecific markers, such as urinary hydroxyproline and serum type I cross-linked C telopeptide [155, 157, 160, 162, 163, 172, 174]. Despite these limitations, GCs do appear to increase resorption in concert with limiting formation, as evidenced by a study of 18 eugonadal female CS patients who were compared to eugonadal healthy controls. This study demonstrated decreased osteoblastic function, increased bone resorption, and reduced bone mineral density (BMD) at the forearm, femur, and spine in CS patients versus healthy controls [164].

An increase in bone resorption is likely responsible for the initial bone loss observed following GC exposure [143]. Previous studies proposed that this was caused by secondary hyperparathyroidism [152, 154–157, 159, 175–178]. GCs are known to decrease calcium absorption in the gastrointestinal system and increase urinary excretion of calcium, resulting in elevated PTH levels [156]. Chiodini et al. identified secondary hyperparathyroidism, indicated by high PTH levels in the presence of normal plasma calcium levels, in a series of eugonadal CS patients, but noted no correlation between bone resorption markers and PTH levels [164]. The specific finding of trabecular bone loss in the setting of hyperparathyroidism, an entity known to typically affect cortical bone [179], further points to direct GC effects as the central cause of bone loss in patients with CS, and not secondary hyperparathyroidism [164]. Furthermore, patients exposed to GCs develop bone disease essentially characterized by decreased bone remodeling, whereas this is increased in patients with hyperparathyroidism [143].

Another mechanism contributing to increased bone resorption in patients with CS is decreased gonadotropin production. In estrogen deficiency, T-cell tumor necrosis factor (TNF-α) increases, stimulating bone resorption [156]. However, it is unclear whether TNF-α is elevated specifically in GC-induced hypogonadism [180]. As a final point, GC-induced bone resorption may involve RANK-L and OPG [143, 181, 182]; GCs increase RANK-L, while decreasing OPG expression, resulting in enhanced osteoclastogenesis and bone resorption [143]. The abovementioned factors are thought to contribute to the initial bone loss seen in GC-induced osteoporosis. Eventually, bone remodeling will be decreased because of the inhibitory effects of GCs on osteoblastogenesis resulting in reduced osteoblasts number and function, which subsequently leads to reduced signals for osteoclastogenesis and increased osteoclast apoptosis [143, 183].

Along with the effects on bone resorption, GCs stimulate collagenases, or matrix metalloproteinases (MMPs), by osteoblasts, which lead to matrix breakdown [156]. Specifically, osteoblasts exposed to GCs have increased collagenase 3 expression [143]. Of the three collagenases which have been described, collagenase 1 and 3 are responsible for the breakdown of type I collagen fibrils, the major component of the bone matrix [143]. Collagenase inhibition decreases bone resorption, as demonstrated by mice with mutations of the collagenase 3 cleavage site in type I collagen that fail to resorb bone after exposure to PTH [143]. GC exposure also results in decreased degradation of collagenases, and when combined with an increased collagenase level, contributes to type I collagen breakdown.

The effects of GCs on osteoblasts are complex and depend upon the stage of osteoblast growth and differentiation. GCs decrease the number of osteoblasts by decreasing cell replication, preventing differentiation of cells into mature osteoblasts [184] and enhancing mature osteoblast cell death [143]. Furthermore, GCs alter the function of osteoblasts; there is an associated decrease in type I collagen synthesis, which is likely secondary to transcriptional and posttranscriptional mechanisms [143] and leads to a decrease in available bone matrix for mineralization. CS patients are noted to have reduced serum levels of alkaline phosphatase and osteocalcin, which further demonstrates the inhibitory effect of GCs on osteoblastic function and parallels the changes seen by bone histomorphometry [143].

GC Effects on Skin

A variety of skin manifestations are seen in patients with CS, including violaceous striae, acne, hirsutism, acanthosis nigricans, superficial fungal infections, thinning skin, and easy bruisability [185]. GCs enhance the metabolism of proteinaceous tissues such as collagen, resulting in skin atrophy and fragility, and leading to striae and bruising [186]. Striae are dermal scars resulting from tears in the dermis that can occur with, but are not limited to, hypercortisolism [185]. GCs affect collagen formation in the dermis, and the cell type most likely to be affected is the skin fibroblast, which is responsible for collagen production and tissue repair [2]. Other skin features may depend upon the etiology of CS. In CD, elevated ACTH levels lead to increased adrenal androgens, which may cause hirsutism, male pattern alopecia, and acne in women [186]. In ectopic ACTH syndrome, excessive circulating ACTH and POMC precursors can result in skin hyper pigmentation; in vitro, ACTH and Melanocyte-stimulating hormone (MSH) are similarly potent stimulators of melanogenesis [187] through binding to the human melanocortin-1 receptor [188, 189].

One study that investigated the frequency and course of remission of skin manifestations in children and adolescents with CD treated with transsphenoidal surgery found variability in skin presentation. Pre-operative dermatologic findings included purple striae (77%), hirsutism (64%), acne (58%), acanthosis nigricans (28%), ecchymoses (28%), hyperpigmentation (17%), and fungal infections (11%) [185], but no correlation was found between circulating GC levels and severity of skin findings. The frequency of all signs decreased significantly within the first 3 months postoperatively, and by one year, all of the skin findings had progressively disappeared, with the exception of striae, which were lighter in color [185]. The persistence of striae for over 1 year after CD remission highlights the significant effects of GCs on skin structure and physiology.

Conclusion

This review highlights the critical role of GCs in regulating body composition and metabolism. Chronic exposure to even subtle GC excess results in the development of excess abdominal and ectopic adipose tissue, myopathy, hepatic steatosis, impaired β-cell function, and insulin resistance. Prevalent forms of GC excess include exogenous exposure, as GCs are widely used in the treatment of autoimmune and rheumatologic diseases, and chronic stress with resultant activation of the HPA axis. In many instances, the extent of these effects depends on the chronicity, dose, and type of GC exposure. Even common obesity and the metabolic syndrome have been proposed as models of excess endogenous GCs, due to enhanced HPA axis activation, altered metabolism, and/ or a flattened cortisol circadian rhythm. Further knowledge of the underpinnings GC effects on adipose tissue and metabolism could provide rationale for new GC therapeutic agents with reduced adverse (e.g. diabetogenic and adipogenic) effects.

References

1. Saenger AK. Discovery of the wonder drug: from cows to cortisone. The effects of the adrenal cortical hormone 17-hydroxy-11-dehydrocorticosterone (Compound E) on the acute phase of rheumatic fever; preliminary report. Mayo Clin Proc 1949;24:277-97. Clin Chem. 2010;56(8):1349–50.
2. Fernandez-Rodriguez E, Stewart PM, Cooper MS. The pituitary-adrenal axis and body composition. Pituitary. 2009;12(2):105–15.
3. Geer EB, Islam J, Buettner C. Mechanisms of glucocorticoid-induced insulin resistance: focus on adipose tissue function and lipid metabolism. Endocrinol Metab Clin North Am. 2014;43(1):75–102.
4. Overman RA, Yeh JY, Deal CL. Prevalence of oral glucocorticoid usage in the United States: a general population perspective. Arthritis Care Res (Hoboken). 2013;65(2):294–8.
5. Geer EB et al. MRI assessment of lean and adipose tissue distribution in female patients with Cushing's disease. Clin Endocrinol (Oxf). 2010;73(4):469–75.
6. Geer EB et al. Body composition and cardiovascular risk markers after remission of Cushing's disease: a prospective study using whole-body MRI. J Clin Endocrinol Metab. 2012;97(5): 1702–11.
7. Greenberg AS, Obin MS. Obesity and the role of adipose tissue in inflammation and metabolism. Am J Clin Nutr. 2006;83(2):461s–5s.
8. Goossens GH. The role of adipose tissue dysfunction in the pathogenesis of obesity-related insulin resistance. Physiol Behav. 2008;94(2):206–18.
9. Flegal KM et al. Association of all-cause mortality with overweight and obesity using standard body mass index categories: A systematic review and meta-analysis. JAMA. 2013;309(1):71–82.
10. Etxabe J, Vazquez JA. Morbidity and mortality in Cushing's disease: an epidemiological approach. Clin Endocrinol (Oxf). 1994;40(4):479–84.
11. Shen W et al. Adipose tissue quantification by imaging methods: a proposed classification. Obes Res. 2003;11(1):5–16.
12. Lee MJ, Wu Y, Fried SK. Adipose tissue heterogeneity: implication of depot differences in adipose tissue for obesity complications. Mol Aspects Med. 2013;34(1):1–11.
13. Schipper HS et al. Adipose tissue-resident immune cells: key players in immunometabolism. Trends Endocrinol Metab. 2012;23(8):407–15.
14. Olefsky JM, Glass CK. Macrophages, inflammation, and insulin resistance. Annu Rev Physiol. 2010;72:219–46.
15. Vazquez-Vela ME, Torres N, Tovar AR. White adipose tissue as endocrine organ and its role in obesity. Arch Med Res. 2008;39(8):715–28.
16. Rebuffe-Scrive M et al. Muscle and adipose tissue morphology and metabolism in Cushing's syndrome. J Clin Endocrinol Metab. 1988;67(6):1122–8.
17. Morton NM, Seckl JR. 11beta-hydroxysteroid dehydrogenase type 1 and obesity. Front Horm Res. 2008;36:146–64.
18. Wajchenberg BL et al. Estimation of body fat and lean tissue distribution by dual energy X-ray absorptiometry and abdominal body fat evaluation by computed tomography in Cushing's disease. J Clin Endocrinol Metab. 1995;80(9):2791–4.
19. Peckett AJ, Wright DC, Riddell MC. The effects of glucocorticoids on adipose tissue lipid metabolism. Metabolism. 2011;60(11):1500–10.
20. Walker BR, Andrew R. Tissue production of cortisol by 11beta-hydroxysteroid dehydrogenase type 1 and metabolic disease. Ann N Y Acad Sci. 2006;1083:165–84.
21. Masuzaki H et al. A transgenic model of visceral obesity and the metabolic syndrome. Science. 2001;294(5549):2166–70.
22. Desbriere R et al. 11beta-hydroxysteroid dehydrogenase type 1 mRNA is increased in both visceral and subcutaneous adipose tissue of obese patients. Obesity (Silver Spring). 2006;14(5):794–8.

23. Sjogren J et al. Glucocorticoid hormone binding to rat adipocytes. Biochim Biophys Acta. 1994;1224(1):17–21.
24. Rebuffe-Scrive M et al. Steroid hormone receptors in human adipose tissues. J Clin Endocrinol Metab. 1990;71(5):1215–9.
25. Bujalska IJ, Kumar S, Stewart PM. Does central obesity reflect "Cushing's disease of the omentum"? Lancet. 1997;349(9060):1210–3.
26. Ottosson M et al. The effects of cortisol on the regulation of lipoprotein lipase activity in human adipose tissue. J Clin Endocrinol Metab. 1994;79(3):820–5.
27. Appel B, Fried SK. Effects of insulin and dexamethasone on lipoprotein lipase in human adipose tissue. Am J Physiol. 1992;262(5 Pt 1):E695–9.
28. Fried SK et al. Lipoprotein lipase regulation by insulin and glucocorticoid in subcutaneous and omental adipose tissues of obese women and men. J Clin Invest. 1993;92(5):2191–8.
29. Bujalska IJ et al. Differentiation of adipose stromal cells: the roles of glucocorticoids and 11beta-hydroxysteroid dehydrogenase. Endocrinology. 1999;140(7):3188–96.
30. Hauner H et al. Promoting effect of glucocorticoids on the differentiation of human adipocyte precursor cells cultured in a chemically defined medium. J Clin Invest. 1989;84(5):1663–70.
31. Kershaw EE, Flier JS. Adipose tissue as an endocrine organ. J Clin Endocrinol Metab. 2004;89(6):2548–56.
32. Ahmed A et al. A switch in hepatic cortisol metabolism across the spectrum of non alcoholic fatty liver disease. PLoS One. 2012;7(2), e29531.
33. Lee MJ et al. Depot-specific regulation of the conversion of cortisone to cortisol in human adipose tissue. Obesity (Silver Spring). 2008;16(6):1178–85.
34. Campbell JE et al. Adipogenic and lipolytic effects of chronic glucocorticoid exposure. Am J Physiol Cell Physiol. 2011;300(1):C198–209.
35. Slavin BG, Ong JM, Kern PA. Hormonal regulation of hormone-sensitive lipase activity and mRNA levels in isolated rat adipocytes. J Lipid Res. 1994;35(9):1535–41.
36. Samra JS et al. Effects of physiological hypercortisolemia on the regulation of lipolysis in subcutaneous adipose tissue. J Clin Endocrinol Metab. 1998;83(2):626–31.
37. Djurhuus CB et al. Effects of cortisol on lipolysis and regional interstitial glycerol levels in humans. Am J Physiol Endocrinol Metab. 2002;283(1):E172–7.
38. Krsek M et al. Increased lipolysis of subcutaneous abdominal adipose tissue and altered noradrenergic activity in patients with Cushing's syndrome: an in-vivo microdialysis study. Physiol Res. 2006;55(4):421–8.
39. Gathercole LL et al. Regulation of lipogenesis by glucocorticoids and insulin in human adipose tissue. PLoS One. 2011;6(10), e26223.
40. Gathercole LL et al. Glucocorticoid modulation of insulin signaling in human subcutaneous adipose tissue. J Clin Endocrinol Metab. 2007;92(11):4332–9.
41. Tomlinson JJ et al. Insulin sensitization of human preadipocytes through glucocorticoid hormone induction of forkhead transcription factors. Mol Endocrinol. 2010;24(1):104–13.
42. Gathercole LL et al. Short- and long-term glucocorticoid treatment enhances insulin signalling in human subcutaneous adipose tissue. Nutr Diabetes. 2011;1, e3.
43. Hazlehurst JM et al. Glucocorticoids fail to cause insulin resistance in human subcutaneous adipose tissue in vivo. J Clin Endocrinol Metab. 2013;98(4):1631–40.
44. Christ-Crain M et al. AMP-activated protein kinase mediates glucocorticoid-induced metabolic changes: a novel mechanism in Cushing's syndrome. FASEB J. 2008;22(6):1672–83.
45. Kola B et al. Changes in adenosine 5'-monophosphate-activated protein kinase as a mechanism of visceral obesity in Cushing's syndrome. J Clin Endocrinol Metab. 2008;93(12):4969–73.
46. Cushing H. The basophil adenomas of the pituitary body and their clinical manifestations (pituitary basophilism). Obes Res. 1994;2(5):486–508.
47. Muller R, Kugelberg E. Myopathy in Cushing's syndrome. J Neurol Neurosurg Psychiatry. 1959;22:314–9.
48. Lane RJ, Mastaglia FL. Drug-induced myopathies in man. Lancet. 1978;2(8089):562–6.
49. Mills GH et al. Respiratory muscle strength in Cushing's syndrome. Am J Respir Crit Care Med. 1999;160(5 Pt 1):1762–5.

50. Khaleeli AA et al. Corticosteroid myopathy: a clinical and pathological study. Clin Endocrinol (Oxf). 1983;18(2):155–66.
51. Scott W, Stevens J, Binder-Macleod SA. Human skeletal muscle fiber type classifications. Phys Ther. 2001;81(11):1810–6.
52. Kendall-Taylor P, Turnbull DM. Endocrine myopathies. Br Med J (Clin Res Ed). 1983;287(6394):705–8.
53. Djaldetti M, Gafter U, Fishman P. Ultrastructural observations in myopathy complicating Cushing's disease. Am J Med Sci. 1977;273(3):273–7.
54. Tomlinson JW et al. Regulation of expression of 11beta-hydroxysteroid dehydrogenase type 1 in adipose tissue: tissue-specific induction by cytokines. Endocrinology. 2001;142(5):1982–9.
55. Schakman O, Gilson H, Thissen JP. Mechanisms of glucocorticoid-induced myopathy. J Endocrinol. 2008;197(1):1–10.
56. Kostyo JL, Redmond AF. Role of protein synthesis in the inhibitory action of adrenal steroid hormones on amino acid transport by muscle. Endocrinology. 1966;79(3):531–40.
57. Shah OJ, Kimball SR, Jefferson LS. Acute attenuation of translation initiation and protein synthesis by glucocorticoids in skeletal muscle. Am J Physiol Endocrinol Metab. 2000;278(1):E76–82.
58. te Pas MF, de Jong PR, Verburg FJ. Glucocorticoid inhibition of C2C12 proliferation rate and differentiation capacity in relation to mRNA levels of the MRF gene family. Mol Biol Rep. 2000;27(2):87–98.
59. Hasselgren PO. Glucocorticoids and muscle catabolism. Curr Opin Clin Nutr Metab Care. 1999;2(3):201–5.
60. Gayan-Ramirez G et al. Acute treatment with corticosteroids decreases IGF-1 and IGF-2 expression in the rat diaphragm and gastrocnemius. Am J Respir Crit Care Med. 1999;159(1):283–9.
61. Florini JR, Ewton DZ, Coolican SA. Growth hormone and the insulin-like growth factor system in myogenesis. Endocr Rev. 1996;17(5):481–517.
62. Frost RA, Lang CH. Regulation of insulin-like growth factor-I in skeletal muscle and muscle cells. Minerva Endocrinol. 2003;28(1):53–73.
63. Dehoux M et al. Role of the insulin-like growth factor I decline in the induction of atrogin-1/MAFbx during fasting and diabetes. Endocrinology. 2004;145(11):4806–12.
64. Latres E et al. Insulin-like growth factor-1 (IGF-1) inversely regulates atrophy-induced genes via the phosphatidylinositol 3-kinase/Akt/mammalian target of rapamycin (PI3K/Akt/mTOR) pathway. J Biol Chem. 2005;280(4):2737–44.
65. Li BG et al. Insulin-like growth factor-I blocks dexamethasone-induced protein degradation in cultured myotubes by inhibiting multiple proteolytic pathways: 2002 ABA paper. J Burn Care Rehabil. 2004;25(1):112–8.
66. Sacheck JM et al. IGF-I stimulates muscle growth by suppressing protein breakdown and expression of atrophy-related ubiquitin ligases, atrogin-1 and MuRF1. Am J Physiol Endocrinol Metab. 2004;287(4):E591–601.
67. Schakman O et al. Insulin-like growth factor-I gene transfer by electroporation prevents skeletal muscle atrophy in glucocorticoid-treated rats. Endocrinology. 2005;146(4):1789–97.
68. Thomas M et al. Myostatin, a negative regulator of muscle growth, functions by inhibiting myoblast proliferation. J Biol Chem. 2000;275(51):40235–43.
69. McPherron AC, Lawler AM, Lee SJ. Regulation of skeletal muscle mass in mice by a new TGF-beta superfamily member. Nature. 1997;387(6628):83–90.
70. Grobet L et al. Modulating skeletal muscle mass by postnatal, muscle-specific inactivation of the myostatin gene. Genesis. 2003;35(4):227–38.
71. Schuelke M et al. Myostatin mutation associated with gross muscle hypertrophy in a child. N Engl J Med. 2004;350(26):2682–8.
72. Reisz-Porszasz S et al. Lower skeletal muscle mass in male transgenic mice with muscle-specific overexpression of myostatin. Am J Physiol Endocrinol Metab. 2003;285(4):E876–88.
73. Durieux AC et al. Ectopic expression of myostatin induces atrophy of adult skeletal muscle by decreasing muscle gene expression. Endocrinology. 2007;148(7):3140–7.
74. Hong DH, Forsberg NE. Effects of dexamethasone on protein degradation and protease gene expression in rat L8 myotube cultures. Mol Cell Endocrinol. 1995;108(1-2):199–209.

75. Gilson H et al. Myostatin gene deletion prevents glucocorticoid-induced muscle atrophy. Endocrinology. 2007;148(1):452–60.
76. DeFronzo RA et al. The effect of insulin on the disposal of intravenous glucose. Results from indirect calorimetry and hepatic and femoral venous catheterization. Diabetes. 1981;30(12): 1000–7.
77. DeFronzo RA, Tripathy D. Skeletal muscle insulin resistance is the primary defect in type 2 diabetes. Diabetes Care. 2009;32 Suppl 2:S157–63.
78. Krebs M et al. Mechanism of amino acid-induced skeletal muscle insulin resistance in humans. Diabetes. 2002;51(3):599–605.
79. Krebs M et al. Free fatty acids inhibit the glucose-stimulated increase of intramuscular glucose-6-phosphate concentration in humans. J Clin Endocrinol Metab. 2001;86(5):2153–60.
80. Perseghin G, Petersen K, Shulman GI. Cellular mechanism of insulin resistance: potential links with inflammation. Int J Obes Relat Metab Disord. 2003;27 Suppl 3:S6–11.
81. Dimitriadis G et al. Effects of glucocorticoid excess on the sensitivity of glucose transport and metabolism to insulin in rat skeletal muscle. Biochem J. 1997;321(Pt 3):707–12.
82. Weinstein SP et al. Dexamethasone inhibits insulin-stimulated recruitment of GLUT4 to the cell surface in rat skeletal muscle. Metabolism. 1998;47(1):3–6.
83. Morgan SA et al. 11beta-hydroxysteroid dehydrogenase type 1 regulates glucocorticoid-induced insulin resistance in skeletal muscle. Diabetes. 2009;58(11):2506–15.
84. Saad MJ et al. Modulation of insulin receptor, insulin receptor substrate-1, and phosphatidylinositol 3-kinase in liver and muscle of dexamethasone-treated rats. J Clin Invest. 1993;92(4):2065–72.
85. Ewart HS, Somwar R, Klip A. Dexamethasone stimulates the expression of GLUT1 and GLUT4 proteins via different signalling pathways in L6 skeletal muscle cells. FEBS Lett. 1998;421(2):120–4.
86. Giorgino F et al. Glucocorticoid regulation of insulin receptor and substrate IRS-1 tyrosine phosphorylation in rat skeletal muscle in vivo. J Clin Invest. 1993;91(5):2020–30.
87. Kuo T, Harris CA, Wang JC. Metabolic functions of glucocorticoid receptor in skeletal muscle. Mol Cell Endocrinol. 2013;380(1-2):79–88.
88. Ruzzin J, Wagman AS, Jensen J. Glucocorticoid-induced insulin resistance in skeletal muscles: defects in insulin signalling and the effects of a selective glycogen synthase kinase-3 inhibitor. Diabetologia. 2005;48(10):2119–30.
89. Short KR, Bigelow ML, Nair KS. Short-term prednisone use antagonizes insulin's anabolic effect on muscle protein and glucose metabolism in young healthy people. Am J Physiol Endocrinol Metab. 2009;297(6):E1260–8.
90. Dinneen S et al. Effects of the normal nocturnal rise in cortisol on carbohydrate and fat metabolism in IDDM. Am J Physiol. 1995;268(4 Pt 1):E595–603.
91. Taskinen MR et al. Plasma lipoproteins, lipolytic enzymes, and very low density lipoprotein triglyceride turnover in Cushing's syndrome. J Clin Endocrinol Metab. 1983;57(3):619–26.
92. Boden G, Shulman GI. Free fatty acids in obesity and type 2 diabetes: defining their role in the development of insulin resistance and beta-cell dysfunction. Eur J Clin Invest. 2002;32 Suppl 3:14–23.
93. Jacob S et al. Association of increased intramyocellular lipid content with insulin resistance in lean nondiabetic offspring of type 2 diabetic subjects. Diabetes. 1999;48(5):1113–9.
94. Goodpaster BH, Thaete FL, Kelley DE. Thigh adipose tissue distribution is associated with insulin resistance in obesity and in type 2 diabetes mellitus. Am J Clin Nutr. 2000;71(4):885–92.
95. Lofberg E et al. Effects of high doses of glucocorticoids on free amino acids, ribosomes and protein turnover in human muscle. Eur J Clin Invest. 2002;32(5):345–53.
96. Patti ME et al. Bidirectional modulation of insulin action by amino acids. J Clin Invest. 1998;101(7):1519–29.
97. Olefsky JM et al. The effects of acute and chronic dexamethasone administration on insulin binding to isolated rat hepatocytes and adipocytes. Metabolism. 1975;24(4):517–27.

98. Caro JF, Amatruda JM. Glucocorticoid-induced insulin resistance: the importance of postbinding events in the regulation of insulin binding, action, and degradation in freshly isolated and primary cultures of rat hepatocytes. J Clin Invest. 1982;69(4):866–75.

99. Jin JY et al. Receptor/gene-mediated pharmacodynamic effects of methylprednisolone on phosphoenolpyruvate carboxykinase regulation in rat liver. J Pharmacol Exp Ther. 2004; 309(1):328–39.

100. Vander Kooi BT et al. The glucose-6-phosphatase catalytic subunit gene promoter contains both positive and negative glucocorticoid response elements. Mol Endocrinol. 2005; 19(12):3001–22.

101. Patel R et al. LXRbeta is required for glucocorticoid-induced hyperglycemia and hepatosteatosis in mice. J Clin Invest. 2011;121(1):431–41.

102. Divertie GD, Jensen MD, Miles JM. Stimulation of lipolysis in humans by physiological hypercortisolemia. Diabetes. 1991;40(10):1228–32.

103. van Raalte DH, Ouwens DM, Diamant M. Novel insights into glucocorticoid-mediated diabetogenic effects: towards expansion of therapeutic options? Eur J Clin Invest. 2009;39(2):81–93.

104. Zimmerman T et al. Contribution of insulin resistance to catabolic effect of prednisone on leucine metabolism in humans. Diabetes. 1989;38(10):1238–44.

105. Bernal-Mizrachi C et al. Dexamethasone induction of hypertension and diabetes is PPAR-alpha dependent in LDL receptor-null mice. Nat Med. 2003;9(8):1069–75.

106. Magomedova L, Cummins CL. Glucocorticoids and Metabolic Control. 2015. Handb Exp Pharmacol.

107. Lim S et al. Fat in liver/muscle correlates more strongly with insulin sensitivity in rats than abdominal fat. Obesity (Silver Spring). 2009;17(1):188–95.

108. Cole TG, Wilcox HG, Heimberg M. Effects of adrenalectomy and dexamethasone on hepatic lipid metabolism. J Lipid Res. 1982;23(1):81–91.

109. Schwarz JM et al. Hepatic de novo lipogenesis in normoinsulinemic and hyperinsulinemic subjects consuming high-fat, low-carbohydrate and low-fat, high-carbohydrate isoenergetic diets. Am J Clin Nutr. 2003;77(1):43–50.

110. Lemke U et al. The glucocorticoid receptor controls hepatic dyslipidemia through Hes1. Cell Metab. 2008;8(3):212–23.

111. Dolinsky VW et al. Regulation of the enzymes of hepatic microsomal triacylglycerol lipolysis and re-esterification by the glucocorticoid dexamethasone. Biochem J. 2004;378(Pt 3):967–74.

112. Ettinger Jr WH, Hazzard WR. Prednisone increases very low density lipoprotein and high density lipoprotein in healthy men. Metabolism. 1988;37(11):1055–8.

113. Zoppini G et al. Relationship of nonalcoholic hepatic steatosis to overnight low-dose dexamethasone suppression test in obese individuals. Clin Endocrinol (Oxf). 2004;61(6):711–5.

114. Westerbacka J et al. Body fat distribution and cortisol metabolism in healthy men: enhanced 5beta-reductase and lower cortisol/cortisone metabolite ratios in men with fatty liver. J Clin Endocrinol Metab. 2003;88(10):4924–31.

115. Rockall AG et al. Hepatic steatosis in Cushing's syndrome: a radiological assessment using computed tomography. Eur J Endocrinol. 2003;149(6):543–8.

116. Ruhl CE, Everhart JE. Determinants of the association of overweight with elevated serum alanine aminotransferase activity in the United States. Gastroenterology. 2003;124(1):71–9.

117. Nomura H et al. Prevalence of fatty liver in a general population of Okinawa, Japan. Jpn J Med. 1988;27(2):142–9.

118. Hilden M et al. Liver histology in a 'normal' population--examinations of 503 consecutive fatal traffic casualties. Scand J Gastroenterol. 1977;12(5):593–7.

119. el-Hassan AY et al. Fatty infiltration of the liver: analysis of prevalence, radiological and clinical features and influence on patient management. Br J Radiol. 1992;65(777):774–8.

120. Szczepaniak LS et al. Magnetic resonance spectroscopy to measure hepatic triglyceride content: prevalence of hepatic steatosis in the general population. Am J Physiol Endocrinol Metab. 2005;288(2):E462–8.

121. Dourakis SP, Sevastianos VA, Kaliopi P. Acute severe steatohepatitis related to prednisolone therapy. Am J Gastroenterol. 2002;97(4):1074–5.
122. Nanki T, Koike R, Miyasaka N. Subacute severe steatohepatitis during prednisolone therapy for systemic lupus erythematosis. Am J Gastroenterol. 1999;94(11):3379.
123. Nesher R et al. Beta-cell protein kinases and the dynamics of the insulin response to glucose. Diabetes. 2002;51 Suppl 1:S68–73.
124. Ogawa A et al. Roles of insulin resistance and beta-cell dysfunction in dexamethasone-induced diabetes. J Clin Invest. 1992;90(2):497–504.
125. Borboni P et al. Quantitative analysis of pancreatic glucokinase gene expression in cultured beta cells by competitive polymerase chain reaction. Mol Cell Endocrinol. 1996;117(2):175–81.
126. Khan A et al. Glucocorticoid increases glucose cycling and inhibits insulin release in pancreatic islets of ob/ob mice. Am J Physiol. 1992;263(4 Pt 1):E663–6.
127. Billaudel B et al. Inhibition by corticosterone of calcium inflow and insulin release in rat pancreatic islets. J Endocrinol. 1984;100(2):227–33.
128. Billaudel B, Sutter BC. Direct effect of corticosterone upon insulin secretion studied by three different techniques. Horm Metab Res. 1979;11(10):555–60.
129. Zawalich WS et al. Dexamethasone suppresses phospholipase C activation and insulin secretion from isolated rat islets. Metabolism. 2006;55(1):35–42.
130. Ullrich S et al. Serum- and glucocorticoid-inducible kinase 1 (SGK1) mediates glucocorticoid-induced inhibition of insulin secretion. Diabetes. 2005;54(4):1090–9.
131. Shao J, Qiao L, Friedman JE. Prolactin, progesterone, and dexamethasone coordinately and adversely regulate glucokinase and cAMP/PDE cascades in MIN6 beta-cells. Am J Physiol Endocrinol Metab. 2004;286(2):E304–10.
132. van Raalte DH et al. Acute and 2-week exposure to prednisolone impair different aspects of beta-cell function in healthy men. Eur J Endocrinol. 2010;162(4):729–35.
133. Schneiter P, Tappy L. Kinetics of dexamethasone-induced alterations of glucose metabolism in healthy humans. Am J Physiol. 1998;275(5 Pt 1):E806–13.
134. Vila G et al. Acute effects of hydrocortisone on the metabolic response to a glucose load: increase in the first-phase insulin secretion. Eur J Endocrinol. 2010;163(2):225–31.
135. Longano CA, Fletcher HP. Insulin release after acute hydrocortisone treatment in mice. Metabolism. 1983;32(6):603–8.
136. Barseghian G, Levine R. Effect of corticosterone on insulin and glucagon secretion by the isolated perfused rat pancreas. Endocrinology. 1980;106(2):547–52.
137. Matsumoto K et al. High-dose but not low-dose dexamethasone impairs glucose tolerance by inducing compensatory failure of pancreatic beta-cells in normal men. J Clin Endocrinol Metab. 1996;81(7):2621–6.
138. Rafacho A et al. High doses of dexamethasone induce increased beta-cell proliferation in pancreatic rat islets. Am J Physiol Endocrinol Metab. 2009;296(4):E681–9.
139. Rafacho A et al. Morphofunctional alterations in endocrine pancreas of short- and long-term dexamethasone-treated rats. Horm Metab Res. 2011;43(4):275–81.
140. Jensen DH et al. Steroid-induced insulin resistance and impaired glucose tolerance are both associated with a progressive decline of incretin effect in first-degree relatives of patients with type 2 diabetes. Diabetologia. 2012;55(5):1406–16.
141. Besse C, Nicod N, Tappy L. Changes in insulin secretion and glucose metabolism induced by dexamethasone in lean and obese females. Obes Res. 2005;13(2):306–11.
142. Larsson H, Ahren B. Insulin resistant subjects lack islet adaptation to short-term dexamethasone-induced reduction in insulin sensitivity. Diabetologia. 1999;42(8):936–43.
143. Canalis E, Delany AM. Mechanisms of glucocorticoid action in bone. Ann N Y Acad Sci. 2002;966:73–81.
144. Ohmori N et al. Osteoporosis is more prevalent in adrenal than in pituitary Cushing's syndrome. Endocr J. 2003;50(1):1–7.
145. Van Staa TP et al. Use of oral corticosteroids and risk of fractures. June, 2000. J Bone Miner Res, 2005. J Bone Miner Res. 2000;20(8):1487–94. discussion 1486.

146. Kristo C et al. Restoration of the coupling process and normalization of bone mass following successful treatment of endogenous Cushing's syndrome: a prospective, long-term study. Eur J Endocrinol. 2006;154(1):109–18.
147. Tauchmanova L et al. Bone loss determined by quantitative ultrasonometry correlates inversely with disease activity in patients with endogenous glucocorticoid excess due to adrenal mass. Eur J Endocrinol. 2001;145(3):241–7.
148. Hadjidakis DJ, Androulakis II. Bone remodeling. Ann N Y Acad Sci. 2006;1092:385–96.
149. Rehman Q, Lane NE. Effect of glucocorticoids on bone density. Med Pediatr Oncol. 2003;41(3):212–6.
150. Karsenty G. Transcriptional control of skeletogenesis. Annu Rev Genomics Hum Genet. 2008;9:183–96.
151. Raggatt LJ, Partridge NC. Cellular and molecular mechanisms of bone remodeling. J Biol Chem. 2010;285(33):25103–8.
152. Lukert BP, Raisz LG. Glucocorticoid-induced osteoporosis: pathogenesis and management. Ann Intern Med. 1990;112(5):352–64.
153. Jowsey J, Riggs BL. Bone formation in hypercortisonism. Acta Endocrinol (Copenh). 1970;63(1):21–8.
154. Hahn TJ et al. Altered mineral metabolism in glucocorticoid-induced osteopenia. Effect of 25-hydroxyvitamin D administration. J Clin Invest. 1979;64(2):655–65.
155. Eastell R. Management of corticosteroid-induced osteoporosis. UK Consensus Group Meeting on Osteoporosis. J Intern Med. 1995;237(5):439–47.
156. Canalis E. Clinical review 83: Mechanisms of glucocorticoid action in bone: implications to glucocorticoid-induced osteoporosis. J Clin Endocrinol Metab. 1996;81(10):3441–7.
157. Sambrook PN, Jones G. Corticosteroid osteoporosis. Br J Rheumatol. 1995;34(1):8–12.
158. Reid IR et al. Low serum osteocalcin levels in glucocorticoid-treated asthmatics. J Clin Endocrinol Metab. 1986;62(2):379–83.
159. Reid IR. Pathogenesis and treatment of steroid osteoporosis. Clin Endocrinol (Oxf). 1989;30(1):83–103.
160. Prummel MF et al. The course of biochemical parameters of bone turnover during treatment with corticosteroids. J Clin Endocrinol Metab. 1991;72(2):382–6.
161. Sartorio A et al. Osteocalcin levels in Cushing's disease before and after treatment. Horm Metab Res. 1988;20(1):70.
162. Sartorio A et al. Serum bone Gla protein and carboxyterminal cross-linked telopeptide of type I collagen in patients with Cushing's syndrome. Postgrad Med J. 1996;72(849):419–22.
163. Osella G et al. Serum markers of bone and collagen turnover in patients with Cushing's syndrome and in subjects with adrenal incidentalomas. J Clin Endocrinol Metab. 1997;82(10):3303–7.
164. Chiodini I et al. Alterations of bone turnover and bone mass at different skeletal sites due to pure glucocorticoid excess: study in eumenorrheic patients with Cushing's syndrome. J Clin Endocrinol Metab. 1998;83(6):1863–7.
165. Laan RF et al. Differential effects of glucocorticoids on cortical appendicular and cortical vertebral bone mineral content. Calcif Tissue Int. 1993;52(1):5–9.
166. Hall GM et al. The effect of rheumatoid arthritis and steroid therapy on bone density in postmenopausal women. Arthritis Rheum. 1993;36(11):1510–6.
167. Verstraeten A, Dequeker J. Vertebral and peripheral bone mineral content and fracture incidence in postmenopausal patients with rheumatoid arthritis: effect of low dose corticosteroids. Ann Rheum Dis. 1986;45(10):852–7.
168. Butler RC et al. Bone mineral content in patients with rheumatoid arthritis: relationship to low-dose steroid therapy. Br J Rheumatol. 1991;30(2):86–90.
169. Kalla AA et al. Loss of trabecular bone mineral density in systemic lupus erythematosus. Arthritis Rheum. 1993;36(12):1726–34.
170. Compston JE et al. Spinal trabecular bone mineral content in patients with non-steroid treated rheumatoid arthritis. Ann Rheum Dis. 1988;47(8):660–4.

171. Sambrook PN et al. Effects of low dose corticosteroids on bone mass in rheumatoid arthritis: a longitudinal study. Ann Rheum Dis. 1989;48(7):535–8.
172. Hermus AR et al. Bone mineral density and bone turnover before and after surgical cure of Cushing's syndrome. J Clin Endocrinol Metab. 1995;80(10):2859–65.
173. Manning PJ, Evans MC, Reid IR. Normal bone mineral density following cure of Cushing's syndrome. Clin Endocrinol (Oxf). 1992;36(3):229–34.
174. Conti A et al. Modifications of biochemical markers of bone and collagen turnover during corticosteroid therapy. J Endocrinol Invest. 1996;19(2):127–30.
175. Suzuki Y et al. Importance of increased urinary calcium excretion in the development of secondary hyperparathyroidism of patients under glucocorticoid therapy. Metabolism. 1983;32(2):151–6.
176. Lukert BP, Adams JS. Calcium and phosphorus homeostasis in man. Effect of corticosteroids. Arch Intern Med. 1976;136(11):1249–53.
177. Fucik RF et al. Effect of glucocorticoids on function of the parathyroid glands in man. J Clin Endocrinol Metab. 1975;40(1):152–5.
178. Adams JS, Wahl TO, Lukert BP. Effects of hydrochlorothiazide and dietary sodium restriction on calcium metabolism in corticosteroid treated patients. Metabolism. 1981;30(3):217–21.
179. Silverberg SJ et al. Skeletal disease in primary hyperparathyroidism. J Bone Miner Res. 1989;4(3):283–91.
180. Cenci S et al. Estrogen deficiency induces bone loss by enhancing T-cell production of TNF-alpha. J Clin Invest. 2000;106(10):1229–37.
181. Hofbauer LC et al. Stimulation of osteoprotegerin ligand and inhibition of osteoprotegerin production by glucocorticoids in human osteoblastic lineage cells: potential paracrine mechanisms of glucocorticoid-induced osteoporosis. Endocrinology. 1999;140(10):4382–9.
182. Rubin J et al. Dexamethasone promotes expression of membrane-bound macrophage colony-stimulating factor in murine osteoblast-like cells. Endocrinology. 1998;139(3):1006–12.
183. Dempster DW et al. Glucocorticoids inhibit bone resorption by isolated rat osteoclasts by enhancing apoptosis. J Endocrinol. 1997;154(3):397–406.
184. Pereira RM, Delany AM, Canalis E. Cortisol inhibits the differentiation and apoptosis of osteoblasts in culture. Bone. 2001;28(5):484–90.
185. Stratakis CA et al. Skin manifestations of Cushing disease in children and adolescents before and after the resolution of hypercortisolemia. Pediatr Dermatol. 1998;15(4):253–8.
186. Phillips PJ, Weightman W. Skin and Cushing syndrome. Aust Fam Physician. 2007;36(7):545–7.
187. Tsatmali M et al. Skin POMC peptides: their actions at the human MC-1 receptor and roles in the tanning response. Pigment Cell Res. 2000;13 Suppl 8:125–9.
188. Kadekaro AL et al. Significance of the melanocortin 1 receptor in regulating human melanocyte pigmentation, proliferation, and survival. Ann N Y Acad Sci. 2003;994:359–65.
189. Suzuki I et al. Binding of melanotropic hormones to the melanocortin receptor MC1R on human melanocytes stimulates proliferation and melanogenesis. Endocrinology. 1996;137(5):1627–33.

Glucocorticoid Regulation of Neurocognitive and Neuropsychiatric Function

Alberto M. Pereira and Onno C. Meijer

Abstract The evolutionary conserved control of behaviour by glucocorticoids translates into a key role for glucocorticoids in the control of neuropsychological functioning. In accordance, both animal and human models of uncontrolled exposure to glucocorticoids show impaired stress responsiveness, cognitive dysfunction, and a broad spectrum of neuropsychiatric disorders, ranging from severe depression and anxiety disorders to acute psychosis and delirium. Importantly, exogenous glucocorticoid administration can induce the same phenotype, proving the causal role of glucocorticoids per se on neurocognitive and neuropsychiatric functioning. Recent findings now indicate that these effects may be long-lasting and even may not be completely reversible because cognitive dysfunction and maladaptive personality traits persist in patients long-term after successful correction of glucocorticoid excess in the presence of altered coping strategies and affected illness perceptions. This implies that long-term care for both patients with pituitary and adrenal disorders and patients using glucocorticoids should incorporate self-management interventions that help to improve quality of life

Keywords Glucocorticoids • Brain • Cortisol • Adrenal Insufficiency • Cushing's syndrome • Animal models • Neurocognitive function • Neuropsychiatric function • Coping strategies • Illness perceptions • Quality of life

Introduction: Regulation of the Stress Response (From an Evolutionary Perspective)

Evolution has provided us with powerful tools to ensure survival, and an adequate response to a stressor in this respect is fundamental. A normal stress response is a prerequisite for a normal behavioural and metabolic adaptation to the stressor. When

A.M. Pereira (✉) • O.C. Meijer
Department of Medicine, Division of Endocrinology, Leiden University Medical Center, Leiden, The Netherlands

Center for Endocrine Tumors, Leiden University Medical Center, Leiden, The Netherlands
e-mail: a.m.pereira@lumc.nl

© Springer International Publishing Switzerland 2017
E.B. Geer (ed.), *The Hypothalamic-Pituitary-Adrenal Axis in Health and Disease*, DOI 10.1007/978-3-319-45950-9_2

an individual is exposed to a stressor, the response is characterized by stimulation of the sympathetic nervous system (leading to catecholamine release) and activation of the hypothalamus–pituitary–adrenal (HPA) axis. Cortisol, or corticosterone in the rodent, is the main mediator of the adrenocortical stress response that ultimately serves only one purpose: to induce the required behavioural and metabolic adaptations enabling the individual to adequately cope with the stressor. Thus, activation of the HPA axis, and consequently, increased cortisol secretion is fundamental for modelling the stress response [1]. Corticotrophin releasing hormone (CRH), secreted from parvocellular neurons of the paraventricular nucleus (PVN) in the hypothalamus, stimulates the pituitary to release adrenocorticotropin (ACTH) after cleavage from the pro-opiomelanocortin precursor. Subsequently, activation of ACTH receptors in the adrenal cortex leads to the synthesis and secretion of glucocorticoids.

The regulation of stress-induced HPA activation occurs by so-called negative glucocorticoid feedback at the level of the anterior pituitary and hypothalamus. In clinical endocrinology, this negative feedback action exerted at the pituitary by synthetic glucocorticoids is exploited in the diagnostic workup and subsequent treatment of primary and secondary adrenal insufficiency. However, this clinical model of the HPA axis actually is a truncated model from a biological perspective, because higher centres, including brain stem catecholamines, modulate CRH production by the hypothalamus and limbic brain structures such as the amygdala [2]. This activation is of paramount importance in the responses to psychological stressors, which trigger emotional arousal and require cognitive operations for coping and storing the experience in the memory for future use. Glucocorticoids exert a strong feedback and feedforward action on these limbic forebrain areas [3]. Two nuclear receptor types mediate this action exerted by these steroids: the mineralocorticoid (MR) and the glucocorticoid receptor (GR) [1].

In addition to the well-known genomic actions of glucocorticoids, recent evidence suggests that rapid, non-genomic effects of glucocorticoids are mediated via lower affinity MR and GR variants localized in the cell membrane [4, 5]. This so-called fast negative-feedback control of glucocorticoid action appears to be mediated by another pleiotropic physiological system: the endocannabinoid system. Endocannabinoids play a pivotal role in the control of glucocorticoid action, via modulation of the excitatory action of glutamate on CRH neurons in the PVN [6]. Glutamate activation is a crucial step in the activation of the HPA axis and the inhibition of glutamate release appears to be specifically mediated by cannabinoids in the hypothalamic PVN.

Dysregulation of the activity of the HPA axis occurs when the glucocorticoid response is either inadequate, or too extreme and prolonged. This aberrant glucocorticoid response to stressors can have deleterious consequences for the organism. The inability to effectively terminate the stress response may lead to continued hypersecretion of glucocorticoids, which eventually leads to wear and tear of tissues and organs with an increased risk for metabolic and cardiovascular diseases, compromised immune responses, and psychopathology. Alternatively, an inadequate cortisol response is unable to restrain the initial stress reactions, as is the case for instance in inflammatory disorders and autoimmune diseases.

The Regulation of Emotion and Cognition by the HPA Axis (For Coping and Storing Experience in the Memory for Future use)

As stated in the introduction, the action of cortisol in the central nervous system is mediated by two steroid receptors, the mineralocorticoid (MR) and glucocorticoid receptor (GR). An appropriate balance of MR and GR activation is key for optimal control of emotion and cognition that is regulated by the limbic system. In accordance, MR and GR expression is high, especially in the hippocampus, amygdala, and prefrontal cortex [7, 8]. Basal levels of cortisol via MR stimulate neuronal excitation and determine the initial defence against the stressor, a finding that translates to vulnerability and resilience to psychiatric disease [9]. In contrast, stress-induced activation of GR coordinates the recovery, processing of information, and storage of the experience in the memory through reduction of neuronal excitation. In a general sense, these effects on excitability affect the overall activity of brain regions and circuits in ways that bias emotional and behavioural responses towards survival (e.g. by increasing likelihood of habitual rather than goal-directed responses [10]).

MR and GR activation depends foremost on binding of cortisol. High-affinity MRs are already occupied by low, basal levels of hormone, whereas GR affinity is such that substantial activation takes place during the circadian peak and after stress. Thus, mildly elevated trough levels may bias receptor activation towards the MR [11]. Intracellular prereceptor metabolism and differential tissue access are two other factors that determine cortisol levels 'seen' by the receptors [12, 13].

Next to hormone levels, absolute and relative MR/GR activation depends on expression and posttranslational modifications. Expression can vary as a consequence of genetic variation [14], early life programming effects [15], and regulation during adult life (see below). Because MRs can be considered tonically activated even at relatively low levels of cortisol, it has been argued that regulation of receptor amount is an important level of regulation. However, receptor regulation of expression is also a relevant variable for GR, for example, in view of its homologous down-regulation upon chronic hormone exposure [16].

The MR- and GR-dependent effects are not autonomous, but occur in conjunction with central stress-responsive transmitters such as noradrenalin, corticotrophin-releasing hormone (CRH), and urocortins. A prime example is the interaction between noradrenalin and glucocorticoid hormones in the amygdala and hippocampus that underlies stress-induced facilitation of memory consolidation [17]. The effects of cortisol interact with those of other neurotransmitters in two ways.

First, because cortisol affects neuronal excitability rather than neuronal firing per se [18], the effects are *permissive*: they bias how the brain responds depends on the current state of activity and demands on the system. For example, neuropsychiatric symptoms that can be induced by cortisol and its synthetic homologues include psychosis [19]. It can be expected that this particular vulnerability is highest in subjects that—in absence of any psychopathology—have high basal activity of dopaminergic signalling, or other pathways that can be causal to psychotic states. Permissive effects imply that 'moving parts' of the circuit are affected most strongly. A hypothesis that is testable is that this vulnerability becomes manifest in an interaction between high cortisol and variation in psychosis-related genes.

A second context-dependence lies in effects of neurotransmitters on functionality of the MR and GR. Animal studies have shown that activation of brain-derived neurotropic factor (BDNF) increases the phosphorylation of the GR in the hypothalamus. This in turn potentiates many effects of GR on gene expression [20]. Likewise, a prior history of stressful circumstances led to a dramatic change in the genes that were regulated in the rat hippocampus upon treatment with a single dose of corticosterone. Genome-wide analysis revealed that corticosterone could regulate the expression of around 600 genes in the hippocampus both in naïve and in chronically stressed rats. Strikingly, only 50 % of these genes were common to both groups. This implies that previous, recent history substantially remodels—via unknown mechanisms—the way in which the neuronal circuits respond to glucocorticoid exposure [21].

Animal Models of HPA Axis Disturbances

Animal studies have been indispensable to gain insight in the many effects of corticosteroids and their underlying mechanisms [22]. Of note, the sole glucocorticoid in rodents is corticosterone, which does differ from cortisol in some aspects, most notably in relation to transport into tissues [12]. Such species differences become even more pronounced when studying cortisol in the context of stress-related brain circuitry, as readouts of psychological state are necessarily indirect in rodents. A prime example has been the Porsolt forced swim test, in which active swimming/ struggling is compared to passive floating. This behaviour is surely strongly dependent on glucocorticoids, but the interpretation of these effects has been given very differently, either as inducing a depression-like state or rather as adaptive memory processes [23].

Nevertheless, animal models do give insights on the brain effects of glucocorticoids per se and on their roles as mediators of the consequences of physical and psychological stress. Classic models of glucocorticoid exposure include treatment via implanted pellets and drinking water. Such studies—in absence of stressors— have revealed many principles of feedback regulation [24] and genomic targets predominantly in the hippocampus. Many of these targets are evolutionary conserved [25]. Such studies have also shown the consequences of chronic hypercortisolemia for the morphology of neurons and size of brain areas. Earlier studies revealed the vulnerability of the hippocampus to glucocorticoid exposure, including shrinking of dendrites of the principal cells in the CA3 area and effects on adult neurogenesis in the dentate gyrus.

Of note, it is not only the overall amount of cortisol that is important, but also the pattern of exposure over the day—as is clear from the imperfections of current replacement therapies. An elegant approach to studying the importance of circadian variation has been to treat animals with low, constant levels of corticosterone, which leads to suppression of the endogenous secretion at the time of circadian peak. This regimen ensures flattened diurnal rhythms in absence of hypercorticism [26]. It has been useful to study both negative feedback and corticosterone effects on hippocampal gene expression [11, 16]. Also, the importance of the ultradian rhythm of glucocorticoid rhythms was revealed in rats, showing marked effects on behavioural and endocrine stress responsiveness that correlated with changes in neuronal activation in the amygdala . Twelve hours of constant low, rather than absence of a corticosterone rhythm led to a blunted neuronal response to an acute stressor stressor, in conjunction with a blunted ACTH response to the stressor. In this setting, also the timing of the stressor relative to the phase of ultradian peaks was of consequence, suggesting rapid feedback effects from these one-hour corticosterone peaks [27].

A last approach to study the effects of glucocorticoids on the brain makes use of the fact that dexamethasone strongly suppresses ACTH secretion at the level of the pituitary, but at low doses do not penetrate into the brain [28, 29]. In this way, a state of selective central hypocorticism can be created [30]. This approach was used to demonstrate the importance of glucocorticoid rhythmicity for the plasticity of dendritic spines—the contact points for synaptic contacts that form the structural basis for plasticity of the brain. Circadian glucocorticoid peaks allowed the formation of dendritic spine, while troughs were required for stabilizing newly formed spines, which are important for long-term memory retention [31].

The role of MR and GR in individual cell types of the brain has also been approached using transgenic methodologies, using either advanced transgenic mice [32] or local manipulation of expression in adult mice [33, 34].

There is a plethora of models for glucocorticoids as mediators of the effects of stress. Steroids in general can have either long-term programming effects, or more adaptive activational effects. In line, there are models that focus on early life stressors, stressors during adolescence, and stressors during adult life. The latter have a logical extension to any animal model for disease that is available.

Early life experience—even in utero—can have major consequences for the development of emotional reactivity in later life [35]. Consequences of early life stress often include the development of anxiety and reprogramming of the HPA axis [36, 37]. This type of programming was recognized in animal studies as early as the 1950s [38]. Many types of early life stressors have been used, ranging from 24 h separation between mother and pup to creating 'disorganized mothers' by limiting the amount of bedding material that is available to the dam [39]. The direct contribution of glucocorticoids in the development of later life changes has mainly been studied in the prenatal models, also in relationship to the barrier function of the placenta for maternal cortisol [40].

The effects of stress-induced corticosterone have also been extensively studied using rodent models. The different types of stressors differ in physical and psychological components, intensity, duration, predictability, and controllability. Much is known on the role of glucocorticoids in models for single traumatic events, based on fear-conditioning paradigms [41]. However, many clinical issues involve more chronic exposure to stress and cortisol. The often-used stressor of repeated restraint can lead to substantial habituation of at least the HPA-axis response [42], and while this is accompanied by strong changes in the brain reactivity [21], it does not model chronically elevated cortisol. Therefore, many recent studies have taken to the non-habituating models of chronic unpredictable stress [43]. Certainly, many effects observed in these models depend on elevation of glucocorticoid levels [44].

However, even if stress and glucocorticoids predispose to disease, a stress-model per se may not suffice to study particular pathologies. In this respect, there is more direct information in combining existing disease models and treatment with MR and GR agonists or antagonists. A case in point is a recent impressive study where the GR antagonist mifepristone was efficacious both in a rat model of alcohol abuse and in a group of addicted human subjects [45]. In particular, such studies using receptor antagonists (or cortisol-lowering agents [46]) point to involvement of cortisol in pathogenic processes, even in situations without an obvious or dominant stress-related component.

Human Models for the Effects of Glucocorticoids on Neuropsychological Function

Cushing's Syndrome

Cushing's syndrome is a rare endocrine disorder characterized by long-term exposure to elevated endogenous glucocorticoid levels. Cushing's syndrome is caused by either an ACTH secreting pituitary adenoma (70 % of cases), ectopic ACTH secretion (mostly bronchial carcinoids), or by autonomous cortisol hyper-secretion secondary to an adrenal adenoma/carcinoma, or adrenal hyperplasia. Cushing's syndrome can also be induced by long-term administration of supraphysiological doses of synthetic corticosteroids, as is prescribed in clinical practice for a variety of inflammatory conditions and autoimmune diseases. This so-called exogenous Cushing's syndrome is highly prevalent and insufficiently recognized in routine clinical practice, especially in the milder cases.

In accordance with the earlier described biological effects of glucocorticoids, the vast majority of patients with Cushing's syndrome have both physical and psychological morbidity [47]. In patients with active or uncontrolled disease, neurocognitive function (that includes cognition, mood, and personality) is affected, and psychopathology is also often observed. In active Cushing's syndrome, the frequency of psychiatric symptoms was reported starting in the early 1980s, demonstrating that symptoms like irritability, depressed mood, and anxiety were present in the majority of the patients [48]. In accordance, depression was present in more than 50 % of patients in a large cohort of patients with Cushing's disease reported by

Sonino and colleagues, and was significantly associated with older age, female sex, higher pretreatment urinary cortisol levels, a more severe clinical condition, and no pituitary adenoma on pituitary imaging [49]. Intriguingly, an increased overall psychiatric disability score was associated with increased cortisol secretion. In addition, patients with active Cushing's syndrome report cognitive impairments, like memory problems and lack of concentration [50, 51]. Thus, the most common comorbid disorder is major depression, and a severe clinical presentation of Cushing's often also includes depression (though to a lesser extent mania and anxiety disorders have also been reported). These observations are in line with the pivotal evolutionary role ascribed to cortisol in the control of mood and behaviour. Because limbic structures like the hippocampus and the prefrontal cortex are rich in glucocorticoid-receptors, these clinical observations suggest that these structures are particularly vulnerable to the cortisol excess as is present in Cushing's syndrome.

The limited numbers of patients who have been reported after treatment indicate that a significant improvement occurs within the first year after treatment [52, 53]. In addition, reduction of glucocorticoid synthesis or action, either with metyrapone, ketoconazole, or mifepristone, rather than treatment with antidepressant drugs, is generally successful in relieving depressive symptoms, as well as other disabling symptoms [54, 55]. Thus, following successful correction of hypercortisolism, both physical and psychiatric signs and symptoms improve substantially. In the long-term, however, it now becomes evident from an accumulating number of studies that patients do not completely return to their premorbid level of functioning. These studies demonstrated residual physical and psychopathological morbidity despite long-term biochemical remission [56–59]. In addition, patients with long-term remission of CD reported persistent impairments in cognitive functioning [58, 60] and a reduced quality of life [61]. To which extent psychopathology still affects general well-being after long-term cure of CS is still, however, not clear.

An emerging topic of interest in this respect is the relation between glucocorticoid excess and changes in brain structure and function, and consequently, its relation with neuropsychological dysfunction.

The first observations in the human indicating that long-term exposure to elevated glucocorticoids may affect the brain were reported by Lupien and colleagues [62]. In that particular study, exposure to prolonged elevated cortisol levels in aged humans led to reduced hippocampal volumes as well as memory deficits (when compared to controls with normal cortisol levels). In later studies, however (in healthy young men), a larger hippocampal volume got associated with a greater cortisol response both in a social stress test (Trier social stress test) and in the cortisol awakening response, questioning the relevance of the former finding in aged individuals for younger individuals [63]. Many psychiatric diseases, like major depressive and bipolar disorder, have been linked to alterations in the HPA axis [64, 65], and GC receptor polymorphisms that alter glucocorticoid sensitivity have been associated with depression (reviewed in [66]). In addition, other studies in patients with psychiatric diseases indicate that limbic structure volumes, like the hippocampus and the amygdala, are smaller

[67, 68], though these changes may also be associated with brain aging and interact with the progression of the disorder [69].

The effects of Cushing's syndrome on the brain, reflecting long-term excessive overexposure to endogenous cortisol, were recently reported in a systematic review [52]. This review systematically evaluated all studies in patients with active and remitted Cushing's disease or syndrome using MRI ($n = 19$). These studies demonstrated that structural abnormalities in the grey matter were present in patients with active disease, which were characterized by smaller hippocampal volumes, enlarged ventricles, and cerebral atrophy (see also: [70]). In addition, functional changes occurred, characterized by alterations in neurochemical concentrations and functional activity. Intriguingly, the reversibility of structural and neurochemical alterations after correction of cortisol excess was incomplete, even when patients were evaluated after long-term remission. The structural alterations after long-term remission included smaller grey matter volumes of the anterior cingulate cortex, greater grey matter volume of the left posterior lobe of the cerebellum [71], and widespread reductions in white matter integrity [72, 73]. Long-lasting functional alterations included increased resting state functional connectivity between the limbic network and the subgenual subregion of the anterior cingulate cortex [74] and altered neural processing of emotional faces [75]. Some findings as obtained using MRI were related to the severity of the cortisol excess, and others also to neuropsychological functioning (as reflected by mood, cognition, and emotional functioning) and quality of life. This points towards persistent changes in brain function after previous exposure to hypercortisolism.

Adrenal Insufficiency

Adrenal insufficiency per se, by definition, will result in impaired stress responsiveness. In the human, this is best exemplified by the clinical application of the insulin tolerance test that is considered the golden standard for the diagnosis of adrenal insufficiency. The test is based upon induction of the stress response by insulin-induced hypoglycaemia, which from an evolutionary perspective is one of the most potent physiological stressors because it is potentially lethal. In accordance, the response to severe hypoglycaemia is characterized both by a sympathetic noradrenergic response (tachycardia, agitation, sweating, etc.) and stimulation of cortisol secretion through activation of the HPA axis. Patients with adrenal insufficiency (regardless the cause) are not able to secrete sufficient cortisol after hypoglycaemia (and fail this test). The subsequent metabolic and behavioural adaptations orchestrated by cortisol via the mineralo- and glucocorticoid receptor are not or insufficiently induced. Thus, by definition, these patients exhibit impaired stress responsiveness, and in accordance, even patients with adrenal insufficiency that were on stable hydrocortisone replacement reported impairments in quality of life [76–78].

Cognitive function in patients with adrenal insufficiency on hydrocortisone replacement has been reported only in seven studies involving a total of 195 patients [79–85]. These studies indicate that mild cognitive deficits may persist, especially in memory and executive functioning tasks. Intriguingly, patients performed better on concentration and attentional tasks when compared with controls [83], and cognitive function was neither affected by the dose used (high vs. low daily dose) [85], nor by postponement of the first daily dose by a few hours [83].

Besides cognition, neurocognitive functioning also includes mood and personality. Adrenal insufficiency may present solely with psychiatric manifestations [86, 87] and epidemiological studies indicate that patients with adrenal insufficiency may be at increased risk of developing severe affective disorders. When hospitalized patients with Addison's disease were compared to hospitalized patients with osteoarthritis, the former had a more than two times greater rate of affective disorders and 1.7 times greater rate of depressive disorders [88]. In the Leiden cohort, we observed more psychosocial morbidity (irritability and somatic arousal) in the presence of impairments in quality of life when patients with adrenal insufficiency were compared with controls. Patients and controls did not differ regarding maladaptive personality traits; however, the daily hydrocortisone dose proved to be strongly associated both with the prevalence of maladaptive personality traits and with depression [78].

Patients Using Glucocorticoids

Glucocorticoids are frequently prescribed for various conditions like chronic obstructive pulmonary diseases and autoimmune diseases to inhibit the inflammatory response. Soon after their introduction in the 1950s, the first cases were reported on severe neuropsychiatric manifestations after the initiation of glucocorticoid therapy [89, 90]. In agreement with the studies in endogenous CS reported by Sonino and colleagues, more than 50 % of patients exposed to glucocorticoids for more than 3 months developed neuropsychiatric symptoms/manifestations [91]. A recent review beautifully summarized the topic of the adverse neuropsychological consequences of glucocorticoid therapy [19]. The acute and long-term effects on both mood and cognition have been studied in prospective studies, and the severe neuropsychiatric effects in case studies and with the use of epidemiological databases [92]. The observed rates and spectrum of manifestations of depression, anxiety disorders, and cognitive dysfunction are similar to those as observed in endogenous Cushing' syndrome and exemplifies that glucocorticoids can induce the same neuropsychological phenotype (in pre-disposed individuals). The most prominent risk factors identified were gender (male patients being more prone to develop mania and delirium, and female patients being more prone for depression), a past history for psychiatric disorders, and the initial daily glucocorticoid dose (in general above 40 mg of prednisone daily equivalent). Finally, withdrawal from long-term glucocorticoid therapy also

Table 1 Functional domains affecting cognition, mood, and behaviour related to abnormal glucocorticoid exposure

	Disease state	Cognitive function	Psychopathology	Coping Strategies and illness perceptions
Cushing's syndrome or Cushing's disease	Active or uncontrolled disease	Impaired	Major depression, anxiety disorders, acute psychosis, delirium	Not studied
	Long-term remission	Subtle cognitive impairments	Maladaptive personality traits, especially: Apathy, irritability, anxiety, negative affect, and lack of positive affect	Less effective coping More negative illness perceptions
Adrenal insufficiency	Disease manifestation	Not studied	Depression, anxiety During Addisonian crisis: agitation, delirium, visual- and auditory hallucinations	Not studied
	After long-term glucocorticoid replacement	Mild cognitive deficits	Irritability and somatic arousal, strong relation with daily hydrocortisone dose	Less effective coping More negative illness perceptions, associated with concerns and stronger beliefs about the necessity of hydrocortisone intake
Glucocorticoid use		Not studied	Depression Delirium/confusion/disorientation Mania Panic disorder Suicidal behaviour	Not studied

Cognitive function: various tests, which evaluated global cognitive functioning, memory, and executive functioning
Psychopathology: evaluated using questionaires, focussing on frequently occurring psychiatric symptoms in somatic illness
Coping strategies: compared with the normal population
Illness perceptions: evaluated using the Illness Perception Questionnaire (IPQ) Revised. Illness perceptions showed a strong correlation with quality of life

increases the risk for severe psychiatric manifestations. Again, a past history of psychiatric disease and also the use of long-acting glucocorticoids (especially dexamethasone) increased the risk for depression and delirium following discontinuation of glucocorticoid therapy [93].

Summary and Conclusions

Glucocorticoids play a key role in the control of neuropsychological functioning, which is exemplified by the evolutionary conserved control of behaviour in the 'fight or flight response'. In accordance, both animal and human models of uncontrolled (and therefore abnormal) exposure to glucocorticoids show impaired stress responsiveness, cognitive dysfunction, and a broad spectrum of neuropsychiatric disorders, ranging from severe depression and anxiety disorders to acute psychosis and delirium (for a summary, see Table 1). The fact that the same phenotype can be induced by exogenous glucocorticoid administration proves the causal role of glucocorticoids per se on neurocognitive and neuropsychiatric functioning. Finally, it now becomes clear that these effects may be long-lasting and even may not be completely reversible because cognitive dysfunction and maladaptive personality traits persist in the presence of altered coping strategies and affected illness perceptions despite long-term optimal treatment. This implies that long-term care for both patients with pituitary and adrenal disorders and patients using glucocorticoids should incorporate self-management interventions that help to improve quality of life.

References

1. de Kloet ER, Joels M, Holsboer F. Stress and the brain: from adaptation to disease. Nat Rev Neurosci. 2005;6(6):463–75.
2. McCall JG, et al. CRH engagement of the locus coeruleus noradrenergic system mediates stress-induced anxiety. Neuron. 2015;87(3):605–20.
3. Laugero KD, et al. Corticosterone infused intracerebroventricularly inhibits energy storage and stimulates the hypothalamo-pituitary axis in adrenalectomized rats drinking sucrose. Endocrinology. 2002;143(12):4552–62.
4. Jiang CL, Liu L, Tasker JG. Why do we need nongenomic glucocorticoid mechanisms? Front Neuroendocrinol. 2014;35(1):72–5.
5. Karst H, et al. Mineralocorticoid receptors are indispensable for nongenomic modulation of hippocampal glutamate transmission by corticosterone. Proc Natl Acad Sci U S A. 2005;102(52):19204–7.
6. Evanson NK, et al. Fast feedback inhibition of the HPA axis by glucocorticoids is mediated by endocannabinoid signaling. Endocrinology. 2010;151(10):4811–9.
7. Datson NA, et al. Identification of corticosteroid-responsive genes in rat hippocampus using serial analysis of gene expression. Eur J Neurosci. 2001;14(4):675–89.
8. de Kloet ER, et al. Brain mineralocorticoid receptors and centrally regulated functions. Kidney Int. 2000;57(4):1329–36.
9. Klok MD, et al. A common and functional mineralocorticoid receptor haplotype enhances optimism and protects against depression in females. Transl Psychiatry. 2011;1(12), e62.

10. Sousa N, Almeida OF. Disconnection and reconnection: the morphological basis of (mal) adaptation to stress. Trends Neurosci. 2012;35(12):742–51.
11. Meijer OC, Van Oosten RV, De Kloet ER. Elevated basal trough levels of corticosterone suppress hippocampal 5-hydroxytryptamine(1A) receptor expression in adrenally intact rats: implication for the pathogenesis of depression. Neuroscience. 1997;80(2):419–26.
12. Karssen AM, et al. Multidrug resistance P-glycoprotein hampers the access of cortisol but not of corticosterone to mouse and human brain. Endocrinology. 2001;142(6):2686–94.
13. Wyrwoll CS, Holmes MC, Seckl JR. 11β-hydroxysteroid dehydrogenases and the brain: from zero to hero, a decade of progress. Front Neuroendocrinol. 2011;32(3):265–86.
14. van Leeuwen N, et al. Human mineralocorticoid receptor (MR) gene haplotypes modulate MR expression and transactivation: implication for the stress response. Psychoneuroendocrinology. 2011;36(5):699–709.
15. Turecki G, Meaney MJ. Effects of the social environment and stress on glucocorticoid receptor gene methylation: a systematic review. Biol Psychiatry. 2016;79(2):87–96.
16. Sarabdjitsingh RA, et al. Disrupted corticosterone pulsatile patterns attenuate responsiveness to glucocorticoid signaling in rat brain. Endocrinology. 2010;151(3):1177–86.
17. Roozendaal B, McGaugh JL. Memory modulation. Behav Neurosci. 2011;125(6):797–824.
18. Joels M, Sarabdjitsingh RA, Karst H. Unraveling the time domains of corticosteroid hormone influences on brain activity: rapid, slow, and chronic modes. Pharmacol Rev. 2012;64(4):901–38.
19. Judd LL, et al. Adverse consequences of glucocorticoid medication: psychological, cognitive, and behavioral effects. Am J Psychiatry. 2014;171(10):1045–51.
20. Lambert WM, et al. Brain-derived neurotrophic factor signaling rewrites the glucocorticoid transcriptome via glucocorticoid receptor phosphorylation. Mol Cell Biol. 2013;33(18):3700–14.
21. Datson NA, et al. Previous history of chronic stress changes the transcriptional response to glucocorticoid challenge in the dentate gyrus region of the male rat hippocampus. Endocrinology. 2013;154(9):3261–72.
22. de Kloet ER, et al. Glucocorticoid signaling and stress-related limbic susceptibility pathway: about receptors, transcription machinery and microRNA. Brain Res. 2009;1293:129–41.
23. Molendijk ML, de Kloet ER. Immobility in the forced swim test is adaptive and does not reflect depression. Psychoneuroendocrinology. 2015;623:89–91.
24. Dallman MF, et al. Regulation of ACTH secretion: variations on a theme of B. Recent Prog Horm Res. 1987;43:113–73.
25. Datson NA, et al. Specific regulatory motifs predict glucocorticoid responsiveness of hippocampal gene expression. Endocrinology. 2011;152(10):3749–57.
26. Akana SF, et al. Feedback sensitivity of the rat hypothalamo-pituitary-adrenal axis and its capacity to adjust to exogenous corticosterone. Endocrinology. 1992;131(2):585–94.
27. Sarabdjitsingh RA, et al. Stress responsiveness varies over the ultradian glucocorticoid cycle in a brain-region-specific manner. Endocrinology. 2010;151(11):5369–79.
28. De Kloet R, Wallach G, McEwen BS. Differences in corticosterone and dexamethasone binding to rat brain and pituitary. Endocrinology. 1975;96(3):598–609.
29. Meijer OC, et al. Penetration of dexamethasone into brain glucocorticoid targets is enhanced in mdr1A P-glycoprotein knockout mice. Endocrinology. 1998;139(4):1789–93.
30. Karssen AM, et al. Low doses of dexamethasone can produce a hypocorticosteroid state in the brain. Endocrinology. 2005;146(12):5587–95.
31. Liston C, et al. Circadian glucocorticoid oscillations promote learning-dependent synapse formation and maintenance. Nat Neurosci. 2013;16(6):698–705.
32. Ambroggi F, et al. Stress and addiction: glucocorticoid receptor in dopaminoceptive neurons facilitates cocaine seeking. Nat Neurosci. 2009;12(3):247–9.
33. Fitzsimons CP, et al. Knockdown of the glucocorticoid receptor alters functional integration of newborn neurons in the adult hippocampus and impairs fear-motivated behavior. Mol Psychiatry. 2013;18(9):993–1005.
34. Kolber BJ, et al. Central amygdala glucocorticoid receptor action promotes fear-associated CRH activation and conditioning. Proc Natl Acad Sci U S A. 2008;105(33):12004–9.

35. Bock J, et al. Stress in utero: prenatal programming of brain plasticity and cognition. Biol Psychiatry. 2015;78(5):315–26.
36. Klengel T, Binder EB. Epigenetics of stress-related psychiatric disorders and gene × environment interactions. Neuron. 2015;86(6):1343–57.
37. Schmidt MV, Wang XD, Meijer OC. Early life stress paradigms in rodents: potential animal models of depression? Psychopharmacology (Berl). 2011;214(1):131–40.
38. Levine S. Infantile experience and resistance to physiological stress. Science. 1957;126(3270):405.
39. Molet J, et al. Naturalistic rodent models of chronic early-life stress. Dev Psychobiol. 2014;56(8):1675–88.
40. Chapman K, Holmes M, Seckl J. 11β-hydroxysteroid dehydrogenases: intracellular gatekeepers of tissue glucocorticoid action. Physiol Rev. 2013;93(3):1139–206.
41. Kaouane N, et al. Glucocorticoids can induce PTSD-like memory impairments in mice. Science. 2012;335(6075):1510–3.
42. Grissom N, Bhatnagar S. Habituation to repeated stress: get used to it. Neurobiol Learn Mem. 2009;92(2):215–24.
43. Willner P, et al. Reduction of sucrose preference by chronic unpredictable mild stress, and its restoration by a tricyclic antidepressant. Psychopharmacology (Berl). 1987;93(3):358–64.
44. Joels M, et al. Chronic stress: implications for neuronal morphology, function and neurogenesis. Front Neuroendocrinol. 2007;28(2-3):72–96.
45. Vendruscolo LF, et al. Glucocorticoid receptor antagonism decreases alcohol seeking in alcohol-dependent individuals. J Clin Invest. 2015;125(8):3193–7.
46. Sooy K, et al. Cognitive and disease-modifying effects of 11β-hydroxysteroid dehydrogenase type 1 inhibition in male Tg2576 mice, a model of Alzheimer's disease. Endocrinology. 2015;156(12):4592–603.
47. Newell-Price J, et al. Cushing's syndrome. Lancet. 2006;367(9522):1605–17.
48. Starkman MN, Schteingart DE. Neuropsychiatric manifestations of patients with Cushing's syndrome. Relationship to cortisol and adrenocorticotropic hormone levels. Arch Intern Med. 1981;141(2):215–9.
49. Sonino N, et al. Clinical correlates of major depression in Cushing's disease. Psychopathology. 1998;31(6):302–6.
50. Starkman MN, et al. Elevated cortisol levels in Cushing's disease are associated with cognitive decrements. Psychosom Med. 2001;63(6):985–93.
51. Webb SM, et al. Evaluation of health-related quality of life in patients with Cushing's syndrome with a new questionnaire. Eur J Endocrinol. 2008;158(5):623–30.
52. Andela CD, et al. Mechanisms in endocrinology: Cushing's syndrome causes irreversible effects on the human brain: a systematic review of structural and functional magnetic resonance imaging studies. Eur J Endocrinol. 2015;173(1):R1–14.
53. Hook JN, et al. Patterns of cognitive change over time and relationship to age following successful treatment of Cushing's disease. J Int Neuropsychol Soc. 2007;13(1):21–9.
54. Jeffcoate WJ, et al. Psychiatric manifestations of Cushing's syndrome: response to lowering of plasma cortisol. Q J Med. 1979;48(191):465–72.
55. Sonino N, Fava GA. Psychiatric disorders associated with Cushing's syndrome. Epidemiology, pathophysiology and treatment. CNS Drugs. 2001;15(5):361–73.
56. Dorn LD, et al. The longitudinal course of psychopathology in Cushing's syndrome after correction of hypercortisolism. J Clin Endocrinol Metab. 1997;82(3):912–9.
57. Milian M, et al. Similar psychopathological profiles in female and male Cushing's disease patients after treatment but differences in the pathogenesis of symptoms. Neuroendocrinology. 2014;100(1):9–16.
58. Resmini E, et al. Verbal and visual memory performance and hippocampal volumes, measured by 3-Tesla magnetic resonance imaging, in patients with Cushing's syndrome. J Clin Endocrinol Metab. 2012;97(2):663–71.
59. Tiemensma J, et al. Increased prevalence of psychopathology and maladaptive personality traits after long-term cure of Cushing's disease. J Clin Endocrinol Metab. 2010;95(10):E129–41.

60. Tiemensma J, et al. Subtle cognitive impairments in patients with long-term cure of Cushing's disease. J Clin Endocrinol Metab. 2010;95(6):2699–714.
61. van Aken MO, et al. Quality of life in patients after long-term biochemical cure of Cushing's disease. J Clin Endocrinol Metab. 2005;90(6):3279–86.
62. Lupien SJ, et al. Cortisol levels during human aging predict hippocampal atrophy and memory deficits. Nat Neurosci. 1998;1(1):69–73.
63. Pruessner M, et al. The associations among hippocampal volume, cortisol reactivity, and memory performance in healthy young men. Psychiatry Res. 2007;155(1):1–10.
64. Antonijevic IA. Depressive disorders—is it time to endorse different pathophysiologies? Psychoneuroendocrinology. 2006;31(1):1–15.
65. Belvederi Murri M, et al. The HPA axis in bipolar disorder: systematic review and meta-analysis. Psychoneuroendocrinology. 2016;63:327–42.
66. Spijker AT, van Rossum EF. Glucocorticoid receptor polymorphisms in major depression. Focus on glucocorticoid sensitivity and neurocognitive functioning. Ann N Y Acad Sci. 2009;1179:199–215.
67. Harrisberger F, et al. BDNF Val66Met polymorphism and hippocampal volume in neuropsychiatric disorders: a systematic review and meta-analysis. Neurosci Biobehav Rev. 2015;55:107–18.
68. Malykhin NV, Coupland NJ. Hippocampal neuroplasticity in major depressive disorder. Neuroscience. 2015;309:200–13.
69. Alves GS, et al. Structural neuroimaging findings in major depressive disorder throughout aging: a critical systematic review of prospective studies. CNS Neurol Disord Drug Targets. 2014;13(10):1846–59.
70. Burkhardt T, et al. Hippocampal and cerebellar atrophy in patients with Cushing's disease. Neurosurg Focus. 2015;39(5), E5.
71. Andela CD, et al. Smaller grey matter volumes in the anterior cingulate cortex and greater cerebellar volumes in patients with long-term remission of Cushing's disease: a case-control study. Eur J Endocrinol. 2013;169(6):811–9.
72. Pires P, et al. White matter alterations in the brains of patients with active, remitted, and cured cushing syndrome: a DTI study. AJNR Am J Neuroradiol. 2015;36(6):1043–8.
73. van der Werff SJ, et al. Widespread reductions of white matter integrity in patients with long-term remission of Cushing's disease. Neuroimage Clin. 2014;46:59–67.
74. van der Werff SJ, et al. Resting-state functional connectivity in patients with long-term remission of Cushing's disease. Neuropsychopharmacology. 2015;40(8):1888–98.
75. Bas-Hoogendam JM, et al. Altered neural processing of emotional faces in remitted Cushing's disease. Psychoneuroendocrinology. 2015;59:134–46.
76. Aulinas A, Webb SM. Health-related quality of life in primary and secondary adrenal insufficiency. Expert Rev Pharmacoecon Outcomes Res. 2014;14(6):873–88.
77. Bancos I, et al. Diagnosis and management of adrenal insufficiency. Lancet Diabetes Endocrinol. 2015;3(3):216–26.
78. Tiemensma J, et al. Psychological morbidity and impaired quality of life in patients with stable treatment for primary adrenal insufficiency: cross-sectional study and review of the literature. Eur J Endocrinol. 2014;171(2):171–82.
79. Harbeck B, Kropp P, Monig H. Effects of short-term nocturnal cortisol replacement on cognitive function and quality of life in patients with primary or secondary adrenal insufficiency: a pilot study. Appl Psychophysiol Biofeedback. 2009;34(2):113–9.
80. Henry M, Thomas KG, Ross IL. Episodic memory impairment in Addison's disease: results from a telephonic cognitive assessment. Metab Brain Dis. 2014;29(2):421–30.
81. Klement J, et al. Effects of glucose infusion on neuroendocrine and cognitive parameters in Addison disease. Metabolism. 2009;58(12):1825–31.
82. Schultebraucks K, et al. Cognitive function in patients with primary adrenal insufficiency (Addison's disease) and the role of mineralocorticoid receptors. Psychoneuroendocrinology. 2015;55:1–7.
83. Tiemensma J, et al. Mild cognitive deficits in patients on stable treatment for primary adrenal insufficiency. Psychoneuroendocrinology. 2015;61:46.

84. Tytherleigh MY, Vedhara K, Lightman SL. Mineralocorticoid and glucocorticoid receptors and their differential effects on memory performance in people with Addison's disease. Psychoneuroendocrinology. 2004;29(6):712–23.
85. Werumeus Buning J, et al. The effects of two different doses of hydrocortisone on cognition in patients with secondary adrenal insufficiency—results from a randomized controlled trial. Psychoneuroendocrinology. 2015;55:36–47.
86. Anglin RE, Rosebush PI, Mazurek MF. The neuropsychiatric profile of Addison's disease: revisiting a forgotten phenomenon. J Neuropsychiatry Clin Neurosci. 2006;18(4):450–9.
87. Pavlovic A. Sivakumar V. Hypoadrenalism presenting as a range of mental disorders. BMJ Case Rep. 2011. pii: bcr0920103305. doi:10.1136/bcr.09.2010.3305.
88. Thomsen AF, et al. The risk of affective disorders in patients with adrenocortical insufficiency. Psychoneuroendocrinology. 2006;31(5):614–22.
89. Manzini B. Psychotic reactions during prednisone therapy. Riv Sper Freniatr Med Leg Alien Ment. 1958;82(2):417–29.
90. Piguet B. Study of attacks of tetany and psychological disorders appearing during adrenal cortex hormone therapy: attacks of tetany and grave psychoses initiated by substitution of delta-cortisone for hydrocortisone and subsequently by ACTH. Rev Rhum Mal Osteoartic. 1958;25(12):814–28.
91. Fardet L, et al. Corticosteroid-induced clinical adverse events: frequency, risk factors and patient's opinion. Br J Dermatol. 2007;157(1):142–8.
92. Fardet L, Petersen I, Nazareth I. Suicidal behavior and severe neuropsychiatric disorders following glucocorticoid therapy in primary care. Am J Psychiatry. 2012;169(5):491–7.
93. Fardet L, et al. Severe neuropsychiatric outcomes following discontinuation of long-term glucocorticoid therapy: a cohort study. J Clin Psychiatry. 2013;74(4):e281–6.

Glucocorticoids: Inflammation and Immunity

Maria G. Petrillo, Carl D. Bortner, and John A. Cidlowski

Abstract Glucocorticoids are universally prescribed as the drug of choice for the treatment of inflammatory and autoimmune disorders. These stress hormones act through their cognate glucocorticoid receptor to regulate transcription of various target genes. The mechanisms of glucocorticoid action are often cell type dependent and involve the regulation of thousands of genes. Glucocorticoids have a tremendous impact on the immune system during inflammation including effects on the plasticity, survival, and function of immune cells. This chapter highlights the dynamic effects of glucocorticoids in regards to both physiological and pathological conditions during inflammation. We address issues involving classical and alternative mechanisms of glucocorticoid inhibition, the effect on innate and adaptive immunity, glucocorticoid tissue-specific actions, and their role in target immune cells.

Keywords Glucocorticoid • Steroids • Stress hormones • Inflammation • Gene expression • Transrepression • Transactivation • Innate immunity • Adaptive immunity • Resistance

Introduction

Glucocorticoids are primary stress hormones that function throughout the body to regulate a diverse array of physiological systems. Glucocorticoids (GCs) derived their name from early observations of their effect in regulating glucose metabolism [1, 2]. Currently, the actions of this class of steroids extend beyond the mobilization of amino acids and gluconeogenesis and are known to play important roles in the control/regulation of various biological processes. In fact, glucocorticoids are required for life, as the absence of these stress hormones results in death prior to or at the time of birth. Glucocorticoids influence a number of physiological systems including the

M.G. Petrillo • C.D. Bortner • J.A. Cidlowski (✉)
Signal Transduction Laboratory, National Institute of Environmental Health Sciences,
Department of Health and Human Services, National Institutes of Health,
111 T. W. Alexander Dr., Research Triangle Park, NC, USA
e-mail: cidlowski@niehs.nih.gov

© Springer International Publishing Switzerland 2017
E.B. Geer (ed.), *The Hypothalamic-Pituitary-Adrenal Axis in Health and Disease*, DOI 10.1007/978-3-319-45950-9_3

43

immune system [3] where in addition to exerting both pro- and anti-inflammatory actions, this stress hormone has a potent role in development and homeostasis of T lymphocytes [4]. Additionally, these stress hormones impact development [5], where glucocorticoids are known to play a key role in the maturation of the fetal lung [6]. Furthermore, glucocorticoids have a role in the brain where they have been shown to regulate arousal and cognitive functions controlled by the hippocampus, amygdala, and the frontal lobes of the brain [7]. The pleiotropic actions of glucocorticoids occur through binding to its cognate receptor, the glucocorticoid receptor (GR), which is expressed in nearly every cell in the body. Glucocorticoids act through its receptor to regulate transcription of various target genes, however as will be discussed later, non-genomic effects have also been described.

Human GR protein is encoded by the *NR3C1* gene located in chromosome 5 (5q31) and is a member of the nuclear receptor superfamily of ligand-dependent transcription factors [8]. Like other steroid receptors, GR is modular in structure containing an N-terminal regulatory domain (NTD), a central DNA-binding domain (DBD), a hinge region, and a C-terminal ligand-binding domain (LBD) [9]. In the absence of hormone, GR resides in the cytoplasm in a complex with other proteins including heat shock protein 90, heat shock protein 70, and FKBP52, the latter being an immunophilin molecule involved in protein folding and trafficking. Upon ligand binding, GR is released from its cytoplasmic complex and translocates into the nucleus where it interacts with specific targeting sequences termed glucocorticoid-response elements (GREs) to regulate thousands of genes. The nature of the GR-occupied GRE results in either induction or repression of target gene expression. Additionally, GR can undergo a conformational change upon binding to the GRE that leads to the recruitment of cofactors and/or coregulators to modulate, and thereby alter the transcriptional rates of target genes [10]. Along with the nature of the GRE and the recruitment of cofactors and/or coregulators, several other factors can also influence GRs ability to regulate gene transcription including chromatin structure, epigenetic regulators, and its physical interaction with other transcription factors such as nuclear factor kappa B (NF-kB) and activator protein 1 (AP-1).

GR, while derived from a single gene, has multiple functionally distinct isoforms due to alternative splicing and translational initiation mechanisms [11]. Alternative splicing accounts for 2 discrete receptor isoforms (GRα and GRβ) that differ at their carboxyl termini, while alternative translation initiation results in 8 additional receptor subtypes, each with a progressively shorter NTD. GRα has been the primary and most extensively studied glucocorticoid receptor, as the GRβ splice variant does not bind GCs [12]. However, expression of GRβ has been associated with glucocorticoid resistance and tissue specificity, as GRβ has been shown to antagonize the activity of GRα [13, 14]. In contrast to GRβ, the 8 unique translational isoforms of GRα have similar binding affinities for glucocorticoids and can interact with GREs. Similar to GRβ, the expression of these various translational GRα isoforms varies widely across tissues. Thus, the existence of numerous GR isoforms is thought to be a major factor contributing to the diverse array of tissue-specific actions of glucocorticoids in the body.

Glucocorticoids are also known to have potent immunosuppressive and anti-inflammatory actions, thus being vital in the treatment of autoimmune and inflammatory diseases and are one of the most widely prescribed drugs in the world. In this review, we will focus on the how glucocorticoids modulate and interact with the immune system, along with its effect on combating inflammation. Additionally, we will discuss how glucocorticoids affect the response and behavior of different immune cells in the management of inflammatory diseases.

Inflammation as a Natural Host Defense Mechanism and Glucocorticoid Regulation

Inflammation is an innate defensive mechanism that protects us from damaging stimuli such as pathogens and harmful irritants. Inflammation is a complex biological process that initially involves increased blood flow and movement of immune cells, especially granulocytes and macrophages along with other molecular mediators, from the blood to the site of injury. This acute response sets the stage for the healing process by combating the initial source of inflammation through the removal of necrotic cells and damaged tissue in a coordinated response involving both the immune and vascular systems. As the inflammation process continues, a shift in the type of cells present at the site of injury results in the repair or healing of the tissue.

Classic signs of inflammation include pain, redness, swelling, warmth, and loss of function or immobility at the site of damage. Additionally, inflammation may result in more global symptoms such as fever, chills, fatigue, and general stiffness. Pain expressly plays an important role in the ability of glucocorticoids to regulate inflammation. Cytokines and inflammatory mediators released into the blood from the damage site activate peripheral pain receptors. Pain signals sent to the brain result in the activation of the hypothalamic–pituitary–adrenal (HPA) axis which, in turn, induces glucocorticoid secretion. GCs inhibit the synthesis of cytokines and inflammatory mediators to counter the extent of the inflammation. Despite the initial unfavorable effects of this condition, inflammation is extremely important and beneficial to human health as infections and wounds, or any damage tissue, would not heal without this homeostatic response.

As with all physiological responses in the body, inflammation needs to be regulated especially in conjunction with other host defense systems. Too little inflammation can result in progressive and detrimental tissue destruction, while excessive inflammation can lead to a host of diseases including allergies, autoimmune disorders, chronic inflammatory diseases, and even cancer. Glucocorticoids regulate and reduce the inflammatory response by entering cells and suppressing the transcription of proteins that promote inflammation. In the absence of glucocorticoids, persistent inflammation can lead to dysregulation of converging pro- and anti-inflammatory factors at the site of injury resulting in abnormalities and pathogenesis [15].

Since glucocorticoids are known to be essential for the regulation of the inflammatory response, they also act to reduce the extent of an overactive immune system. Thus, glucocorticoids are among the most widely prescribed drugs for the treatment of asthma, allergies, and autoimmune diseases. This class of steroid hormones initiates a multitude of diverse signaling pathways that hold inflammation in check and counter a rampant immune system, limiting the excessive damage that can occur to the host cells and surrounding tissue [16]. A major barrier in employing glucocorticoid therapy in the clinic has been our lack of understanding of the molecular mechanisms that resolves inflammation. At one level, the mechanism of glucocorticoid action in counteracting inflammation may appear simplistic as this class of drug (GC) acting on its receptor (GR) modulates gene transcription to inhibit the extent of inflammation. However, the heterogeneity of glucocorticoid receptor isoforms and the cell-type specific biological responses suggest that GR's ability to prevent inflammation is not a simple endeavor but a complex series of events. Thus, there are many ways glucocorticoids exert their anti-inflammatory effects.

Classical Mechanisms of Glucocorticoid Inhibition of Inflammation

Glucocorticoids utilize a variety of processes simultaneously to control inflammation; from the activation of anti-inflammatory genes, to suppressing proinflammatory cytokines and chemokines, to moderating key proinflammatory regulators such as NF-kB and AP-1. Several fundamental mechanisms have been elucidated for these actions of GR. First, direct binding of GR to GREs in DNA can enhance the transcription of anti-inflammatory genes (transactivation). Glucocorticoid-induced transactivation of genes such as IL-10, IL-1 receptor antagonist, and mitogen-activated protein kinase phosphatase-1 (MKP-1) is known to increase gene expression and thus protein expression of these anti-inflammatory molecules [17, 18]. Second, direct binding of GR to negative GREs (nGRE) in DNA can suppress transcription (transrepression) of various proinflammatory cytokines and modulators such as iNOS, COX-2, IL-1β, and TNF [17, 18]. Finally, GR binding directly to transcription factors like the p65 subunit of NF-kB or AP-1 can prevent downstream transcription of proinflammatory mediators to control the extent of inflammation [19]. This latter mechanism of transcriptional repression, known as tethering, where GR does not directly bind DNA response sequences, has been shown to be key in GRs ability to regulate inflammation [20–22]. Additionally, cross-talk between GR and other transcription factors can occur through the binding to composite or overlapping response elements [23]. Interestingly, while the repression of NF-kB by GR has long been considered a crucial determinant in reducing the expression of specific proinflammatory targets, a recent study by Altonsy et al. showed the cooperative association between the GR and NF-kB enhanced the expression of TNFAIP3, a potent anti-inflammatory gene and inhibitor of NF-kB, suggesting a greater complexity in the cross-talk of these two molecules [24].

GR binding directly to either GREs or nGREs that initiate or suppress gene expression of anti- or pro-inflammatory genes, respectively, is not the only mechanism to consider in GRs ability to control inflammation. While it has been suggested that the number of GREs present play a role in the activation or suppression of gene transcription, it has become increasingly clear that their proximity to the TATA box also is an important factor [25]. Additionally, the recruitment of various coactivators such as TIF2 and SRC-1, corepressors such as NCoR and SMRT, along with various other comodulators can interact with the GR-DNA complex enabling an additional level of transcriptional regulation in GR's ability to control inflammation [26, 27]. GR may regulate inflammation by reducing mRNA half-life through the GC responsive gene tristetraprolin (TTP) [28, 29]. GR can also be phosphorylated by various kinases that can affect its stability, its DNA binding capacity, its ability to translocate to the nucleus, and its interactions with other transcription factors and/or molecular chaperones [30]. Thus, the simple concept of one drug (GC) working on its receptor (GR) has evolved to comprise numerous multifaceted mechanisms to control and regulate inflammation.

Furthermore, glucocorticoids have been shown to regulate inflammation in the absence of DNA binding or interactions with other transcription factors. Nongenomic GC–GR interactions were shown to account for the cardioprotective effect of an acute high dose of corticosteroids resulting in the nontranscriptional activation of endothelial nitric oxide synthase (eNOS) [31]. eNOS has been shown to play an important role during inflammation in regulating the expression of proinflammatory molecules such as NF-kB and cyclooxygenase-2 (Cox-2) [32, 33]. Additionally, nongenomic GC–GR mechanisms involving the activation/inhibition of various signaling pathways, including the p42 MAPK and MAPK ERK1/2 pathways, and the activation of proteins with SH3 domains such as Src and Ras that in turn activate the aforementioned kinase pathways, have been shown to occur [34]. Finally, a mechanism of GC anti-inflammatory action still in its infancy involves posttranscriptional gene regulation via RNA-binding proteins and microRNAs [35]. Interestingly, the role of these posttranscriptional gene regulation actions is thought not to function specifically in turning on or off genes, but to act more as a rheostat in controlling the appropriate amplitude and duration of the response [36].

As the modulation of gene transcription is the major consequence of glucocorticoid activity, the changes in gene transcription that occur directly via activation/repression of GC target genes, or through tethering to another transcription factor, are the most well studied means in controlling inflammation. However, GR can also modulate gene transcription indirectly through the consequences of the activation of the initial target gene. A classic example of this mode of GR regulation involves the glucocorticoid-induced leucine zipper (GILZ) gene. GILZ encodes for a potent anti-inflammatory protein with immunosuppressive and cell survival-promoting effects. GILZ was initially identified as a molecule that protected lymphocytes from TCR/CD3-activated cell death [37]. However, subsequently it was shown that GILZ itself did not directly protect lymphocytes from death, but inhibited the ability of the T cell receptor to induce interleukin-2/interleukin-2 receptor expression and NF-kB activity [38]. Specifically, it was shown that GILZ inhibited NF-kB nuclear translocation and DNA binding via direct protein-to-protein interaction of GILZ with NF-kB. Since these initial observations, GILZ has been shown to be a multifunctional protein that can inhibit key immune cell signaling pathways. Recently, GILZ was shown to regulate Th17 responses and to restrain IL-17-mediated skin inflammation [39]. While the anti-inflammatory and immune-modulatory effects of GILZ have been widely described [40, 41], the induction of this protein has also been associated with provoking apoptosis. GILZ expression in human neutrophils promoted apoptosis through the down-regulation of the myeloid leukemia cell differentiation protein Mcl-1, an antiapoptotic protein of the Bcl-2 family [42].

Alternative Mechanisms of Glucocorticoid Inhibition of Inflammation

The ability of glucocorticoids to regulate and control inflammation goes beyond simply regulating gene transcription of pro- and anti-inflammatory genes. Recently, a unique mechanism of action for anti-inflammatory effects of GCs was reported during the early phase of acute lung injury (ALI) [43]. These authors showed that glucocorticoids attenuate inflammation associated with ALI via up-regulation of the *SphK1* gene in macrophages. The up-regulation of sphingosine kinase 1 in the lung resulted in the synthesis of sphingosine 1 (S1P) that in turn binds to the S1P receptor type 1 (S1PR1) and triggers the Rho family-dependent reorganization of the cytoskeleton leading to enhanced barrier function of the endothelium. The protection afforded by glucocorticoids to enhance the barrier function through this mechanism prevents vascular leakage and the massive infiltration of immune cells into the lung as a way of controlling inflammation.

Interestingly, another recent study linked the action of glucocorticoids to the circadian clock to control time-of-day variations and magnitude of pulmonary inflammation [44]. In this study, the authors observed that pulmonary antibacterial responses of neutrophil recruitment via the chemokine and glucocorticoid responsive gene CXCL5 were modulated by a circadian clock mechanism within epithelial club (Clara) cells [45]. Intriguingly, adrenalectomy blocked this circadian neutrophil recruitment and rhythmic inflammatory responses afforded by CXCL5 upon intraperitoneal injection of LPS. Therefore, this study suggests that the therapeutic effects of glucocorticoids can depend on the local circadian circuit regulation of GR function.

Glucocorticoids have also been shown to suppress overactive inflammatory responses by induction of negative feedback regulators such as the interleukin-1 receptor-associated kinase M (IRAK-M; also known as IRAK3). IRAK-M is known to be a critical negative feedback regulator of Toll-like receptor/Interleukin (IL)-1 receptor (TLR/IL-1R) superfamily of signaling molecules that trigger increased expression of multiple inflammatory genes [46]. Miyata et al. have shown that glucocorticoids suppress bacteria-induced inflammation by directly binding to and up-regulating IRAK-M in airway macrophages and epithelial cells [47]. Additionally, these authors show that IRAK-M depletion results in the enhanced expression of proinflammatory mediators. Thus, glucocorticoids can suppress an overactive inflammatory response via negative feedback to tightly control the inflammatory response and maintain homeostasis.

Overview of Glucocorticoids and the Immune System

The first medical use of glucocorticoids some 60 years ago was for the treatment of rheumatoid arthritis [48, 49]. Since then, glucocorticoids have remained the most commonly used anti-inflammatory and immunomodulatory agents. Their therapeutic activity is substantial in a wide spectrum of diseases, including acute and chronic inflammations, autoimmune disorders, organ transplantations, and the treatment of hematologic cancers [50]. Over the years, numerous publications have focused on glucocorticoids effects on the immune system, and much has been discovered about the molecular mechanism by which GCs act. As in other cells, GR is able to regulate gene expression both positively and negatively in immune cells [9, 18] and can control the inflammatory processes either by a direct binding with glucocorticoid-responsive sequences expressed in the promoters of target genes, or by binding other crucial transcription factors, thus inhibiting the propagation of proinflammatory signals [51]. Furthermore, recent studies have described discrepancies between the immunosuppressive and immunostimulatory effects of glucocorticoids [52, 53].

As described earlier, the inflammatory response is the first protective host response elicited by an injury prompting mobilization of the immune system. The inflammatory recruitment of immune cells neutralizes injurious stimuli and restores the function and structure of damaged tissues [54]. The initial manifestation is the release of intracellular contents after cellular necrosis within the inflammatory site that induces the activation of innate immune components. Through invariant pattern-recognition receptors (PRRs), innate immunity is promptly activated upon detection of conserved structures known as pathogen-associated molecular patterns (PAMPs) or damage-associated molecular patterns (DAMPs) [55]. Mast cells and resident macrophages exert different effector functions, one of which is the increased production of proinflammatory molecules such as interleukin-1 and TNF-α, free radicals, histamine, nitric oxide, prostaglandins, and leukotrienes. This increase in proinflammatory molecules results in vasodilatation, capillary permeability, growth of new blood vessels, and leukocyte migration into the inflamed region. The ability of these cells to generate a chemotactic gradient to recruit cells into the injured tissue is rapid; thus, granulocytes and monocyte migrate from blood into tissue within minutes of injury. Among granulocytes, neutrophils are the most important cells at this first stage because of their capacity to destroy invading microorganisms through phagocytosis and microbicidal activity. Pathogenic antigens are engulfed by resident dendritic cells that rapidly differentiate, migrate to lymph nodes, and present antigens to T and B lymphocytes, thus priming and propagating the adaptive immune components including cell-mediated immunity, cytokine production, antigen-specific antibody production, and immunological memory [56].

The resolution of acute inflammation is a dynamic, limited, and finely regulated process that depends upon the crosstalk between innate and adaptive compartments that restore homeostasis after the elimination of harmful agent. An excessive immune response that continues to counteract the persistence of the injurious stimulus triggers a domino effect that leads to chronic inflammation. For this purpose, in the management of numerous mechanisms that control the development and maintenance of inflammation and autoimmune diseases, glucocorticoids have been the most potent drugs of choice, affecting nearly every cell of the immune system, depending on their state of differentiation or activation [57, 58].

Glucocorticoid Effects on Innate Immunity

The innate immune system is the first line of defense that acts rapidly after encountering noxious agents without the reliance on antibody or other acquired responses. For this reason, the effects of glucocorticoids on innate immune cells must be immediate (in terms of minutes) thus contributing to the resolution of inflammation. Among innate immune cells, glucocorticoids strongly influence the plasticity, survival, and function of monocytes and macrophages according to the plasma GC concentration and the state of cell activation [59]. To enhance the clearance of pathogens, dead cells, and toxins, low GC concentrations enhance antigen uptake, scavenger function, and phagocytosis. For this purpose, the induction of the opsonins MFG-E8, Mertk and protein S [60, 61], the up-regulation of mannose receptor MR/CD206 [62], the scavenger receptor CD163 [63], and the increase of IFN-γ-induced FcR [64] have been observed. Glucocorticoids target macrophages to ensure survival in response to LPS-induced sepsis and to suppress inflammation associated with contact allergy [65, 66]. These results suggest that low concentrations of glucocorticoids have an immune-stimulatory effect on macrophage function in the presence of inflammatory stimuli, whereas high concentrations of this stress hormone exert inhibitory functions on macrophages. High dose actions abrogate the production of proinflammatory mediators as numerous cytokines are down-regulated, the secretion of many chemokines is inhibited, the expression of adhesion molecules such as beta-2 integrin is reduced, and antigen presentation and expression of HLA molecules are decreased by GCs [53, 59, 67].

Consistent with the immunomodulatory properties, some studies have shown that steroid hormones induce highly phagocytic monocyte-derived macrophages. Glucocorticoid exposure functions to reprogram monocyte differentiation through changes in intracellular components that regulate cytoskeletal reorganization following adhesion. The enhanced phagocytic activity and increased expression of the anti-inflammatory cytokine IL-10 observed in these cells support the hypothesis that glucocorticoids do not cause a global suppression of macrophages effectors, but result in the differentiation of a specific anti-inflammatory phenotype which seems to be actively involved in the resolution of inflammatory conditions [68–70].

Furthermore, glucocorticoids exert many of their anti-inflammatory effects through the regulation of granulocyte trafficking. GCs can induce apoptosis and degranulation of basophils and eosinophils. However, at the same time GCs promote the survival and expansion of neutrophils increasing the release of bone marrow precursors [71–75]. In the presence of glucocorticoids, the flow and movement of granulocytes appear tightly regulated to that of monocytes. In an inflammatory scenario, endothelial cells increase the expression of adhesion molecules that bind to their cognate receptor on granulocytes thus allowing cellular transmigration into inflammatory sites. To reduce cellular infiltration, GCs promote shedding of L-selectin and E-selectin from neutrophils [76], suppress the synthesis of many chemokines including IL-8, Mip-1β, Mip-3β, Mcp-2, Mcp-3, Mcp-4, RANTES, TARC, and eotaxin, and increase IL-1RII expression, a decoy receptor which limit the deleterious effects of IL-1 [77].

Natural or synthetic glucocorticoids can also alter natural killer (NK) cell activity. Acting as regulatory cells, this homogenous population of innate lymphocytes interacts with various components of the immune system suppressing the immune response [78]. Nevertheless, glucocorticoid treatment is able to reduce NK cell cytolytic activity by the reduction of histone promoter acetylation for perforin and granzyme B. In contrast, glucocorticoids increase histone acetylation in regulatory regions for INF-γ and IL-6. The increase in histone acetylation is associated with increased proinflammatory cytokine mRNA and protein production upon cellular stimulation and epigenetic modifications [79]. These immunologic effects demonstrate how glucocorticoids epigenetically reduce NK cell cytolytic activity, while at the same time prime NK cells for proinflammatory cytokine production that can act as a powerful tool in cancer immunotherapy [80].

Coupling Innate to Adaptive Immunity

While innate immunity provides the first line of defense against pathogens, adaptive immunity, also known as acquired immunity, is also involved during inflammation. Adaptive immunity creates immunological memory after an initial exposure to antigen, resulting in an enhanced antigen-specific response upon subsequent exposure. It is now appreciated that the innate immune response shapes the acquired immune response. The link between the innate and adaptive components involves soluble cytokines and chemokines, and cellular interactions between antigen-specific lymphocytes and antigen-bearing dendritic cells (DCs). In response to a plethora of stimuli, DCs change from immature cells specialized for antigen capture, processing, and presentation, into mature cells that migrate to draining lymph nodes to interact with naive T cell. On this basis, it is evident that innate immune receptors on DCs play a pivotal role in determining the type of adaptive immune response triggered.

Impairment of DC maturation and function is one of the immunosuppressive effects of glucocorticoids [81]. Synthetic GC treatment interferes with the lifecycle of DCs both in vitro and in vivo [82]. After in vitro maturation with LPS and CD40L, DCs treated with the synthetic glucocorticoid methylprednisolone exhibit enhanced antigen uptake, but down-regulated expression of CD80, CD86, and CD54, and decreased production of TNF-α, IL-6, and IL-12, thus inhibiting the induction of primary T cell responses. Similar results were observed when TNF-α was used to activate DCs [83]. In this study, a different synthetic glucocorticoid dexamethasone, inhibited DC expression of MHC I and II, costimulatory molecules (including B7.1 and B7.2), and the ICAM-1/LFA-1 complex, thus promoting the formation of tolerogenic DCs [82, 84–86]. Tolerogenic DCs are able to drive uncommitted T cells toward the Treg subtype [87] and promote the conversion of CD4+ T cells into IL-10 producing type 1 Tregs (Tr1) [88–90]. Moreover, tolerogenic DCs inhibit the proliferation of allospecific T cells [91, 92], preventing acute graft rejection in mice [93], decreasing the number of IFN-γ producing CD4+ T cells, and promoting NK cell function toward an alternative activated phenotype unable to secrete IFN-γ [94].

Besides their capacity to modulate and induce effector T cell responses, tolerogenic DCs are defined based on the expression of various surface markers such as Ig-like transcript (ILT) molecules [95], transcriptional regulators like glucocorticoid-induced leucine zipper (GILZ) [40, 96, 97], and enzymes such as retinaldehyde dehydrogenase (RALDH) [98] or NO synthetase-2 (NOS-2) [99], all contributing to their functional properties. Additionally, dendritic cells can also facilitate communication between the immune system and the endocrine system. A recent study described a novel DC population in the pituitary gland that produces cytokines, controls LPS-dependent ACTH secretion, and expresses factors for glucocorticoid release [100]. These data suggest that pituitary DCs relay an immune challenge (such as LPS) to the HPA axis by secreting proinflammatory cytokines, which stimulates the anterior pituitary gland to release ACTH. Therefore, DCs are distinguished not only by their role in linking the innate and adaptive immune responses but also in directing communication between the immune and endocrine systems.

Glucocorticoid Effects on Adaptive Immunity

Autoimmune diseases are associated with the generation of an adaptive immune response mounted against self-antigens. During development, most lymphocytes bearing high affinity receptors for self-antigens are deleted, but not all self-reactive lymphocytes are eliminated. The activity of self-reactive lymphocytes is regulated by peripheral tolerance, an active immunosuppressive process that involves clonal anergy and clonal suppression. A failure in peripheral tolerance allows the activation of self-reactive T or B cell clones, thus eliciting (or inducing) cell-mediated or humoral responses against self-antigen. As potent immunosuppressors, synthetic GCs are extensively used for the treatment of various autoimmune and chronic inflammatory conditions [101]. GCs target several aspects of adaptive immunity, including thymocyte maturation, as well as T and B cell proliferation, survival, and differentiation.

Glucocorticoids and Thymocyte Development

The thymus is the key organ for T cell maturation, and glucocorticoids play an important role in thymocyte selection and survival. The selection process drives immature CD4⁻CD8⁻ "double negative" thymocytes to CD4⁺CD8⁺TCR^{low} "double positive" thymocytes, which represent about 80% of the cells in the thymus. At this stage double positive thymocytes are extremely sensitive to glucocorticoid-induced apoptosis, but escape apoptosis when both TCR and GR signal simultaneously, according to the "mutual antagonism" model [102]. Thymocytes that are unable to process a functional TCR undergo GC-induced apoptosis, since the TCR signal cannot counteract GR signaling, while only thymocytes expressing a TCR with high affinity for self-peptides undergo negative selection due to the inability of GR signaling to overcome the strong TCR-dependent signal. Finally, only those thymocytes exerting a moderate avidity for self-antigens will survive, suggesting interplay between TCR and GC signals. The grade of affinity between TCR and self-peptides is crucial for the survival of a mature T cell repertoire expressing either the CD4 or CD8 receptor.

A number of in vitro experiments implicate glucocorticoids in regulating T cell number, survival, and TCR repertoire, although the in vivo evidence is still contradictory regarding the correlation between GR expression and thymocyte sensitivity to glucocorticoid-induced death. Adrenalectomy induces thymic hypertrophy [103], mice overexpressing GR in T cells exhibit a reduced number of double positive thymocytes, despite the fact that these cells express lower level of GR compared to thymocytes in other developmental stages. In GILZ-overexpressing transgenic mice, CD4⁺CD8⁺ thymocyte number is significantly decreased and ex vivo thymocyte apoptosis is increased [104]. In contrast, intrathymic T cell development and selection proceed normally in mice expressing antisense GR in the thymus and in fetal mice from GR-KO mice [105, 106]. However, these studies suggest the molecular and cellular mechanisms that regulate thymocyte maturation need further investigation.

Moreover, corticosterone synthesis has been suggested to occur in the thymus and there is debate about how locally produced GCs may regulate thymocyte development as well as may affect the initiation of age-associated thymic involution [107].

Glucocorticoid Function in T Cells

Although immature T cells are extremely sensitive to undergoing apoptosis, cell death can also occur in mature T cells either by a glucocorticoid-directed action, or by the involvement of factors that mediate activation-induced cell death, i.e. inhibiting IL-2-mediated activation. According to the activation state and the timing of hormone exposure, T cells can be sensitive or resistant to glucocorticoid-induce cell death. Moreover, mature T cells are susceptible to mutual antagonism between GR and TCR. Mice lacking GR in T cells (GR^{LckCre}) or carrying a point mutation which inhibits GR dimerization and DNA binding (GR^{dim}) demonstrate that inhibition of activation-induced cell death depends on direct binding of the GR to two nGREs in the CD95 (APO-1/Fas) ligand promoter [108]. In contrast, overexpression of the SWI3-related gene (SRG3) protein in peripheral T cells renders them sensitive to GC-induced apoptosis through the GR–SRG3 complex formation, suggesting that SRG3 may play a critical role in controlling GC-mediated apoptosis of developing thymocytes. Studies have also shown that a dominant negative SRG3 decreases GC sensitivity in thymoma cells. In addition, mice overexpressing the SRG3 protein appear to be more susceptible to stress-induced deletion of peripheral T cells than WT mice, which may result in an immunosuppressive condition [109].

In addition inducing apoptosis, glucocorticoids also affect T cell polarization. Since endogenous or exogenous glucocorticoids attenuate IL-12 synthesis, T cell response is shifted from the Th1 to Th2 phenotype [110]. GCs inhibit both T-bet and GATA-3 transcriptional activity through two different mechanisms. T-bet does not only function as an activator of IFN-γ expression, but also interacts with the GATA-3 transcription factor, inhibiting Th2 cytokine gene expression. Consistent with these results, GCs inhibit both Th1 and Th2 master regulator factors, however long-term treatment favors Th2 expansion [111]. In addition, glucocorticoids induce T polarization toward Th2 phenotype through an increase of Itk expression, a Tec kinase inducing T helper 2 differentiation via negative regulation of T-bet [112, 113].

The effects of GCs on a third subset of IL-17-producing effector T helper cells, called Th17 cells, are still a matter of debate. Dexamethasone inhibits anti-CD3/anti-CD28-stimulated IFN-γ, IL-4, IL-17A, IL-17F, and IL-22 in various cell clones [114]. Interestingly, IL-17A and IL-17F, but not IL-22, lead to resistance of GC-induced apoptosis in in vitro-differentiated Th17 cells despite immunocytochemistry confirming glucocorticoid receptor translocation to the nucleus following treatment [114]. Mice lacking GILZ exhibit severe inflammation and a proinflammatory cytokine expression pattern in the imiquimod model of psoriasis, and DCs lacking GILZ produced greater IL-1, IL-23, and IL-6 in response to imiquimod stimulation in vitro [39]. These studies assessing glucocorticoid-dependent inhibition of IL-17 synthesis are in stark contrast with other studies describing Th17 sensitivity upon GC administration [115–117].

While the role of GC and Th17 is controversial, how glucocorticoids affect Treg function is much clearer. Both in humans and mice, treatment with dexamethasone increases the frequency of Treg cells, suggesting GC-mediated immune suppression is achieved, in part by enhancing Treg cell number or activity [118, 119] and by promoting the development of IL-10-producing T cells, an inducible peripheral Treg subpopulation [120]. There are several mechanisms that have been proposed to explain GC-mediated increase in Treg frequency. First, dexamethasone inhibits IL-2-mediated activation of T effector cells, increasing the proportion of Treg cells [118]; moreover, Treg cells were relatively more resistant to Dex-induced cell death and they were further protected by IL-2 [121]. Second, glucocorticoids synergize with TGF-β in FoxP3 induction [122, 123].

Glucocorticoid Function in B Cells

Circulating B lymphocytes are reduced by GC treatment, however not to the same extent as T cells [124]. This results from a reduction in splenic and lymph node B cell numbers. Furthermore, in vivo administration of dexamethasone to adrenalectomized mice reduced B cell numbers in both the spleen and bone marrow [125]. Studies on human leukemic lymphoblasts have shown that glucocorticoids have preferential apoptotic effects in certain lymphoid cell populations including B cell lymphomas [126]. B-lymphoblastic leukemia/lymphoma has been characterized in having increased expression of Bcl-2 resulting in resistance to glucocorticoid-induced apoptosis. Additionally, it was shown that deletion of GILZ in murine B lymphocytes leads to an accumulation of B cells in the bone marrow, blood, and lymphoid tissues due to impaired glucocorticoid-induced apoptosis. Since GILZ inhibits NF-kB in B cells, increased nuclear translocation of p65 has been shown in GILZ-deficient cells resulting in an increase in Bcl-2 gene transcription [127].

Regarding the humoral immune response, glucocorticoids increase IgE synthesis, which is driven by the synergistic effects of hormones and IL-4 [128]. GC-induced increases in IgE synthesis support why systemic administration of corticosteroids does not interfere in skin prick tests to common allergens [128]. In vivo studies using mice deficient for either CD40L or CD40 lack serum IgE and fail to undergo isotype switching after immunization with T cell-dependent antigens [129, 130]. In addition, patients with X-linked hyper-IgM syndrome are deficient in CD40L and have low serum levels of IgG, IgA, and IgE due to impaired isotype switching [131]. To explain this effect of glucocorticoids, many studies suggest that glucocorticoid- and IL-4-induced IgE production is dependent on CD40L increased transcription and, thereby expression [132]. In vitro experiments demonstrated that agonist antibodies against CD40 mimic CD40L-dependent triggering of IL-4-driven isotype switching to IgE [132], while soluble CD40 inhibits IL-4 dependent IgE synthesis [133]. These results suggest that a rise in IgE production associated with glucocorticoid treatment is not clinically detrimental but presents additional immunomodulatory effects of corticosteroids on the T cell response.

Glucocorticoid Therapy and Resistance During Inflammation

The advantages of GC therapy in controlling and regulating inflammation are many, however due to the numerous signaling pathways glucocorticoids activate, consequences of this class of corticosteroids can also result in harmful side effects. GC-related side effects include musculoskeletal complications such as osteoporosis, hypertension, rapid weight gain, diabetes, glaucoma, peptic ulcer disease, and decelerated wound healing [134, 135]. Additionally, an increased risk of infection can occur resulting from a compromised immune system [136].

The successful resolution of acute inflammation afforded by glucocorticoids occurs by the delicate balance of pro- and anti-inflammatory molecules. However, long-term glucocorticoid therapy for the treatment of chronic inflammation typically results in reduced anti-inflammatory effects. These diminished effects can occur through a variety of ways including down-regulation of the GR itself [137, 138], defective GR binding and translocation (exemplified by GR^{dim}) [139, 140], GR nitrosylation by nitric oxide (NO) donors [141], and/or increased expression of GRβ which competes with GRα for binding to GREs [142]. Recently, it was shown that hypoxia attenuates the anti-inflammatory effects of GCs by down-regulating GR along with inhibiting nuclear translocation [143]. Genetic factors may also contribute to GC resistance [144], such as the occurrence of polymorphisms in GR that may occur within families. A recent study by Mohamed et al. suggested a marked association of the glucocorticoid receptor 646 C>G polymorphism in resistance to GCs, resulting in severe bronchial asthma [145]. Finally, causative factors for glucocorticoid resistance go beyond defects in GR and include cigarette smoke, where in asthmatic patients who smoke, diminished anti-inflammatory actions in response to glucocorticoids were observed [146], viral infections where a study showed that rhinovirus infection can reduce GR nuclear translocation and GC function [147], and hypoxia observed at the site of inflammation that can impair GR transactivation [148].

Despite the well-known adverse side effects of prolonged GC treatment and the occurrence of GC resistance associated with long-term usage of glucocorticoids, these stress hormones remain the most effective treatment and commonly prescribed medication for controlling inflammation. The beneficial effects of GCs in treating anti-inflammatory and immunosuppressive disorders such as rheumatic diseases, allergy, asthma, and sepsis still outweigh their unfavorable consequences. Further research into the practice of GC therapy for combating inflammation to minimize harmful side effects and reduce the resistance associated with chronic treatment will be required to fully understand the pharmacological characteristics and biological actions of these stress hormones.

References

1. Munck A. Studies on the mode of action of glucocorticoids in rats. II. The effects in vivo and in vitro on net glucose uptake by isolated adipose tissue. Biochim Biophys Acta. 1962;57:318–26.
2. Munck A, Koritz SB. Studies on the mode of action of glucocorticoids in rats. I. Early effects of cortisol on blood glucose and on glucose entry into muscle, liver and adipose tissue. Biochim Biophys Acta. 1962;57:310–7.
3. Oppong E, Cato AC. Effects of Glucocorticoids in the Immune System. Adv Exp Med Biol. 2015;872:217–33.
4. Savino W, Mendes-da-Cruz DA, Lepletier A, Dardenne M. Hormonal control of T-cell development in health and disease. Nat Rev Endocrinol. 2016;12(2):77–89.
5. Singh RR, Cuffe JS, Moritz KM. Short- and long-term effects of exposure to natural and synthetic glucocorticoids during development. Clin Exp Pharmacol Physiol. 2012;39(11):979–89.
6. Grier DG, Halliday HL. Effects of glucocorticoids on fetal and neonatal lung development. Treat Respir Med. 2004;3(5):295–306.

7. Bellavance MA, Rivest S. The HPA-immune axis and the immunomodulatory actions of glucocorticoids in the brain. Front Immunol. 2014;5:136.
8. Evans RM. The steroid and thyroid hormone receptor superfamily. Science. 1988;240(4854):889–95.
9. Kumar R, Thompson EB. Gene regulation by the glucocorticoid receptor: structure: function relationship. J Steroid Biochem Mol Biol. 2005;94(5):383–94.
10. Lonard DM, O'Malley BW. Expanding functional diversity of the coactivators. Trends Biochem Sci. 2005;30(3):126–32.
11. Oakley RH, Cidlowski JA. The biology of the glucocorticoid receptor: new signaling mechanisms in health and disease. J Allergy Clin Immunol. 2013;132(5):1033–44.
12. Lewis-Tuffin LJ, Cidlowski JA. The physiology of human glucocorticoid receptor beta (hGRbeta) and glucocorticoid resistance. Ann N Y Acad Sci. 2006;1069:1–9.
13. Oakley RH, Jewell CM, Yudt MR, Bofetiado DM, Cidlowski JA. The dominant negative activity of the human glucocorticoid receptor beta isoform. Specificity and mechanisms of action. J Biol Chem. 1999;274(39):27857–66.
14. Oakley RH, Webster JC, Sar M, Parker Jr CR, Cidlowski JA. Expression and subcellular distribution of the beta-isoform of the human glucocorticoid receptor. Endocrinology. 1997;138(11):5028–38.
15. Coussens LM, Werb Z. Inflammation and cancer. Nature. 2002;420(6917):860–7.
16. Barnes PJ. Mechanisms and resistance in glucocorticoid control of inflammation. J Steroid Biochem Mol Biol. 2010;120(2-3):76–85.
17. Clark AR. Anti-inflammatory functions of glucocorticoid-induced genes. Mol Cell Endocrinol. 2007;275(1-2):79–97.
18. Smoak KA, Cidlowski JA. Mechanisms of glucocorticoid receptor signaling during inflammation. Mech Ageing Dev. 2004;125(10-11):697–706.
19. Gottlicher M, Heck S, Herrlich P. Transcriptional cross-talk, the second mode of steroid hormone receptor action. J Mol Med (Berl). 1998;76(7):480–9.
20. Barnes PJ. Corticosteroid effects on cell signalling. Eur Respir J. 2006;27(2):413–26.
21. De Bosscher K, Vanden Berghe W, Haegeman G. The interplay between the glucocorticoid receptor and nuclear factor-kappaB or activator protein-1: molecular mechanisms for gene repression. Endocr Rev. 2003;24(4):488–522.
22. De Bosscher K, Vanden Berghe W, Haegeman G. Cross-talk between nuclear receptors and nuclear factor kappaB. Oncogene. 2006;25(51):6868–86.
23. Ratman D, Vanden Berghe W, Dejager L, Libert C, Tavernier J, Beck IM, et al. How glucocorticoid receptors modulate the activity of other transcription factors: a scope beyond tethering. Mol Cell Endocrinol. 2013;380(1-2):41–54.
24. Altonsy MO, Sasse SK, Phang TL, Gerber AN. Context-dependent cooperation between nuclear factor kappaB (NF-kappaB) and the glucocorticoid receptor at a TNFAIP3 intronic enhancer: a mechanism to maintain negative feedback control of inflammation. J Biol Chem. 2014;289(12):8231–9.
25. Jantzen HM, Strahle U, Gloss B, Stewart F, Schmid W, Boshart M, et al. Cooperativity of glucocorticoid response elements located far upstream of the tyrosine aminotransferase gene. Cell. 1987;49(1):29–38.
26. Glass CK, Rosenfeld MG. The coregulator exchange in transcriptional functions of nuclear receptors. Genes Dev. 2000;14(2):121–41.
27. Simons Jr SS. Glucocorticoid receptor cofactors as therapeutic targets. Curr Opin Pharmacol. 2010;10(6):613–9.
28. Hahn RT, Hoppstadter J, Hirschfelder K, Hachenthal N, Diesel B, Kessler SM, et al. Downregulation of the glucocorticoid-induced leucine zipper (GILZ) promotes vascular inflammation. Atherosclerosis. 2014;234(2):391–400.
29. Smoak K, Cidlowski JA. Glucocorticoids regulate tristetraprolin synthesis and posttranscriptionally regulate tumor necrosis factor alpha inflammatory signaling. Mol Cell Biol. 2006;26(23):9126–35.
30. Weigel NL, Moore NL. Steroid receptor phosphorylation: a key modulator of multiple receptor functions. Mol Endocrinol. 2007;21(10):2311–9.

31. Hafezi-Moghadam A, Simoncini T, Yang Z, Limbourg FP, Plumier JC, Rebsamen MC, et al. Acute cardiovascular protective effects of corticosteroids are mediated by non-transcriptional activation of endothelial nitric oxide synthase. Nat Med. 2002;8(5):473–9.

32. Connelly L, Jacobs AT, Palacios-Callender M, Moncada S, Hobbs AJ. Macrophage endothelial nitric-oxide synthase autoregulates cellular activation and pro-inflammatory protein expression. J Biol Chem. 2003;278(29):26480–7.

33. Grumbach IM, Chen W, Mertens SA, Harrison DG. A negative feedback mechanism involving nitric oxide and nuclear factor kappa-B modulates endothelial nitric oxide synthase transcription. J Mol Cell Cardiol. 2005;39(4):595–603.

34. Mitre-Aguilar IB, Cabrera-Quintero AJ, Zentella-Dehesa A. Genomic and non-genomic effects of glucocorticoids: implications for breast cancer. Int J Clin Exp Pathol. 2015;8(1):1–10.

35. Stellato C. Posttranscriptional gene regulation: novel pathways for glucocorticoids' anti-inflammatory action. Transl Med UniSa. 2012;3:67–73.

36. Anderson P. Post-transcriptional regulons coordinate the initiation and resolution of inflammation. Nat Rev Immunol. 2010;10(1):24–35.

37. D'Adamio F, Zollo O, Moraca R, Ayroldi E, Bruscoli S, Bartoli A, et al. A new dexamethasone-induced gene of the leucine zipper family protects T lymphocytes from TCR/CD3-activated cell death. Immunity. 1997;7(6):803–12.

38. Ayroldi E, Migliorati G, Bruscoli S, Marchetti C, Zollo O, Cannarile L, et al. Modulation of T-cell activation by the glucocorticoid-induced leucine zipper factor via inhibition of nuclear factor kappaB. Blood. 2001;98(3):743–53.

39. Jones SA, Perera DN, Fan H, Russ BE, Harris J, Morand EF. GILZ regulates Th17 responses and restrains IL-17-mediated skin inflammation. J Autoimmun. 2015;61:73–80.

40. Ayroldi E, Riccardi C. Glucocorticoid-induced leucine zipper (GILZ): a new important mediator of glucocorticoid action. FASEB J. 2009;23(11):3649–58.

41. Ronchetti S, Migliorati G, Riccardi C. GILZ as a mediator of the anti-inflammatory effects of glucocorticoids. Front Endocrinol (Lausanne). 2015;6:170.

42. Espinasse MA, Pepin A, Virault-Rocroy P, Szely N, Chollet-Martin S, Pallardy M, et al. Glucocorticoid-Induced Leucine Zipper Is Expressed in Human Neutrophils and Promotes Apoptosis through Mcl-1 Down-Regulation. J Innate Immun. 2016;8(1):81–96.

43. Vettorazzi S, Bode C, Dejager L, Frappart L, Shelest E, Klassen C, et al. Glucocorticoids limit acute lung inflammation in concert with inflammatory stimuli by induction of SphK1. Nat Commun. 2015;6:7796.

44. Gibbs J, Ince L, Matthews L, Mei J, Bell T, Yang N, et al. An epithelial circadian clock controls pulmonary inflammation and glucocorticoid action. Nat Med. 2014;20(8):919–26.

45. Smith JB, Herschman HR. Glucocorticoid-attenuated response genes encode intercellular mediators, including a new C-X-C chemokine. J Biol Chem. 1995;270(28):16756–65.

46. Loiarro M, Ruggiero V, Sette C. Targeting TLR/IL-1R signalling in human diseases. Mediators Inflamm. 2010;2010:674363.

47. Miyata M, Lee JY, Susuki-Miyata S, Wang WY, Xu H, Kai H, et al. Glucocorticoids suppress inflammation via the upregulation of negative regulator IRAK-M. Nat Commun. 2015;6:6062.

48. Bunim JJ. The clinical effects of cortisone and ACTH on rheumatic diseases. Bull N Y Acad Med. 1951;27(2):75–100.

49. Hench P. Effects of cortisone in the rheumatic diseases. Lancet. 1950;2(6634):483–4.

50. Rhen T, Cidlowski JA. Antiinflammatory action of glucocorticoids--new mechanisms for old drugs. N Engl J Med. 2005;353(16):1711–23.

51. De Bosscher K, Vanden Berghe W, Haegeman G. Mechanisms of anti-inflammatory action and of immunosuppression by glucocorticoids: negative interference of activated glucocorticoid receptor with transcription factors. J Neuroimmunol. 2000;109(1):16–22.

52. Zhong HJ, Wang HY, Yang C, Zhou JY, Jiang JX. Low concentrations of corticosterone exert stimulatory effects on macrophage function in a manner dependent on glucocorticoid receptors. Int J Endocrinol. 2013;2013:405127.

53. Zhou JY, Zhong HJ, Yang C, Yan J, Wang HY, Jiang JX. Corticosterone exerts immunostimulatory effects on macrophages via endoplasmic reticulum stress. Br J Surg. 2010;97(2):281–93.

54. Medzhitov R, Horng T. Transcriptional control of the inflammatory response. Nat Rev Immunol. 2009;9(10):692–703.
55. Akira S. Pathogen recognition by innate immunity and its signaling. Proc Jpn Acad Ser B Phys Biol Sci. 2009;85(4):143–56.
56. Pancer Z, Cooper MD. The evolution of adaptive immunity. Annu Rev Immunol. 2006;24 :497–518.
57. Baschant U, Tuckermann J. The role of the glucocorticoid receptor in inflammation and immunity. J Steroid Biochem Mol Biol. 2010;120(2-3):69–75.
58. Coutinho AE, Chapman KE. The anti-inflammatory and immunosuppressive effects of glucocorticoids, recent developments and mechanistic insights. Mol Cell Endocrinol. 2011;335(1):2–13.
59. Lim HY, Muller N, Herold MJ, van den Brandt J, Reichardt HM. Glucocorticoids exert opposing effects on macrophage function dependent on their concentration. Immunology. 2007;122(1):47–53.
60. McColl A, Bournazos S, Franz S, Perretti M, Morgan BP, Haslett C, et al. Glucocorticoids induce protein S-dependent phagocytosis of apoptotic neutrophils by human macrophages. J Immunol. 2009;183(3):2167–75.
61. McColl A, Michlewska S, Dransfield I, Rossi AG. Effects of glucocorticoids on apoptosis and clearance of apoptotic cells. ScientificWorldJournal. 2007;7:1165–81.
62. Roszer T. Understanding the mysterious M2 macrophage through activation markers and effector mechanisms. Mediators Inflamm. 2015;2015:816460.
63. Tang Z, Niven-Fairchild T, Tadesse S, Norwitz ER, Buhimschi CS, Buhimschi IA, et al. Glucocorticoids enhance CD163 expression in placental Hofbauer cells. Endocrinology. 2013;154(1):471–82.
64. Warren MK, Vogel SN. Opposing effects of glucocorticoids on interferon-gamma-induced murine macrophage Fc receptor and Ia antigen expression. J Immunol. 1985;134(4):2462–9.
65. Bhattacharyya S, Brown DE, Brewer JA, Vogt SK, Muglia LJ. Macrophage glucocorticoid receptors regulate Toll-like receptor 4-mediated inflammatory responses by selective inhibition of p38 MAP kinase. Blood. 2007;109(10):4313–9.
66. Tuckermann JP, Kleiman A, Moriggl R, Spanbroek R, Neumann A, Illing A, et al. Macrophages and neutrophils are the targets for immune suppression by glucocorticoids in contact allergy. J Clin Invest. 2007;117(5):1381–90.
67. Broug-Holub E, Kraal G. Dose- and time-dependent activation of rat alveolar macrophages by glucocorticoids. Clin Exp Immunol. 1996;104(2):332–6.
68. Ehrchen J, Steinmuller L, Barczyk K, Tenbrock K, Nacken W, Eisenacher M, et al. Glucocorticoids induce differentiation of a specifically activated, anti-inflammatory subtype of human monocytes. Blood. 2007;109(3):1265–74.
69. Giles KM, Ross K, Rossi AG, Hotchin NA, Haslett C, Dransfield I. Glucocorticoid augmentation of macrophage capacity for phagocytosis of apoptotic cells is associated with reduced p130Cas expression, loss of paxillin/pyk2 phosphorylation, and high levels of active Rac. J Immunol. 2001;167(2):976–86.
70. Heasman SJ, Giles KM, Ward C, Rossi AG, Haslett C, Dransfield I. Glucocorticoid-mediated regulation of granulocyte apoptosis and macrophage phagocytosis of apoptotic cells: implications for the resolution of inflammation. J Endocrinol. 2003;178(1):29–36.
71. Meagher LC, Cousin JM, Seckl JR, Haslett C. Opposing effects of glucocorticoids on the rate of apoptosis in neutrophilic and eosinophilic granulocytes. J Immunol. 1996;156(11):4422–8.
72. Cox G. Glucocorticoid treatment inhibits apoptosis in human neutrophils. Separation of survival and activation outcomes. J Immunol. 1995;154(9):4719–25.
73. Liles WC, Dale DC, Klebanoff SJ. Glucocorticoids inhibit apoptosis of human neutrophils. Blood. 1995;86(8):3181–8.
74. Yoshimura C, Miyamasu M, Nagase H, Iikura M, Yamaguchi M, Kawanami O, et al. Glucocorticoids induce basophil apoptosis. J Allergy Clin Immunol. 2001;108(2):215–20.

75. Jia W, Wu J, Jia H, Yang Y, Zhang X, Chen K, et al. The peripheral blood neutrophil-to-lymphocyte ratio is superior to the lymphocyte-to-monocyte ratio for predicting the long-term survival of triple-negative breast cancer patients. PLoS One. 2015;10(11), e0143061.
76. Weber PS, Toelboell T, Chang LC, Tirrell JD, Saama PM, Smith GW, et al. Mechanisms of glucocorticoid-induced down-regulation of neutrophil L-selectin in cattle: evidence for effects at the gene-expression level and primarily on blood neutrophils. J Leukoc Biol. 2004;75(5):815–27.
77. Re F, Muzio M, De Rossi M, Polentarutti N, Giri JG, Mantovani A, et al. The type II "receptor" as a decoy target for interleukin 1 in polymorphonuclear leukocytes: characterization of induction by dexamethasone and ligand binding properties of the released decoy receptor. J Exp Med. 1994;179(2):739–43.
78. Lunemann A, Lunemann JD, Munz C. Regulatory NK-cell functions in inflammation and autoimmunity. Mol Med. 2009;15(9-10):352–8.
79. Eddy JL, Krukowski K, Janusek L, Mathews HL. Glucocorticoids regulate natural killer cell function epigenetically. Cell Immunol. 2014;290(1):120–30.
80. Fauci AS, Dale DC, Balow JE. Glucocorticosteroid therapy: mechanisms of action and clinical considerations. Ann Intern Med. 1976;84(3):304–15.
81. Piemonti L, Monti P, Allavena P, Sironi M, Soldini L, Leone BE, et al. Glucocorticoids affect human dendritic cell differentiation and maturation. J Immunol. 1999;162(11):6473–81.
82. Woltman AM, de Fijter JW, Kamerling SW, Paul LC, Daha MR, van Kooten C. The effect of calcineurin inhibitors and corticosteroids on the differentiation of human dendritic cells. Eur J Immunol. 2000;30(7):1807–12.
83. Chabot V, Martin L, Meley D, Sensebe L, Baron C, Lebranchu Y, et al. Unexpected impairment of TNF-alpha-induced maturation of human dendritic cells in vitro by IL-4. J Transl Med. 2016;14(1):93.
84. Bros M, Jahrling F, Renzing A, Wiechmann N, Dang NA, Sutter A, et al. A newly established murine immature dendritic cell line can be differentiated into a mature state, but exerts tolerogenic function upon maturation in the presence of glucocorticoid. Blood. 2007;109(9):3820–9.
85. Zimmer A, Luce S, Gaignier F, Nony E, Naveau M, Biola-Vidamment A, et al. Identification of a new phenotype of tolerogenic human dendritic cells induced by fungal proteases from Aspergillus oryzae. J Immunol. 2011;186(7):3966–76.
86. Bosma BM, Metselaar HJ, Nagtzaam NM, de Haan R, Mancham S, van der Laan LJ, et al. Dexamethasone transforms lipopolysaccharide-stimulated human blood myeloid dendritic cells into myeloid dendritic cells that prime interleukin-10 production in T cells. Immunology. 2008;125(1):91–100.
87. Kapsenberg ML. Dendritic-cell control of pathogen-driven T-cell polarization. Nat Rev Immunol. 2003;3(12):984–93.
88. Calmette J, Ellouze M, Tran T, Karaki S, Ronin E, Capel F, et al. Glucocorticoid-induced leucine zipper enhanced expression in dendritic cells is sufficient to drive regulatory T cells expansion in vivo. J Immunol. 2014;193(12):5863–72.
89. Zimmer A, Bouley J, Le Mignon M, Pliquet E, Horiot S, Turfkruyer M, et al. A regulatory dendritic cell signature correlates with the clinical efficacy of allergen-specific sublingual immunotherapy. J Allergy Clin Immunol. 2012;129(4):1020–30.
90. Schlecht G, Leclerc C, Dadaglio G. Induction of CTL and nonpolarized Th cell responses by CD8alpha(+) and CD8alpha(-) dendritic cells. J Immunol. 2001;167(8):4215–21.
91. Wang Y, Kissenpfennig A, Mingueneau M, Richelme S, Perrin P, Chevrier S, et al. Th2 lympho-proliferative disorder of LatY136F mutant mice unfolds independently of TCR-MHC engagement and is insensitive to the action of Foxp3+ regulatory T cells. J Immunol. 2008;180(3):1565–75.
92. Kamanaka M, Kim ST, Wan YY, Sutterwala FS, Lara-Tejero M, Galan JE, et al. Expression of interleukin-10 in intestinal lymphocytes detected by an interleukin-10 reporter knockin tiger mouse. Immunity. 2006;25(6):941–52.
93. Inaba K, Inaba M, Romani N, Aya H, Deguchi M, Ikehara S, et al. Generation of large numbers of dendritic cells from mouse bone marrow cultures supplemented with granulocyte/macrophage colony-stimulating factor. J Exp Med. 1992;176(6):1693–702.

94. Spallanzani RG, Torres NI, Avila DE, Ziblat A, Iraolagoitia XL, Rossi LE, et al. Regulatory dendritic cells restrain NK Cell IFN-gamma production through mechanisms involving NKp46, IL-10, and MHC Class I-specific inhibitory receptors. J Immunol. 2015;195(5):2141–8.
95. Wu J, Horuzsko A. Expression and function of immunoglobulin-like transcripts on tolerogenic dendritic cells. Hum Immunol. 2009;70(5):353–6.
96. Karaki S, Garcia G, Tcherakian C, Capel F, Tran T, Pallardy M, et al. Enhanced glucocorticoid-induced leucine zipper in dendritic cells induces allergen-specific regulatory CD4(+) T-cells in respiratory allergies. Allergy. 2014;69(5):624–31.
97. Benkhoucha M, Molnarfi N, Dunand-Sauthier I, Merkler D, Schneiter G, Bruscoli S, et al. Hepatocyte growth factor limits autoimmune neuroinflammation via glucocorticoid-induced leucine zipper expression in dendritic cells. J Immunol. 2014;193(6):2743–52.
98. Vicente-Suarez I, Larange A, Reardon C, Matho M, Feau S, Chodaczek G, et al. Unique lamina propria stromal cells imprint the functional phenotype of mucosal dendritic cells. Mucosal Immunol. 2015;8(1):141–51.
99. Kania G, Siegert S, Behnke S, Prados-Rosales R, Casadevall A, Luscher TF, et al. Innate signaling promotes formation of regulatory nitric oxide-producing dendritic cells limiting T-cell expansion in experimental autoimmune myocarditis. Circulation. 2013;127(23):2285–94.
100. Glennon E, Kaunzner UW, Gagnidze K, McEwen BS, Bulloch K. Pituitary dendritic cells communicate immune pathogenic signals. Brain Behav Immun. 2015;50:232–40.
101. Kadmiel M, Cidlowski JA. Glucocorticoid receptor signaling in health and disease. Trends Pharmacol Sci. 2013;34(9):518–30.
102. Erlacher M, Knoflach M, Stec IE, Bock G, Wick G, Wiegers GJ. TCR signaling inhibits glucocorticoid-induced apoptosis in murine thymocytes depending on the stage of development. Eur J Immunol. 2005;35(11):3287–96.
103. Inomata T, Nakamura T. Influence of adrenalectomy on the development of the neonatal thymus in the rat. Biol Neonate. 1989;55(4-5):238–43.
104. Delfino DV, Agostini M, Spinicelli S, Vito P, Riccardi C. Decrease of Bcl-xL and augmentation of thymocyte apoptosis in GILZ overexpressing transgenic mice. Blood. 2004;104(13):4134–41.
105. Purton JF, Boyd RL, Cole TJ, Godfrey DI. Intrathymic T cell development and selection proceeds normally in the absence of glucocorticoid receptor signaling. Immunity. 2000;13(2):179–86.
106. Wiegers GJ, Kaufmann M, Tischner D, Villunger A. Shaping the T-cell repertoire: a matter of life and death. Immunol Cell Biol. 2011;89(1):33–9.
107. van den Brandt J, Luhder F, McPherson KG, de Graaf KL, Tischner D, Wiehr S, et al. Enhanced glucocorticoid receptor signaling in T cells impacts thymocyte apoptosis and adaptive immune responses. Am J Pathol. 2007;170(3):1041–53.
108. Baumann S, Dostert A, Novac N, Bauer A, Schmid W, Fas SC, et al. Glucocorticoids inhibit activation-induced cell death (AICD) via direct DNA-dependent repression of the CD95 ligand gene by a glucocorticoid receptor dimer. Blood. 2005;106(2):617–25.
109. Han S, Choi H, Ko MG, Choi YI, Sohn DH, Kim JK, et al. Peripheral T cells become sensitive to glucocorticoid- and stress-induced apoptosis in transgenic mice overexpressing SRG3. J Immunol. 2001;167(2):805–10.
110. Liberman AC, Antunica-Noguerol M, Ferraz-de-Paula V, Palermo-Neto J, Castro CN, Druker J, et al. Compound A, a dissociated glucocorticoid receptor modulator, inhibits T-bet (Th1) and induces GATA-3 (Th2) activity in immune cells. PLoS One. 2012;7(4), e35155.
111. Liberman AC, Druker J, Refojo D, Holsboer F, Arzt E. Glucocorticoids inhibit GATA-3 phosphorylation and activity in T cells. FASEB J. 2009;23(5):1558–71.
112. Petrillo MG, Fettucciari K, Montuschi P, Ronchetti S, Cari L, Migliorati G, et al. Transcriptional regulation of kinases downstream of the T cell receptor: another immunomodulatory mechanism of glucocorticoids. BMC Pharmacol Toxicol. 2014;15:35.
113. Miller AT, Wilcox HM, Lai Z, Berg LJ. Signaling through Itk promotes T helper 2 differentiation via negative regulation of T-bet. Immunity. 2004;21(1):67–80.
114. Banuelos J, Shin S, Cao Y, Bochner BS, Morales-Nebreda L, Budinger GR, et al. BCL-2 protects human and mouse Th17 cells from glucocorticoid-induced apoptosis. Allergy. 2016;71(5):640–50.

115. Momcilovic M, Miljkovic Z, Popadic D, Markovic M, Savic E, Ramic Z, et al. Methylprednisolone inhibits interleukin-17 and interferon-gamma expression by both naive and primed T cells. BMC Immunol. 2008;9:47.

116. Schewitz-Bowers LP, Lait PJ, Copland DA, Chen P, Wu W, Dhanda AD, et al. Glucocorticoid-resistant Th17 cells are selectively attenuated by cyclosporine A. Proc Natl Acad Sci U S A. 2015;112(13):4080–5.

117. Chambers ES, Nanzer AM, Pfeffer PE, Richards DF, Timms PM, Martineau AR, et al. Distinct endotypes of steroid-resistant asthma characterized by IL-17A(high) and IFN-gamma(high) immunophenotypes: potential benefits of calcitriol. J Allergy Clin Immunol. 2015;136(3):628–37. e4.

118. Chen X, Oppenheim JJ, Winkler-Pickett RT, Ortaldo JR, Howard OM. Glucocorticoid amplifies IL-2-dependent expansion of functional FoxP3(+)CD4(+)CD25(+) T regulatory cells in vivo and enhances their capacity to suppress EAE. Eur J Immunol. 2006;36(8):2139–49.

119. Suarez A, Lopez P, Gomez J, Gutierrez C. Enrichment of CD4+ CD25high T cell population in patients with systemic lupus erythematosus treated with glucocorticoids. Ann Rheum Dis. 2006;65(11):1512–7.

120. Karagiannidis C, Akdis M, Holopainen P, Woolley NJ, Hense G, Ruckert B, et al. Glucocorticoids upregulate FOXP3 expression and regulatory T cells in asthma. J Allergy Clin Immunol. 2004;114(6):1425–33.

121. Chen X, Murakami T, Oppenheim JJ, Howard OM. Differential response of murine CD4+CD25+ and CD4+CD25- T cells to dexamethasone-induced cell death. Eur J Immunol. 2004;34(3):859–69.

122. Bereshchenko O, Coppo M, Bruscoli S, Biagioli M, Cimino M, Frammartino T, et al. GILZ promotes production of peripherally induced Treg cells and mediates the crosstalk between glucocorticoids and TGF-beta signaling. Cell Rep. 2014;7(2):464–75.

123. Prado C, Gomez J, Lopez P, de Paz B, Gutierrez C, Suarez A. Dexamethasone upregulates FOXP3 expression without increasing regulatory activity. Immunobiology. 2011;216(3):386–92.

124. Slade JD, Hepburn B. Prednisone-induced alterations of circulating human lymphocyte subsets. J Lab Clin Med. 1983;101(3):479–87.

125. Gruver-Yates AL, Quinn MA, Cidlowski JA. Analysis of glucocorticoid receptors and their apoptotic response to dexamethasone in male murine B cells during development. Endocrinology. 2014;155(2):463–74.

126. Smith LK, Cidlowski JA. Glucocorticoid-induced apoptosis of healthy and malignant lymphocytes. Prog Brain Res. 2010;182:1–30.

127. Bruscoli S, Biagioli M, Sorcini D, Frammartino T, Cimino M, Sportoletti P, et al. Lack of glucocorticoid-induced leucine zipper (GILZ) deregulates B-cell survival and results in B-cell lymphocytosis in mice. Blood. 2015;126(15):1790–801.

128. Barnes PJ. Corticosteroids, IgE, and atopy. J Clin Invest. 2001;107(3):265–6.

129. Xu J, Foy TM, Laman JD, Elliott EA, Dunn JJ, Waldschmidt TJ, et al. Mice deficient for the CD40 ligand. Immunity. 1994;1(5):423–31.

130. Kawabe T, Naka T, Yoshida K, Tanaka T, Fujiwara H, Suematsu S, et al. The immune responses in CD40-deficient mice: impaired immunoglobulin class switching and germinal center formation. Immunity. 1994;1(3):167–78.

131. Callard RE, Smith SH, Herbert J, Morgan G, Padayachee M, Lederman S, et al. CD40 ligand (CD40L) expression and B cell function in agammaglobulinemia with normal or elevated levels of IgM (HIM). Comparison of X-linked, autosomal recessive, and non-X-linked forms of the disease, and obligate carriers. J Immunol. 1994;153(7):3295–306.

132. Jabara HH, Fu SM, Geha RS, Vercelli D. CD40 and IgE: synergism between anti-CD40 monoclonal antibody and interleukin 4 in the induction of IgE synthesis by highly purified human B cells. J Exp Med. 1990;172(6):1861–4.

133. Fanslow WC, Anderson DM, Grabstein KH, Clark EA, Cosman D, Armitage RJ. Soluble forms of CD40 inhibit biologic responses of human B cells. J Immunol. 1992;149(2):655–60.

134. Ozbakir B, Crielaard BJ, Metselaar JM, Storm G, Lammers T. Liposomal corticosteroids for the treatment of inflammatory disorders and cancer. J Control Release. 2014;190:624–36.

135. Kleiman A, Tuckermann JP. Glucocorticoid receptor action in beneficial and side effects of steroid therapy: lessons from conditional knockout mice. Mol Cell Endocrinol. 2007;275(1-2):98–108.

136. Kulkarni NN, Gunnarsson HI, Yi Z, Gudmundsdottir S, Sigurjonsson OE, Agerberth B, et al. Glucocorticoid dexamethasone down-regulates basal and vitamin D3 induced cathelicidin expression in human monocytes and bronchial epithelial cell line. Immunobiology. 2016;221(2):245–52.

137. Cidlowski JA, Cidlowski NB. Regulation of glucocorticoid receptors by glucocorticoids in cultured HeLa S3 cells. Endocrinology. 1981;109(6):1975–82.

138. Okret S, Poellinger L, Dong Y, Gustafsson JA. Down-regulation of glucocorticoid receptor mRNA by glucocorticoid hormones and recognition by the receptor of a specific binding sequence within a receptor cDNA clone. Proc Natl Acad Sci U S A. 1986;83(16):5899–903.

139. Adams M, Meijer OC, Wang J, Bhargava A, Pearce D. Homodimerization of the glucocorticoid receptor is not essential for response element binding: activation of the phenylethanolamine N-methyltransferase gene by dimerization-defective mutants. Mol Endocrinol. 2003;17(12):2583–92.

140. Jewell CM, Scoltock AB, Hamel BL, Yudt MR, Cidlowski JA. Complex human glucocorticoid receptor dim mutations define glucocorticoid induced apoptotic resistance in bone cells. Mol Endocrinol. 2012;26(2):244–56.

141. Galigniana MD, Piwien-Pilipuk G, Assreuy J. Inhibition of glucocorticoid receptor binding by nitric oxide. Mol Pharmacol. 1999;55(2):317–23.

142. Webster JC, Oakley RH, Jewell CM, Cidlowski JA. Proinflammatory cytokines regulate human glucocorticoid receptor gene expression and lead to the accumulation of the dominant negative beta isoform: a mechanism for the generation of glucocorticoid resistance. Proc Natl Acad Sci U S A. 2001;98(12):6865–70.

143. Zhang P, Fang L, Wu H, Ding P, Shen Q, Liu R. Down-regulation of GRalpha expression and inhibition of its nuclear translocation by hypoxia. Life Sci. 2016;146:92–9.

144. Silverman MN, Sternberg EM. Glucocorticoid regulation of inflammation and its functional correlates: from HPA axis to glucocorticoid receptor dysfunction. Ann N Y Acad Sci. 2012;1261:55–63.

145. Mohamed NA, Abdel-Rehim AS, Farres MN, Muhammed HS. Influence of glucocorticoid receptor gene NR3C1 646 C>G polymorphism on glucocorticoid resistance in asthmatics: a preliminary study. Cent Eur J Immunol. 2015;40(3):325–30.

146. Hew M, Chung KF. Corticosteroid insensitivity in severe asthma: significance, mechanisms and aetiology. Intern Med J. 2010;40(5):323–34.

147. Papi A, Contoli M, Adcock IM, Bellettato C, Padovani A, Casolari P, et al. Rhinovirus infection causes steroid resistance in airway epithelium through nuclear factor kappaB and c-Jun N-terminal kinase activation. J Allergy Clin Immunol. 2013;132(5):1075–85. e6.

148. Kodama T, Shimizu N, Yoshikawa N, Makino Y, Ouchida R, Okamoto K, et al. Role of the glucocorticoid receptor for regulation of hypoxia-dependent gene expression. J Biol Chem. 2003;278(35):33384–91.

Part II
Cushing's Syndrome

Molecular Pathogenesis of Primary Adrenal Cushing's Syndrome

Nada El Ghorayeb, Isabelle Bourdeau, and André Lacroix

Abstract Recent advances in whole genome/exome sequencing have greatly accelerated our understanding of the molecular mechanisms of tumorigenesis in adrenocortical tumors and hyperplasia. Maintenance of hypercortisolism in primary adrenal Cushing's syndrome despite the suppression of ACTH secretion by the pituitary results from germline or somatic mutations in a variety of genes as well as from aberrant expression and function of several hormone receptors. This review focuses on novel genetic alterations involved in the cAMP signaling pathway or in armadillo proteins such as *ARMC5* and *β-catenin* as well as on autocrine/paracrine regulatory secretory loops responsible for the abnormal adrenal steroidogenesis in primary adrenal causes of Cushing's syndrome.

Keywords Cushing's syndrome • Adrenal steroidogenesis • Aberrant hormone receptors • Autocrine/paracrine regulation • *ARMC5* • *PRKACA* • *PRKAR1A* • *β-catenin* gene mutations

Introduction

Cushing's syndrome (CS) comprises all causes of hypercortisolism that are associated with symptoms and signs of prolonged exposure to inappropriately elevated free cortisol concentrations activating glucocorticoid (GC) and mineralocorticoid receptors expressed in most tissues [1]. The median age of diagnosis of endogenous CS is 41.4 years with a female-to-male ratio of 3:1. The incidence of this rare condition is estimated to be about 0.2–5.0 per million persons per year [2, 3]. Patients with CS have a higher risk of mortality compared to the general population particularly if left untreated, mainly from cardiovascular, venous thrombo-embolic, and infectious causes [4–6]. Exogenous administration of supraphysiological doses of GC to treat

N. El Ghorayeb • I. Bourdeau • A. Lacroix, M.D. (✉)
Division of Endocrinology, Department of Medicine, Centre de Recherche du Centre
hospitalier de l'Université de Montréal (CHUM), Université de Montréal,
Montréal, QC H2W 1T8, Canada
e-mail: ghorayebnada85@gmail.com; isabelle.bourdeau@umontreal.ca;
andre.lacroix@umontreal.ca

© Springer International Publishing Switzerland 2017
E.B. Geer (ed.), *The Hypothalamic-Pituitary-Adrenal Axis in Health
and Disease*, DOI 10.1007/978-3-319-45950-9_4

various inflammatory or oncologic conditions is the most frequent cause of CS. Endogenous etiologies are less frequent and are divided into corticotropin-dependent and corticotropin-independent causes [1]. Primary adrenal causes account for 20–30 % of overt endogenous hypercortisolism and include unilateral adrenal adenomas (10–20 %), carcinomas (5–7 %), or rarely bilateral adrenal hyperplasias (BAH) (<2 %). BAH is classified in two subtypes: macronodular (nodules >1 cm) and micronodular (nodules <1 cm) [7]. Macronodular disease, which was previously known as ACTH-independent macronodular adrenal hyperplasia was renamed as primary bilateral macronodular adrenal hyperplasia (BMAH) after the description of cortisol regulation by intraadrenal paracrine ACTH production in macronodular adrenals [8, 9]. Micronodular subtype includes the pigmented form of primary pigmented nodular adrenocortical disease (PPNAD) and the nonpigmented form of micronodular adrenocortical disease (MAD) [1, 10]. PPNAD presents either as isolated disease or as part of Carney complex (CNC).

In a patient with suspected CS, it is important to exclude a pseudo-Cushing's state, which is defined by the presence of clinical features of CS with some biochemical evidence of hypercortisolism. It could result from alcohol abuse, depression, or obesity. Its main feature resides in the disappearance of the Cushingoid state with the resolution of the underlying cause [1]. Manifestations of GC excess could be permanent or cyclical with mild-insidious or rapid-severe onset; they range from classic features as centripetal obesity, moon plethoric face, hirsutism, proximal myopathy, and easy bruising to more subtle features sometimes difficult to uncover, yet with major consequences on metabolism, bone, skin, eye, cardiovascular, neuropsychiatric, inflammatory, and reproductive systems [1]. Subclinical CS, which is most often discovered during evaluation of a unilateral or bilateral adrenal incidentalomas, refers to the presence of mild hypercortisolism (abnormal suppression to dexamethasone) in a patient who does not display overt signs of CS; "dysregulated hypercortisolism" seems to be more appropriate in describing this entity because patients with subclinical CS could present with nonspecific features of CS such as weight gain, hypertension, diabetes, or osteopenia, with considerable impact on their morbidity and mortality [11].

Normal Physiology of the Hypothalamic–Pituitary–Adrenal Axis

CRH, first identified in 1981 [12] is secreted into the hypophyseal portal blood, where it binds to specific type I CRH receptors on anterior pituitary corticotrophs to stimulate pro-opiomelanocortin (POMC) gene transcription through a process that includes activation of adenylate cyclase [13]. POMC, the precursor of ACTH, is a 241-amino-acid synthesized within the anterior pituitary. POMC is cleaved in a tissue-specific fashion to yield the secretion of β-lipoprotein (β-LPH) and pro-ACTH, the latter being further cleaved to an amino-terminal peptide, joining peptide, and ACTH itself in pituitary corticotroph cells [14–16]. The enzymes which specifically participate in the proteolysis of polypeptide hormone precursors have been identified as a superfamily of homologous subtilisin-like enzymes, called prohormone convertases and include PC1 (also called PC3) and PC2 [17, 18]. Although CRH is the principal stimulator for ACTH secretion, arginin-vasopressin (AVP) is able to potentiate CRH-mediated secretion by acting through the V1B receptor to activate protein kinase C [19]. Other factors such as stress, food ingestion, and circadian rhythm can modulate POMC secretion in addition to angiotensin II, cholecystokinin, atrial natriuretic factor, and vasoactive peptides [20]. ACTH is a 39-amino-acid peptide and its first 24 amino acids are common to all species. Pituitary control on adrenocortical function was described in the 1920s, but it was not until ACTH was isolated from sheep that it was shown to stimulate adrenal GC biosynthesis and secretion [21]. The precursor of γ-melanocyte stimulating hormone (pro-γ-MSH) is cleaved by a serine protease, which is expressed in the outer adrenal cortex and it is thought to mediate the trophic action of "ACTH" on the adrenal cortex [22]. The adult pyramidal-shaped adrenal gland weighs approximately 4 g; it is located on the posteromedial surface of the kidney. Cortisol secreting cells in the zona fasciculata (ZF), which comprises 75 % of the cortex are large and lipid laden and form radial cords within the fibrovascular radial network; in contrast, the small aldosterone-secreting cells are clustered in spherical nests under the adrenal capsule and the irregular androgen-secreting cells containing fewer lipid droplets and localized on the inner portion of the adrenal cortex. Adrenal cell renewal is thought to occur through the amplification, centripetal migration, and differentiation of initially undifferentiated subcapsular mesenchymal progenitor cells [23]. Cellular proliferation from a progenitor population occurs in a zone lying between the ZG and ZF; then cells migrate to ZF where they will undergo differentiation. Adrenal steroidogenesis from the common cholesterol precursor occurs in a specific "zonal" manner and involves the synchronized action of several cytochromes P450, which are classified according to cellular localization into mitochondrial (type I) and micrososomal (type II) segments [24, 25]. ACTH can result in reversible changes in adrenal cortex with glomerulosa cells adopting a fasciculate phenotype, whereas fasciculate cells adopt a reticularis phenotype. An important aspect of CRH and ACTH secretion is the classic endocrine negative feedback control exerted by cortisol. It is principally mediated via the GC receptor which inhibits POMC gene transcription in the anterior pituitary [26, 27] as well as CRH/AVP mRNA synthesis in the hypothalamus [28, 29]. The synthesis and release of annexin 1 (formerly known as lipocortin 1), from the folliculo-stellate cells of the anterior pituitary gland is induced by the binding of GC to its receptor; it participates in the negative feedback of GC on ACTH and CRH release which is particularly pertinent for the early onset actions of steroids that are mediated via a nongenomic mechanism [30].

The different diagnostic and therapeutic strategies to identify the etiologies of CS are beyond the scope of this review and are found elsewhere [1, 31]. A giant step forward to uncover the pathogenesis of adrenocortical tumors was made possible in recent years by major advances in genetic technologies. In this section, we will review the progress in molecular mechanisms regulating steroidogenesis in CS, despite suppression of ACTH, which includes germline or somatic mutations in a variety of genes as well as aberrant protein expression and function [1] (Table 1).

Table 1 Molecular mechanisms implicated in adrenal Cushing's syndrome

	Adrenocortical adenoma	Micronodular disease PPNAD/iMAD/CNC	Macronodular disease BMAH	MAS
A. Genetic alterations				
1. cAMP/ PKA signaling pathway	–	–	*MC2R^a (missense)*	–
	GNAS^a	–	*GNAS^a*	*GNAS^a (postzygotic)*
	PRKAR1A^b (allelic losses)	**PRKAR1A^b (LOH)**	–	–
	PRKACAa *(missense or insertion)*	*PRKACA^c*	*PRKACA^c*	–
	PDE8B^b	*PDE8B^b*	*PDE8B^b*	–
	PDE11A^b		*PDE11A^b*	–
2. Armadillo proteins	–	–	**ARMC5^b(LOH,** *nonsense or missense)*	–
	CTNNB1^a AXIN2^a	*CTNNB1^a*	–	–
3. Other	–	–	*MEN1^b, FAP^b, FH^b, EDNRA, DOTL1, HDAC9, PRUNE2*	–
B. Abnormal protein expression				
	GPCR	–	**GPCR ACTH Serotonin, vasopressin**	–
	PRKAR1A	–	PRKAR1A	–
	–	PRKACA	–	–
	–	Glucocorticoid receptor	–	–
	–	Estrogen receptor	–	–

The most frequent mechanisms are highlighted in bold
PPNAD = primary pigmented nodular adrenocortical disease, iMAD = isolated micronodular adrenocortical disease, CNC = Carney Complex, BMAH = bilateral macronodular adrenal hyperplasia, MAS = McCune–Albright syndrome
[a]Activating mutation
[b]Inactivating mutation
[c]Gene duplication (complex genomic rearrangements resulting in copy number gain leading either to micronodular or macronodular hyperplasia depending on the extent of gene amplification)

Genetic Alterations Leading to Abnormal Steroidogenesis

Role of cAMP/PKA Signaling Pathway in Adrenal Steroidogenesis and Proliferation

In primary adrenal causes of CS, the production of corticotropin-releasing hormone (CRH) in hypothalamus and of adrenocorticotropin (ACTH) by the corticotroph cells is suppressed by excess secretion of cortisol. The binding of ACTH to its specific melanocortin type 2 receptor (MC2R) regulates cortisol secretion; MC2R is a seven transmembrane domain receptor that belongs to the family of G-protein-coupled hormone receptor (GPCR) [32, 33]; it is expressed on zona fasciculata cells that interacts with MC2R-associated proteins [34] and induces the dissociation of Gs-α subunit, which generates cAMP from ATP by activation of adenylate cyclase (AC) [35]. The second messenger cAMP and its effector protein kinase A (PKA) are key regulators of adrenocortical cells. PKA is a prototypical serine/threonine kinase consisting of a dimer of two regulatory (with four known isoforms RIα, RIβ, RIIα, RIIβ) and two catalytic subunits (with four isoforms Cα, Cβ, Cγ, Prk) [36]. They constitute a tetramer in its inactive holoenzyme form [37] where two cAMP molecules are needed to bind to specific domains of the R subunits of PKA thereby dissociating the tetramer and releasing the C subunit (PRKACA) from its inactivating regulatory subunits; activated PRKACA phosphorylates different intracellular targets, including the transcription factor c-AMP-responsive element-binding protein (CREB). The latter activates the transcription of cAMP-responsive element containing genes in the nucleus including cholesterol transporters and steroidogenic enzymes, which stimulates acutely cortisol synthesis and chronically cellular proliferation [38, 39]. Specific phosphodiesterases (PDEs) are responsible of the degradation of the intracellular cAMP in order for the two R and C subunits of PKA to be reassembled to return to their inactive state [10] (Fig. 1a). Therefore, the cAMP signaling pathway appears to play a fundamental role in regulation of metabolism, cell replication, differentiation, and apoptosis in adrenal tissues; this implies that any defect in this pathway leading to its constitutive activation would be expected to result in cell proliferation and excess hormone production [40] (Table 1).

MC2R Mutations

MC2R mutations are extremely rare causes of adrenal hyperplasia or tumor formation [41, 42]. In only two patients with BMAH, constitutive activation of the *MC2R* with consequent enhanced basal receptor activity resulted either from impaired desensitization of a C-terminal *MC2R* mutation (F278C) [43] or from synergistic interaction between two naturally occurring missense mutations in the same allele of the *MC2R*: substitution of Cys 21 by Arg (C21R) and of Ser 247 by Gly (S247G) [44].

Gs-α Subunit Mutations

Activating mutations of the *Gs-α* subunit of heterotrimeric G protein also termed *gsp* mutations *(GNAS)* were the first identified in primary adrenal CS [45, 46]. It occurred in a mosaic pattern in some fetal adrenal cells during early embryogenesis resulting in the local constitutive activation of the cAMP pathway. This mutation was identified initially in the McCune–Albright Syndrome (MAS) where a minority of patients develops nodular adrenal hyperplasia and CS among other more common manifestations such as *café au lait* spots and bone fibrous dysplasia or other endocrine tumors causing ovarian precocious puberty, acromegaly, or

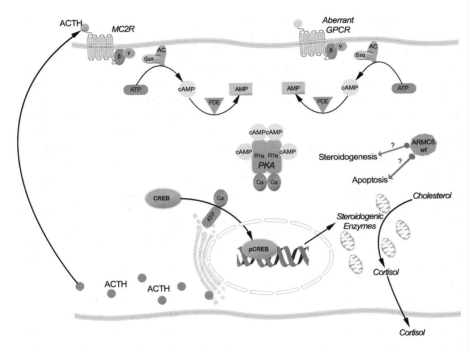

Fig. 1 (**a**) Schematic representation of the cAMP signaling pathway involved in the control of cortisol secretion in primary adrenal Cushing's syndrome. The binding of corticotropin (ACTH) to the melanocortin type 2 receptor (MC2R) leads to dissociation of Gs-α subunit and activation of adenylate cyclase (AC) generating cAMP from ATP. The binding of cAMP to specific domains of the regulatory subunits of protein kinase A (R1 α) dissociates the tetramer thereby releasing the catalytic subunit (C α), which phosphorylates different intracellular targets, including the transcription factor CREB; the latter activates the transcription of cAMP-responsive element-containing genes in the nucleus including cholesterol transporters and steroidogenic enzymes. Specific phosphodiesterases (PDEs) are responsible of the degradation of the intracellular cAMP in order for the two R1 α and C α subunits of PKA to be reassembled to return to their inactive state. Genetic defect in this pathway leading to its constitutive activation can underlie tumor development and excess hormone production. (**b**) Bilateral macronodular adrenal hyperplasia cells can express several functional aberrant G protein-coupled hormone receptors (GPCR). Activation of these receptors by their natural ligands induces the activation of intracellular cascade similar to the one activated normally by the binding of ACTH to MC2R thereby stimulating the release of both cortisol and locally produced ACTH which also triggers cortisol production through autocrine and paracrine mechanisms involving the MC2R. (**c**) Armadillo repeat-containing 5 (ARMC5), a new indirect or direct regulator of steroidogenesis and apoptosis. ARMC5 inactivating mutations induce a decreased steroidogenic capacity and a protection against cell death

hyperthyroidism [45, 47]. In MAS patients with CS, *GNAS* mutations are found in the cortisol-secreting nodules, whereas the internodular adrenal cortex which is not affected by the mutation becomes atrophic as ACTH becomes suppressed. Isolated somatic *GNAS* mutations can also occur in 5–17 % of cortisol-secreting adenomas

[48–50] and in rare cases of BMAH [51, 52] without any other manifestations of MAS. This suggests that the somatic mutation in MAS occurs at an early stage of embryogenesis in cells which are precursors of several tissues. In isolated BMAH, the somatic mutation probably occurs in mosaic pattern in more differentiated adrenocortical progenitor cells only which will migrate to generate bilateral macronodular adrenal glands; a somatic *GNAS* mutation giving rise to a unilateral adenoma occurs later in life in a single committed zona fasciculata cell.

PRKAR1A Mutations

PRKAR1A is an adrenocortical tumor suppressor gene according to in vitro and transgenic mouse studies. Its inactivation leads to ACTH-independent cortisol secretion [36, 53]. Constitutive PKA activation due to *PRKAR1A* mutations results either from reduced expression of the RIα subunits or from impaired binding to C subunits [54]. Loss of RIα is sufficient to induce autonomous adrenal hyperactivity and bilateral hyperplasia and was demonstrated for the first time in vivo in an adrenal cortex-specific *PRKAR1A* KO mouse model referred to as AdKO. Pituitary-independent CS with increased PKA activity developed in AdKO mice with evidence of deregulated adrenocortical cells differentiation, increased proliferation, and resistance to apoptosis. Moreover, RIα loss led to regression of adult cortex and emergence of a new cell population with fetal characteristics [53]. In vitro and in vivo models of PPNAD (AdKO mice) showed that PKA signaling increased mTOR complex 1, leading to increased cell survival and possibly tumor formation [55]. Tumor-specific loss of heterozygosity (LOH) involving the 17q22-24 chromosomal region harboring PRKAR1A and inactivating mutations of *PRKAR1A* are responsible for CS in isolated or familial PPNAD and CNC [54, 56–58]. They are found in more than 60 % of patients with CNC and in up to 80 % of CNC patients who develop CS from PPNAD [57, 59]. Furthermore, somatic allelic losses of the17q22–24 region and inactivating mutations in *PRKAR1A* were identified in 23 and 20 % of adrenocortical tumors, respectively [60]. Although, *PRKAR1A* mutations are not found in BMAH, somatic losses of the 17q22–24 region and PKA subunit and enzymatic activity changes show that PKA signaling is altered in BMAH similarly to what is found in adrenal tumors with 17q losses or *PRKAR1A* mutations [61]. CS presenting in persons younger than 30 years of age with bilateral, small (usually 2–4 mm in diameter), black-pigmented adrenal nodules are all characteristics of PPNAD. A distinctive feature of PPNAD compared to BMAH is the presence of atrophy in the internodular adrenal tissue. CNC is a familial autosomal variant that includes PPNAD among other tumors such atrial myxomas, peripheral nerve tumors, breast/testicular tumors, and GH-secreting pituitary tumors along with skin manifestations [62]. Patients with CS due to *PRKAR1A* mutations tend to have a lower BMI with evidence of increased PKA signaling in periadrenal adipose tissue, which is in concordance with the role of PKA enzyme in the regulation of adiposity and fat distribution [63].

PRKACA Mutations

The most frequent mechanism of adrenal CS secondary to unilateral adrenal adenoma involves somatic mutations in the gene encoding the catalytic subunit of PKA *(PRKACA)*. They occur in patients diagnosed with CS at a younger age (45.3 ± 13.5 vs. 52.5 ± 11.9 years) [49] with a female predominance [64]. The first two mutations identified in a cohort of ten cortisol-producing adrenal adenomas were shown to inhibit the binding of the R subunit making the Cα subunit constitutively active [65]. A combination of biochemical and optical assays, including fluorescence resonance energy transfer in living cells showed that neither mutant can form a stable PKA complex, due to the location of the mutations at the interface between the catalytic and the regulatory subunits [66]. The most common mutation p.Leu206Arg was present in 37 % of these adrenal tumors [65]. It consists of substitution of a small hydrophobic leucine with a large positively charge hydrophilic arginine at position 206. It is located in the active cleft of the C subunit and it inactivates the site where the regulatory subunit RIIβ binds leading to cAMP-independent PKA activation. The second mutation (Leu199_Cys200insTrp) entails the insertion of a tryptophan residue between the amino acid 199 and 200 and was present in one case only. Later, two novel mutations were identified in a study of 22 adrenal adenomas with CS with p.Cys200_Gly201insVal and p.Ser213Arg+p.Leu212_Lys214insIle-Ile-Leu-Arg being found in three and one adenomas, respectively. They indirectly interfere with the formation of a stable PKA holoenzyme by impairing the association between C and R subunits [67]. Other groups confirmed the presence of these mutations in unilateral adrenal adenomas with overt hypercortisolism at a rate of 23–65 % [48, 49, 64, 67, 68]. However, they are seldom present in adenomas with mild cortisol secretion, which might justify why subclinical CS rarely becomes overt CS [65, 67, 68]. These observations suggest that subclinical CS has a different genetic etiology than overt CS rather than being a part of the same pathophysiological spectrum [69]. In contrast to somatic mutations causing cortisol-secreting adenomas, germline complex genomic rearrangements in the chromosome 19p13.2p13.12 locus, resulting in copy number gains that includes *PRKACA* gene rarely lead either to micronodular or macronodular hyperplasia depending on the extent of gene amplification [65, 70, 71]. Finally, Bimpaki et al. demonstrated that adrenal adenomas of patients with CS could have functional abnormalities of cAMP signaling, independently of their *GNAS, PRKAR1A, PDE11A*, and *PDE8B* mutation status most probably due to epigenetic events or other gene defects [72].

PDE Mutations

PDE play a role in the hydrolysis of cAMP. There are two types of PDE8 enzymes coded by two distinct genes, *PDE8A* and *PDE8B*, which are highly expressed in steroidogenic tissues such as the adrenal, ovaries, and the testis as well as in the pituitary, thyroid, and pancreas [73, 74]. Genetic ablation of *PDE8B* in mouse models or long-term pharmacological inhibition of PDE8s in adrenocortical cell lines was shown to increase the expression of steroidogenic enzymes such as StAR and p450scc (CYP11A); furthermore, they potentiated ACTH stimulation of steroidogenesis by increasing cAMP-dependent PKA activity [75]. A *PDE8B* missense mutation (p.H305P) was described in a young girl with isolated micronodular adrenocortical disease (iMAD), which is a nonpigmented micronodular hyperplasia without *PRKAR1A* [76]. HEK293 cells transfected with the *PDE8B* mutant gene exhibited higher cAMP levels than with wild-type *PDE8B*, indicating an impaired ability of the mutant protein to degrade cAMP [76]. Other inactivating mutations in phosphodiesterase 11A isoform 4 gene *(PDE11A)* and 8B *(PDE8B)* have been also described in adrenal adenomas, carcinomas, and BMAH [1, 50, 72, 75, 77–79].

Role of Armadillo Proteins in Adrenal Tumorigenesis

Armadillo Proteins form a large family of proteins that are characterized by the presence of tandem repeats of a 42 amino acid motif with each single ARM-repeat unit consisting of 3 α-helices [80]. The most well-known protein of this family is β-catenin, which is crucial in the regulation of development and adult tissue homeostasis through its two independent functions, acting in cellular adhesion in addition to being a transcriptional coactivator (Fig. 2b). Deregulation in the Wnt/ β-catenin signaling pathway is involved in the pathogenesis of adrenocortical adenomas and carcinomas (Fig. 2c). Armadillo repeat containing five *(ARMC5)* is a novel Armadillo (ARM)-repeat-containing gene and encodes a protein of 935 amino acids; its peptide sequence reveals two distinctive domains: ARM domain in the N-terminal and a BTB/POZ in the C-terminal (Bric-a-Brac, Tramtrack, Broad-complex/Pox virus, and Zinc finger) [81]. *ARMC5* mutations were recently identified to be related to primary bilateral macronodular adrenal hyperplasia [80] (Fig. 1c).

β-Catenin Mutations

Regulatory mechanisms of cortisol production in adrenocortical carcinomas remain not fully elucidated. Decreased activity of steroidogenic enzymes translates into elevated urinary metabolites of several androgens or glucocorticoid precursors [82]. However, the Wnt/β-catenin signaling pathway appears to play a key role in the pathogenesis of both adrenal adenomas and carcinomas. β-catenin forms a complex with other proteins (adenomatous polyposis coli and axin, which facilitates its phosphorylation, making it available for degradation in the absence of Wnt signaling [83] (Fig. 2a). Adrenocortical carcinomas can harbor among other mutations in conserved serine/threonine phosphorylation sites at the amino terminus of β-catenin that block its phosphorylation within the destruction complex, thereby preventing its ubiquitinylation and proteasomal degradation; consequently, it accumulates in the nucleus and forms active transcription factor complexes with T cell factor/lymphoid enhancer factor proteins [83] (Fig. 2c, Table 1). Although *CTNNB1* mutations are

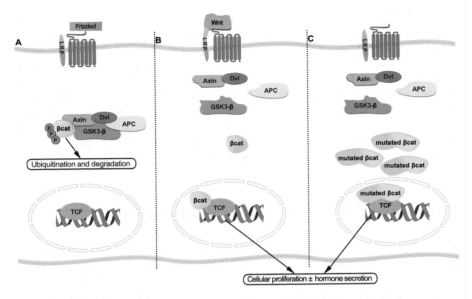

Fig. 2 (**a**) Schematic representation of deregulation in the Wnt/β-catenin signaling pathway involved in the pathogenesis of adrenocortical adenomas and carcinomas. In the absence of a Wnt signal, β-catenin is captured by adenomatous polyposis coli (APC) and axin within the destruction complex, facilitating its phosphorylation by casein kinases 1 α (CK1 α) and glycogen synthase kinase-3 β (GSK3β) through the ubiquitin pathway involving interactions with β-transducin repeat-containing protein (β-TrCP). (**b**) The presence of Wnt ligand inhibits the destruction complex activity. Therefore, β-catenin is accumulated in the cytoplasm; it may migrate to the nucleus where it activates the transcription of target genes, upon its activation. (**c**) *β-catenin* mutations block its phosphorylation within the destruction complex preventing its ubiquitinylation and proteasomal degradation; consequently, β-catenin accumulates in the nucleus and forms active transcription factor complexes with T cell factor/lymphoid enhancer factor proteins (TCF/LEF)

mainly observed in larger and nonsecreting adrenocortical adenomas, suggesting that the Wnt/β-catenin pathway activation is associated with the development of less differentiated tumors, Bonnet et al. described *β-catenin* mutations in 6 and 8 out of 19 and 46 subclinical and overt cortisol-producing tumors, respectively [84]. Recently, Goh et al. identified *β-catenin (CTNNB1)* mutations as responsible for 16 % of the cortisol-secreting adenomas [49]; they were also noted by other groups in some cases of adrenal adenomas with CS or SCS [64, 85, 86]. Somatic *β-catenin* mutations were also found in 2 out of 18 patients with PPNAD (11 %). In both cases, the mutations occurred in relatively larger adenomas that had formed in the background of PPNAD [87] (Table 1). Somatic *CTNNB1* mutations may explain only about 50 % of β-catenin accumulation observed in adrenocortical tumors, indicating that other components of the Wnt pathway may be involved; in fact, genetic alterations of the negative regulator of Wnt signaling, "*AXIN2* gene" were identified in adrenocortical adenomas and carcinomas yet at a low prevalence, 7 and 17 %, respectively [88] (Fig. 2c, Table 1).

ARMC5 Mutations

Inactivating germline mutations in *ARMC5* gene were first described in apparently sporadic cases of BMAH [89] (Table 1). The bilateral nature of macronodular hyperplasia as well as its long and insidious onset motivated the search for a genetic predisposition that could result in earlier diagnosis and better management to avoid bilateral adrenalectomy. Single-nucleotide polymorphism arrays, microsatellite markers, whole-genome and Sanger sequencing were applied to genotype leucocyte and tumor DNA obtained from patients with BMAH. The search for the responsible genes was conducted in apparently sporadic and familial cases [89–94]. The initial germline mutation in the *ARMC5* gene, located at 16p11.2 was detected in 18 out of 33 apparently sporadic tumors 55% of cases of BMAH with Cushing's syndrome [89]. Further studies in sporadic cases found that the prevalence of germline ARMC5 mutations was closer to 25% [90, 92, 93]. Inactivation of *ARMC5* is biallelic, one mutated allele being germline and the second allele being a somatic secondary event that occurs in a macronodule; these findings are consistent with its role as a potential tumor suppressor gene according to Knudson's 2-hits model [89, 92]. Correa et al. demonstrated that *ARMC5* has an extensive genetic variance by Sanger sequencing 20 different adrenal nodules in the same patient with BMAH [95]. They found the same germline mutation in the 20 nodules (p.Trp476* sequence change) but uncovered 16 other mutation variants in 16 of the nodules. This suggests that the germline mutation is responsible for the diffuse hyperplasia but second somatic hits are required to enhance adrenal macronodular formation [89, 95]. In the first large BMAH family studied, a heterozygous germline variant in the *ARMC5* gene (p.Leu365Pro) was identified in all 16 affected Brazilian family members as well as other mutations in two of three other families [92]. Interestingly, only two mutation carriers had overt CS and the majority had subclinical disease and one carrier had no manifestations despite being 72 year old. In addition, in one-third of the affected individuals only unilateral adrenal lesion was present as progression of the full-blown disease, needing many years and requiring the occurrence of additional somatic mutations in several macronodules. This raises the question of the prevalence of *ARMC5* mutation in apparently unilateral incidentalomas in the general population. Other families with BMAH have also been identified with *ARMC5* mutations or alterations [91, 94, 96, 97]. A germline deletion rather than mutation of *ARMC5* was reported in a family presenting with vasopressin-responsive SCS and BMAH [96]. By applying droplet digital polymerase chain reaction, the mother and her son had germline deletion in exon 1–5 of *ARMC5* gene locus. Furthermore, Sanger sequencing of DNA from the right and left adrenal nodules as well as peripheral blood of the son revealed the presence of another germline, missense mutation in *ARMC5* exon 3 (p.P347S) [96].

The presence of *ARMC5* mutation in patients with BMAH and aberrant GPCR has been reported, but the relationship has not been well established yet. The most frequent aberrant responses were to upright posture, isoproterenol, vasopressin, and metoclopramide tests [89, 90, 97]. In contrast, none of the patients with food-dependent CS carried *ARMC5* mutations [89, 90]. *ARMC5* inactivation decreases steroidogenesis, and its overexpression alters cell survival, which could argue why relatively inefficient cortisol overproduction is seen despite massive adrenal enlargement [7, 8, 98]. Despite this, the index cases operated for Cushing's syndrome and carrying *ARMC5* mutations carriers presented more severe CS than cases operated for Cushing's syndrome without *ARMC5* mutation; carrier patients had a more severe clinical phenotype and biochemical profile as well as larger adrenals on imaging with a higher number of nodules [90, 93]. *ARMC5* mutations appear to be the most frequent genetic alteration in BMAH with 61 different mutations, 27 germinal and 30 somatic, found all along the protein in different domains. Thus, genetic counseling and screening for these mutations are highly encouraged in family members of patients with BMAH even without the evidence of a clinical disease [8, 81, 92]. As *ARMC5* appears to be a tumor suppressor gene and is widely expressed in many tissues other than the adrenal, it was of interest to examine whether mutation carriers could develop other tumors. In a few families with BMAH, the occurrence of intracranial meningiomas was described and a somatic *ARMC5* mutation was found in a meningioma of a patient with familial BMAH with a germline *ARMC5* mutation suggesting the possibility of a new multiple neoplasia syndrome [94]. Finally, *ARMC5* mutations have been identified in primary hyperaldosteronism where 6 patients of 56 (10.7 %, all Afro-Americans) had germline mutations in the *ARMC5* gene; among these 6 patients, 2 suffered from BMAH [99].

Other Genetic Defects Associated with Abnormal Steroidogenesis

Several other gene mutations have been reported in patients with CS mainly presenting with BMAH including the multiple endocrine neoplasia type 1 (*MEN1*), familial adenomatous polyposis (*APC*), type A endothelin receptor (*EDNRA*) [52, 98, 100, 101]. Furthermore, somatic mutations other than *ARMC5* have also been found in patients with BMAH such as the *DOT1L* (DOT1-like histone H3K79 methyltransferase) and *HDAC9* (histone deacetylase 9) genes; these two nuclear proteins are involved in the transcriptional regulation; however, their mutations were found at a much lower frequency than *ARMC5* [64] (Table 1).

In a Carney Complex patient without Cushing's syndrome but with skin pigmentation, acromegaly and myxomas, gene triplication of chromosome 1p31.1, including *PRKACB*, which codes for the catalytic subunit beta (Cβ) resulted in increased PKA activity. It is likely that whereas the loss of RIα leads to the full Carney complex phenotype, the gain of function in Cα leads to adrenal adenomas and Cushing's syndrome, while in this case, amplification of Cβ resulted in certain nonadrenal manifestations of CNC [102].

Major Molecular Mechanisms Involved in Adrenal CS Other than Genetic Mutations

Independently of circulating ACTH, many bioactive signals released in the vicinity of adrenocortical cells by chromaffin cells, neurons, cells of the immune system, adipocytes, and endothelial cells can influence the secretory activity of the normal adrenal cortex [103, 104]. In contrast to the mechanisms that mainly lead to constitutive activation of the cAMP system or deregulation in the Wnt/β-catenin signaling pathway, abnormal regulation of steroidogenesis can result from the aberrant adrenal expression of several hormone receptors, particularly GPCR [105–108] and from aberrant autocrine/paracrine loops [9] (Fig. 1). These concepts offer the possibility of targeted therapy using specific receptor-targeted peptide antagonists [108]. These mechanisms are implicated in the pathogenesis of adrenal CS (Table 1) as well as in other endocrine tumors such as primary hyperaldosteronism and pituitary tumors; yet they are the most frequently described mechanism of regulation of hypercortisolism in BMAH [108]. Despite being a rare disease representing less than 1 % of all causes of CS [1], the prevalence of incidentally discovered BMAH due to extensive development and use of abdominal imaging has markedly increased [109]. It is diagnosed in the fifth and sixth decades and occurs more frequently in women [7]. The adrenal glands are hypertrophied with a mass reaching 10–100 times the normal weight of an adrenal gland [110]; however, most of the patients have subclinical hypercortisolism [98]. This discrepancy might be explained by the unequal distribution of steroidogenic enzymes among the different adrenocortical cell types leading to inefficient steroidogenesis [110, 111] as well as to decreased expression of steroidogenic enzymes [89, 112].

Aberrant Expression and Function of GPCR

The expression of ectopic receptors that are not expressed at significant levels in normal zona fasciculata cells and the increased expression or coupling to steroidogenesis of eutopic receptors can lead to abnormal cortisol production by mimicking the cellular events that are triggered normally by MC2R [7] (Fig. 1b). There are ectopic receptors such as those for glucose-dependent insulinotropic peptide (GIPR), β-adrenergic receptors (β-AR), vasopressin AVP (V_2-V_3 R), serotonin (5-HT$_7$R), glucagon (GCGR), and angiotensin II (AT1R). Among eutopic receptors are those for vasopressin (V_1R), luteinizing hormone/human chorionic gonadotropin (LHCGR), or serotonin (5-HT$_4$R) [108]. Five systematic studies have screened for aberrant expression of GPCR in overt and SCS; they demonstrated abnormal expression of more than one type of GPCR with 80 % showing aberrant cortisol responses to at least one stimulus. Multiple responses within individual patients occurred with up to four stimuli in 50 % of the patients; AVP and 5-HTR$_4$ agonists were the most prevalent hormonal stimuli triggering aberrant responses in vivo [52, 113–116].

The percentage of aberrant responses in patients with unilateral adenoma and mild CS or SCS was similar to those in BMAH patients [114]. However, it was less frequent in patients with unilateral adenomas and overt CS [113] most probably due to higher prevalence of *PRKACA* mutations in these patients [65].

Food-Dependent CS

Initially described by Hamet et al. in a case of unilateral adenoma and CS [117], food-dependent CS was identified to be GIP dependent by two different groups in cases of BMAH [118, 119]. To date, more than 30 cases of ectopic GIPR expression were published, being the most extensively studied GPCR in BMAH and unilateral adenomas, though it is not the most prevalent [107]. The transfection of bovine adrenal cells with the GIPR and its injection under the renal capsule in mice led to the development of hyperplastic adrenals and hypercortisolism which supports the role of the GIPR in steroidogenesis and cell proliferation [120]. Low fasting plasma cortisol levels in the morning due to suppressed pituitary ACTH, which increases following meals and its physiological elevation of GIP, is the hallmark of GIP-dependent CS. Since other aberrant GPCR can be expressed with GIPR in the same tissue such as LHCGR and 5-HT$_4$R, fasting cortisol levels may not always be suppressed [121–123]. Short-term control of hypercortisolism in BMAH patients with aberrant expression of GIPR was achieved by octreotide or pasireotide presumably because GIP suppression escapes as downregulation of somatostatin receptors in K cells occurs during chronic administration of the long-acting agonists [124, 125]. In vitro, ACTH-receptor antagonists were able to significantly inhibit cortisol secretion in perifused GIP-dependent BMAH tissues because GIP stimulated ACTH secretion and this effect was reduced by blocking ACTH binding to its own receptor [9].

Posture-Dependent CS

Upright posture induces abnormal steroidogenesis in BMAH with aberrant expression of either β-adrenergic receptors (β-AR), vasopressin AVP (V$_2$–V$_3$-vasopressin receptor), or angiotensin II (AT1R). Further, in vivo testing with β-agonists, vasopressin, and angiotensin II can identify each of these aberrant receptors, respectively [107]. Administration of antagonists of V1aR, AT1R, or β-AR was effective in reducing cortisol levels in patients with posture-related CS [126–129] [130]. Posture and specifically AVP were the most prevalent hormonal stimulus triggering aberrant responses in vivo [108].

LH/hCG-Dependent CS

Cases of transient CS during sequential pregnancies with spontaneous resolution following delivery were reported, while persistent CS occurred only after menopause, as a consequence of aberrant adrenal expression of LH/hCGR;

coexpression of LH/CGR with GIPR and 5-HT$_4$R was found in some patients [121–123, 131, 132]. Some cases of CS outside of pregnancy were also found with aberrant cortisol response to injection of GnRH and hCG [123, 133]. A heterozygous mutation of Gsα at codon 201 was found in addition to the aberrant LH/CG receptors [134]. Chronically elevated serum LH following gonadectomy induced functional LH receptor expression in mouse adrenal cortex, leading to adrenal hyperplasia and LH-dependent hypercortisolism [135]. Leuprolide acetate, a GnRH analog, was able to achieve long-term control of hypercortisolism in LH/hCG-dependent CS [136]. Aberrant expression of LH/CGR and GNRHR was described in pregnant and postmenopausal patients with primary hyperaldosteronism [137]. Recently, activating *β-catenin* mutations (*CTNNB1*) were identified in aldosteronomas which largely overexpressed GNRHR and LHCGR, suggesting that these mutations stimulate Wnt activation and cause adrenocortical cells to de-differentiate toward their common adrenal-gonadal precursor cell type [138].

Serotonin-Dependent CS

In patients with aberrant expression of 5-HT$_4$ R which is also among the most frequent aberrant responses [108], metoclopramide and cisapride (5-HT$_4$ agonists) can stimulate abnormal cortisol production on one hand and 5-HT$_4$ antagonists (GR113808) can inhibit steroidogenesis on the other hand [139, 140]. Ectopic expression of 5-HT$_7$R in an adrenocortical carcinoma cosecreting renin and cortisol as well in BMAH was also reported [141, 142].

Other GPCR

The presence of ectopic **glucagon** receptors was demonstrated in patients with subclinical or overt CS [115, 143–145] as well as the expression of **somatostatin** receptors SSTRs (particularly of SSTR1-3) was increased in PPNAD tissues carrying a *PRKAR1A* mutation compared to normal adrenal and to tissues from other adrenal diseases. Somatostatin analogs such as octreotide were not able to reduce cortisol significantly yet they remain a potential therapeutic tool in PPNAD [146]. Overexpression of receptors for **motilin** (MLNR), **gamma-aminobutyric acid** (GABBR1), and **α2 adrenergic** (ADRA2A) was identified in BMAH tissues by using transcriptome approach [147, 148]. Leptin, which normally inhibits stimulated cortisol secretion in humans, participated in cortisol hypersecretion in a case of BMAH with aberrant response to GIP [149].

Finally, some in vitro studies revealed the expression of GPCRs for **thyrotropin, follicle stimulating hormone** and **interleukin-1** in addition to those clearly confirmed in vivo [105, 106, 150].

Abnormal Autocrine/Paracrine Regulation of Cortisol Secretion

Autocrine Role of Intradrenal ACTH Produced in BMAH

Chromaffin cells of the adrenal gland can express the gene encoding POMC and therefore synthesize ACTH [151]. These chromaffin ACTH-producing cells have been described in BMAH tissues [152] (Fig. 1b). In two patients undergoing adrenal vein catheterization, a significant ACTH concentration gradient between the adrenal and the peripheral vein indicated that BMAH tissues are able to produce ACTH [9]. In a large recent series of 30 cases of BMAH, POMC mRNA and ACTH were expressed in the adrenocortical hyperplastic tissues along with the proconvertase 1, which converts POMC into ACTH. In addition, a positive correlation was observed between ACTH and cortisol levels in culture medium during perifusion of BMAH samples; another positive correlation was found between MC2R mRNA levels and POMC mRNA [9]. In fact, MC2R was upregulated by ACTH in BMAH tissues although it is normally underexpressed [147, 153]. The MC2R antagonist cortico-statin significantly inhibited the production of cortisol in vitro in contrast to dexamethasone and RU486 or CRH that failed to affect ACTH release indicating that intraadrenal ACTH is not regulated negatively by cortisol or stimulated by CRH [9]. Furthermore, it was demonstrated by the same group that ACTH synthesis might originate from abnormal gonadal-like differentiation of some adrenocortical cells since ACTH-producing cells were labeled by antibodies directed against the Leydig cell marker insulin-like 3 (INSL3) [9]. In vitro studies revealed that hyperplastic adrenal tissues secrete ACTH in a pulsatile manner in concordance with previous studies that demonstrated that cortisol production in patients with BMAH was pulsatile (67). Finally, 5-HT, LH/hCG, and GIP were found to stimulate ACTH release from BMAH tissues whereas MC2R antagonists were able to partially reduce the response of cortisol response to GIP [9]. Hence we can summarize that activation of GPCR in BMAH may stimulate cortisol production via two mechanisms including a direct effect on steroidogenesis [123], and an indirect action via ACTH secretion, which amplifies the action of these aberrant GPCR [9] (Fig. 1a, b).

Amplification of the Serotonin Paracrine Pathways in BMAH

Perivascular mast cells, located in the subcapsular region of the cortex, produce serotonin in the normal adrenal gland [154], which activates glucocorticoid synthesis through activation of the cAMP/PKA pathway [154–156]. However in BMAH, molecular and cellular defects reinforce the stimulatory effect of the intraadrenal serotonergic tone on cortisol production mainly due to the aberrant expression of the eutopic 5-HT$_4$R and ectopic 5HT$_7$R which are positively coupled with adenylyl cyclase [157] (Table 1). In the same context, PKA inhibitor H89 was found to inhibit the stimulatory action of serotonin on steroidogenesis in BMAH [142].

Regulation by Steroid Hormone Receptors

A distinctive feature of PPNAD is the paradoxical increase in urinary free cortisol during the 6-day dexamethasone suppression test (Liddle test), which was found in 69–75% of two small series of patients with PPNAD [158, 159]. Conversely, no paradoxical increase in cortisol was seen in nine patients with BMAH, but it was observed in three of 15 patients with a unilateral adenoma [159]. The glucocorticoid receptor (GR) was largely overexpressed in PPNAD nodules [160] (Table 1). In these cases, cortisol secretion was regulated by a **glucocorticoid receptor**-mediated effect on PKA catalytic subunits: the PKA inhibitor and RU486 inhibited the cortisol response to dexamethasone. The stimulatory effect of dexamethasone on cortisol release was not reduced by an adenylyl cyclase inhibitor or potentiated by a phosphodiesterase inhibitor or a cAMP analog [158]. Independently of the presence or absence of PRKAR1A mutation, dexamethasone was found to increase glucocorticoid synthesis in vitro and in vivo, suggesting a direct effect on adrenocortical tissue of PPNAD/CNC patients. Furthermore, in a patient with PPNAD, who had increased cortisol secretion during pregnancy and oral contraceptive use (and dexamethasone), β-estradiol (E_2) stimulated cortisol secretion in a dose–response manner in the absence of ACTH [161]. In PPNAD tissues associated with CS, E_2 abnormally stimulated cortisol secretion through activation of overexpressed **estrogen receptors** ERα and G protein-coupled receptor 30 (GRP30) [162]. This finding may explain why the CS of PPNAD is more frequent after puberty in female patients with CNC [162] (Table 1).

Conclusion

In summary, tumorigenesis and abnormal regulation of steroidogenesis in primary adrenal CS can result from complex interactions between various mechanisms including aberrant expression and function of hormone receptors together with other genetic alterations in several signaling pathways mainly cAMP/PKA and Wnt/β-catenin activating cascades. Whether each isolated mechanism can constitute the initiating event or is the consequence of another genetic alteration leading to adrenal dedifferentiation, hyperplasia, or tumorigenesis remains to be clarified. Further research is needed to uncover the link between aberrant receptors and germline mutations such as *ARMC5*, *PRKACA*, or *β-catenin* in CS as well as in other steroid secreting syndromes. Specific alterations in the cAMP pathway, in autocrine/paracrine secretion loops and aberrant receptors may offer promising specific targeted medical therapies for bilateral diseases as well as targets for PET imaging with specific ligands in the future [9, 163].

References

1. Lacroix A, Feelders RA, Stratakis CA, Nieman LK. Cushing's syndrome. Lancet. 2015;386(9996):913–27.
2. Lindholm J, Juul S, Jorgensen JO, Astrup J, Bjerre P, Feldt-Rasmussen U, et al. Incidence and late prognosis of Cushing's syndrome: a population-based study. J Clin Endocrinol Metab. 2001;86(1):117–23.
3. Steffensen C, Bak AM, Rubeck KZ, Jorgensen JO. Epidemiology of Cushing's syndrome. Neuroendocrinology. 2010;92 Suppl 1:1–5.
4. Hammer GD, Tyrrell JB, Lamborn KR, Applebury CB, Hannegan ET, Bell S, et al. Transsphenoidal microsurgery for Cushing's disease: initial outcome and long-term results. J Clin Endocrinol Metab. 2004;89(12):6348–57.
5. Bolland MJ, Holdaway IM, Berkeley JE, Lim S, Dransfield WJ, Conaglen JV, et al. Mortality and morbidity in Cushing's syndrome in New Zealand. Clin Endocrinol (Oxf). 2011;75(4):436–42.
6. Dekkers OM, Horvath-Puho E, Jorgensen JO, Cannegieter SC, Ehrenstein V, Vandenbroucke JP, et al. Multisystem morbidity and mortality in Cushing's syndrome: a cohort study. J Clin Endocrinol Metab. 2013;98(6):2277–84.
7. Lacroix A. ACTH-independent macronodular adrenal hyperplasia. Best Pract Res Clin Endocrinol Metab. 2009;23(2):245–59.
8. Lacroix A. Heredity and cortisol regulation in bilateral macronodular adrenal hyperplasia. N Engl J Med. 2013;369(22):2147–9.
9. Louiset E, Duparc C, Young J, Renouf S, Tetsi Nomigni M, Boutelet I, et al. Intraadrenal corticotropin in bilateral macronodular adrenal hyperplasia. N Engl J Med. 2013;369(22):2115–25.
10. Stratakis CA, Boikos SA. Genetics of adrenal tumors associated with Cushing's syndrome: a new classification for bilateral adrenocortical hyperplasias. Nat Clin Pract Endocrinol Metab. 2007;3(11):748–57.
11. Nieman LK. Update on subclinical Cushing's syndrome. Curr Opin Endocrinol Diabetes Obes. 2015;22(3):180–4.
12. Vale W, Spiess J, Rivier C, Rivier J. Characterization of a 41-residue ovine hypothalamic peptide that stimulates secretion of corticotropin and beta-endorphin. Science. 1981;213(4514):1394–7.
13. Chen R, Lewis KA, Perrin MH, Vale WW. Expression cloning of a human corticotropin-releasing-factor receptor. Proc Natl Acad Sci U S A. 1993;90(19):8967–71.
14. Mains RE, Eipper BA, Ling N. Common precursor to corticotropins and endorphins. Proc Natl Acad Sci U S A. 1977;74(7):3014–8.
15. Bertagna XY, Nicholson WE, Sorenson GD, Pettengill OS, Mount CD, Orth DN. Corticotropin, lipotropin, and beta-endorphin production by a human nonpituitary tumor in culture: evidence for a common precursor. Proc Natl Acad Sci U S A. 1978;75(10):5160–4.
16. Nakanishi S, Inoue A, Kita T, Nakamura M, Chang AC, Cohen SN, et al. Nucleotide sequence of cloned cDNA for bovine corticotropin-beta-lipotropin precursor. Nature. 1979;278(5703):423–7.
17. Seidah NG, Marcinkiewicz M, Benjannet S, Gaspar L, Beaubien G, Mattei MG, et al. Cloning and primary sequence of a mouse candidate prohormone convertase PC1 homologous to PC2, Furin, and Kex2: distinct chromosomal localization and messenger RNA distribution in brain and pituitary compared to PC2. Mol Endocrinol. 1991;5(1):111–22.
18. Smeekens SP, Steiner DF. Identification of a human insulinoma cDNA encoding a novel mammalian protein structurally related to the yeast dibasic processing protease Kex2. J Biol Chem. 1990;265(6):2997–3000.
19. Hauger RL, Aguilera G. Regulation of pituitary corticotropin releasing hormone (CRH) receptors by CRH: interaction with vasopressin. Endocrinology. 1993;133(4):1708–14.
20. Watanabe T, Oki Y, Orth DN. Kinetic actions and interactions of arginine vasopressin, angiotensin-II, and oxytocin on adrenocorticotropin secretion by rat anterior pituitary cells in the microperifusion system. Endocrinology. 1989;125(4):1921–31.

Molecular Pathogenesis of Primary Adrenal Cushing's Syndrome

21. Li CH, Evans HM, Simpson ME. Adrenocorticotropic hormone. J Biol Chem. 1943;149: 413–24.

22. Bicknell AB, Lomthaisong K, Woods RJ, Hutchinson EG, Bennett HP, Gladwell RT, et al. Characterization of a serine protease that cleaves pro-gamma-melanotropin at the adrenal to stimulate growth. Cell. 2001;105(7):903–12.

23. Kim AC, Barlaskar FM, Heaton JH, Else T, Kelly VR, Krill KT, et al. In search of adrenocortical stem and progenitor cells. Endocr Rev. 2009;30(3):241–63.

24. Miller WL. Minireview: regulation of steroidogenesis by electron transfer. Endocrinology. 2005;146(6):2544–50.

25. Payne AH, Hales DB. Overview of steroidogenic enzymes in the pathway from cholesterol to active steroid hormones. Endocr Rev. 2004;25(6):947–70.

26. Lundblad JR, Roberts JL. Regulation of proopiomelanocortin gene expression in pituitary. Endocr Rev. 1988;9(1):135–58.

27. Gagner JP, Drouin J. Opposite regulation of pro-opiomelanocortin gene transcription by glucocorticoids and CRH. Mol Cell Endocrinol. 1985;40(1):25–32.

28. Davis LG, Arentzen R, Reid JM, Manning RW, Wolfson B, Lawrence KL, et al. Glucocorticoid sensitivity of vasopressin mRNA levels in the paraventricular nucleus of the rat. Proc Natl Acad Sci U S A. 1986;83(4):1145–9.

29. Keller-Wood ME, Dallman MF. Corticosteroid inhibition of ACTH secretion. Endocr Rev. 1984;5(1):1–24.

30. Buckingham JC, John CD, Solito E, Tierney T, Flower RJ, Christian H, et al. Annexin 1, glucocorticoids, and the neuroendocrine-immune interface. Ann N Y Acad Sci. 2006;1088:396–409.

31. Pivonello R, De Leo M, Cozzolino A, Colao A. The treatment of Cushing's disease. Endocr Rev. 2015;36(4):385–486.

32. Mountjoy KG, Robbins LS, Mortrud MT, Cone RD. The cloning of a family of genes that encode the melanocortin receptors. Science. 1992;257(5074):1248–51.

33. Cone RD, Mountjoy KG, Robbins LS, Nadeau JH, Johnson KR, Roselli-Rehfuss L, et al. Cloning and functional characterization of a family of receptors for the melanotropic peptides. Ann N Y Acad Sci. 1993;680:342–63.

34. Chan LF, Metherell LA, Clark AJ. Effects of melanocortins on adrenal gland physiology. Eur J Pharmacol. 2011;660(1):171–80.

35. de Joussineau C, Sahut-Barnola I, Levy I, Saloustros E, Val P, Stratakis CA, et al. The cAMP pathway and the control of adrenocortical development and growth. Mol Cell Endocrinol. 2012;351(1):28–36.

36. Almeida MQ, Stratakis CA. How does cAMP/protein kinase A signaling lead to tumors in the adrenal cortex and other tissues? Mol Cell Endocrinol. 2011;336(1-2):162–8.

37. Taylor SS, Ilouz R, Zhang P, Kornev AP. Assembly of allosteric macromolecular switches: lessons from PKA. Nat Rev Mol Cell Biol. 2012;13(10):646–58.

38. Pearce LR, Komander D, Alessi DR. The nuts and bolts of AGC protein kinases. Nat Rev Mol Cell Biol. 2010;11(1):9–22.

39. Sewer MB, Waterman MR. cAMP-dependent transcription of steroidogenic genes in the human adrenal cortex requires a dual-specificity phosphatase in addition to protein kinase A. J Mol Endocrinol. 2002;29(1):163–74.

40. Rosenberg D, Groussin L, Jullian E, Perlemoine K, Bertagna X, Bertherat J. Role of the PKA-regulated transcription factor CREB in development and tumorigenesis of endocrine tissues. Ann N Y Acad Sci. 2002;968:65–74.

41. Latronico AC, Reincke M, Mendonca BB, Arai K, Mora P, Allolio B, et al. No evidence for oncogenic mutations in the adrenocorticotropin receptor gene in human adrenocortical neoplasms. J Clin Endocrinol Metab. 1995;80(3):875–7.

42. Light K, Jenkins PJ, Weber A, Perrett C, Grossman A, Pistorello M, et al. Are activating mutations of the adrenocorticotropin receptor involved in adrenal cortical neoplasia? Life Sci. 1995;56(18):1523–7.

43. Swords FM, Baig A, Malchoff DM, Malchoff CD, Thorner MO, King PJ, et al. Impaired desensitization of a mutant adrenocorticotropin receptor associated with apparent constitutive activity. Mol Endocrinol. 2002;16(12):2746–53.
44. Swords FM, Noon LA, King PJ, Clark AJ. Constitutive activation of the human ACTH receptor resulting from a synergistic interaction between two naturally occurring missense mutations in the MC2R gene. Mol Cell Endocrinol. 2004;213(2):149–54.
45. Mauras N, Blizzard RM. The McCune-Albright syndrome. Acta Endocrinol Suppl. 1986;279:207–17.
46. Weinstein LS, Shenker A, Gejman PV, Merino MJ, Friedman E, Spiegel AM. Activating mutations of the stimulatory G protein in the McCune-Albright syndrome. N Engl J Med. 1991;325(24):1688–95.
47. Brown RJ, Kelly MH, Collins MT. Cushing syndrome in the McCune-Albright syndrome. J Clin Endocrinol Metab. 2010;95(4):1508–15.
48. Sato Y, Maekawa S, Ishii R, Sanada M, Morikawa T, Shiraishi Y, et al. Recurrent somatic mutations underlie corticotropin-independent Cushing's syndrome. Science. 2014;344(6186):917–20.
49. Goh G, Scholl UI, Healy JM, Choi M, Prasad ML, Nelson-Williams C, et al. Recurrent activating mutation in PRKACA in cortisol-producing adrenal tumors. Nat Genet. 2014;46(6):613–7.
50. Libe R, Fratticci A, Coste J, Tissier F, Horvath A, Ragazzon B, et al. Phosphodiesterase 11A (PDE11A) and genetic predisposition to adrenocortical tumors. Clin Cancer Res. 2008;14(12):4016–24.
51. Fragoso MC, Domenice S, Latronico AC, Martin RM, Pereira MA, Zerbini MC, et al. Cushing's syndrome secondary to adrenocorticotropin-independent macronodular adrenocortical hyperplasia due to activating mutations of GNAS1 gene. J Clin Endocrinol Metab. 2003;88(5):2147–51.
52. Hsiao HP, Kirschner LS, Bourdeau I, Keil MF, Boikos SA, Verma S, et al. Clinical and genetic heterogeneity, overlap with other tumor syndromes, and atypical glucocorticoid hormone secretion in adrenocorticotropin-independent macronodular adrenal hyperplasia compared with other adrenocortical tumors. J Clin Endocrinol Metab. 2009;94(8):2930–7.
53. Sahut-Barnola I, de Joussineau C, Val P, Lambert-Langlais S, Damon C, Lefrancois-Martinez AM, et al. Cushing's syndrome and fetal features resurgence in adrenal cortex-specific Prkar1a knockout mice. PLoS Genet. 2010;6(6), e1000980.
54. Kirschner LS, Carney JA, Pack SD, Taymans SE, Giatzakis C, Cho YS, et al. Mutations of the gene encoding the protein kinase A type I-alpha regulatory subunit in patients with the Carney complex. Nat Genet. 2000;26(1):89–92.
55. de Joussineau C, Sahut-Barnola I, Tissier F, Dumontet T, Drelon C, Batisse-Lignier M, et al. mTOR pathway is activated by PKA in adrenocortical cells and participates in vivo to apoptosis resistance in primary pigmented nodular adrenocortical disease (PPNAD). Hum Mol Genet. 2014;23(20):5418–28.
56. Groussin L, Jullian E, Perlemoine K, Louvel A, Leheup B, Luton JP, et al. Mutations of the PRKAR1A gene in Cushing's syndrome due to sporadic primary pigmented nodular adrenocortical disease. J Clin Endocrinol Metab. 2002;87(9):4324–9.
57. Groussin L, Kirschner LS, Vincent-Dejean C, Perlemoine K, Jullian E, Delemer B, et al. Molecular analysis of the cyclic AMP-dependent protein kinase A (PKA) regulatory subunit 1A (PRKAR1A) gene in patients with Carney complex and primary pigmented nodular adrenocortical disease (PPNAD) reveals novel mutations and clues for pathophysiology: augmented PKA signaling is associated with adrenal tumorigenesis in PPNAD. Am J Hum Genet. 2002;71(6):1433–42.
58. Groussin L, Horvath A, Jullian E, Boikos S, Rene-Corail F, Lefebvre H, et al. A PRKAR1A mutation associated with primary pigmented nodular adrenocortical disease in 12 kindreds. J Clin Endocrinol Metab. 2006;91(5):1943–9.
59. Cazabat L, Ragazzon B, Groussin L, Bertherat J. PRKAR1A mutations in primary pigmented nodular adrenocortical disease. Pituitary. 2006;9(3):211–9.

60. Bertherat J, Groussin L, Sandrini F, Matyakhina L, Bei T, Stergiopoulos S, et al. Molecular and functional analysis of PRKAR1A and its locus (17q22-24) in sporadic adrenocortical tumors: 17q losses, somatic mutations, and protein kinase A expression and activity. Cancer Res. 2003;63(17):5308–19.
61. Bourdeau I, Matyakhina L, Stergiopoulos SG, Sandrini F, Boikos S, Stratakis CA. 17q22-24 chromosomal losses and alterations of protein kinase a subunit expression and activity in adrenocorticotropin-independent macronodular adrenal hyperplasia. J Clin Endocrinol Metab. 2006;91(9):3626–32.
62. Boikos SA, Stratakis CA. Carney complex: the first 20 years. Curr Opin Oncol. 2007;19(1):24–9.
63. London E, Rothenbuhler A, Lodish M, Gourgari E, Keil M, Lyssikatos C, et al. Differences in adiposity in Cushing syndrome caused by PRKAR1A mutations: clues for the role of cyclic AMP signaling in obesity and diagnostic implications. J Clin Endocrinol Metab. 2014;99(2):E303–10.
64. Cao Y, He M, Gao Z, Peng Y, Li Y, Li L, et al. Activating hotspot L205R mutation in PRKACA and adrenal Cushing's syndrome. Science. 2014;344(6186):913–7.
65. Beuschlein F, Fassnacht M, Assie G, Calebiro D, Stratakis CA, Osswald A, et al. Constitutive activation of PKA catalytic subunit in adrenal Cushing's syndrome. N Engl J Med. 2014;370(11):1019–28.
66. Calebiro D, Hannawacker A, Lyga S, Bathon K, Zabel U, Ronchi C, et al. PKA catalytic subunit mutations in adrenocortical Cushing's adenoma impair association with the regulatory subunit. Nat Commun. 2014;5:5680.
67. Di Dalmazi G, Kisker C, Calebiro D, Mannelli M, Canu L, Arnaldi G, et al. Novel somatic mutations in the catalytic subunit of the protein kinase A as a cause of adrenal Cushing's syndrome: a European multicentric study. J Clin Endocrinol Metab. 2014;99(10): E2093–100.
68. Nakajima Y, Okamura T, Gohko T, Satoh T, Hashimoto K, Shibusawa N, et al. Somatic mutations of the catalytic subunit of cyclic AMP-dependent protein kinase (PRKACA) gene in Japanese patients with several adrenal adenomas secreting cortisol [Rapid Communication]. Endocr J. 2014;61(8):825–32.
69. Calebiro D, Di Dalmazi G, Bathon K, Ronchi CL, Beuschlein F. cAMP signaling in cortisol-producing adrenal adenoma. Eur J Endocrinol. 2015;173(4):M99–106.
70. Carney JA, Lyssikatos C, Lodish MB, Stratakis CA. Germline PRKACA amplification leads to Cushing syndrome caused by 3 adrenocortical pathologic phenotypes. Hum Pathol. 2015;46(1):40–9.
71. Lodish MB, Yuan B, Levy I, Braunstein GD, Lyssikatos C, Salpea P, et al. Germline PRKACA amplification causes variable phenotypes that may depend on the extent of the genomic defect: molecular mechanisms and clinical presentations. Eur J Endocrinol. 2015;172(6):803–11.
72. Bimpaki EI, Nesterova M, Stratakis CA. Abnormalities of cAMP signaling are present in adrenocortical lesions associated with ACTH-independent Cushing syndrome despite the absence of mutations in known genes. Eur J Endocrinol. 2009;161(1):153–61.
73. Tsai LC, Beavo JA. The roles of cyclic nucleotide phosphodiesterases (PDEs) in steroidogenesis. Curr Opin Pharmacol. 2011;11(6):670–5.
74. Chen C, Wickenheisser J, Ewens KG, Ankener W, Legro RS, Dunaif A, et al. PDE8A genetic variation, polycystic ovary syndrome and androgen levels in women. Mol Hum Reprod. 2009;15(8):459–69.
75. Tsai LC, Shimizu-Albergine M, Beavo JA. The high-affinity cAMP-specific phosphodiesterase 8B controls steroidogenesis in the mouse adrenal gland. Mol Pharmacol. 2011;79(4):639–48.
76. Horvath A, Mericq V, Stratakis CA. Mutation in PDE8B, a cyclic AMP-specific phosphodiesterase in adrenal hyperplasia. N Engl J Med. 2008;358(7):750–2.
77. Horvath A, Giatzakis C, Robinson-White A, Boikos S, Levine E, Griffin K, et al. Adrenal hyperplasia and adenomas are associated with inhibition of phosphodiesterase 11A in

carriers of PDE11A sequence variants that are frequent in the population. Cancer Res. 2006;66(24):11571–5.

78. Horvath A, Boikos S, Giatzakis C, Robinson-White A, Groussin L, Griffin KJ, et al. A genome-wide scan identifies mutations in the gene encoding phosphodiesterase 11A4 (PDE11A) in individuals with adrenocortical hyperplasia. Nat Genet. 2006;38(7):794–800.

79. Rothenbuhler A, Horvath A, Libe R, Faucz FR, Fratticci A, Raffin Sanson ML, et al. Identification of novel genetic variants in phosphodiesterase 8B (PDE8B), a cAMP-specific phosphodiesterase highly expressed in the adrenal cortex, in a cohort of patients with adrenal tumours. Clin Endocrinol (Oxf). 2012;77(2):195–9.

80. Berthon A, Stratakis CA. From beta-catenin to ARM-repeat proteins in adrenocortical disorders. Horm Metab Res. 2014;46(12):889–96.

81. Drougat L, Omeiri H, Lefevre L, Ragazzon B. Novel Insights into the Genetics and Pathophysiology of Adrenocortical Tumors. Front Endocrinol. 2015;6:96.

82. Arlt W, Biehl M, Taylor AE, Hahner S, Libe R, Hughes BA, et al. Urine steroid metabolomics as a biomarker tool for detecting malignancy in adrenal tumors. J Clin Endocrinol Metab. 2011;96(12):3775–84.

83. Mazzuco TL, Durand J, Chapman A, Crespigio J, Bourdeau I. Genetic aspects of adrenocortical tumours and hyperplasias. Clin Endocrinol (Oxf). 2012;77(1):1–10.

84. Bonnet S, Gaujoux S, Launay P, Baudry C, Chokri I, Ragazzon B, et al. Wnt/beta-catenin pathway activation in adrenocortical adenomas is frequently due to somatic CTNNB1-activating mutations, which are associated with larger and nonsecreting tumors: a study in cortisol-secreting and -nonsecreting tumors. J Clin Endocrinol Metab. 2011;96(2):E419–26.

85. Tadjine M, Lampron A, Ouadi L, Bourdeau I. Frequent mutations of beta-catenin gene in sporadic secreting adrenocortical adenomas. Clin Endocrinol (Oxf). 2008;68(2):264–70.

86. Masi G, Lavezzo E, Iacobone M, Favia G, Palu G, Barzon L. Investigation of BRAF and CTNNB1 activating mutations in adrenocortical tumors. J Endocrinol Invest. 2009;32(7):597–600.

87. Tadjine M, Lampron A, Ouadi L, Horvath A, Stratakis CA, Bourdeau I. Detection of somatic beta-catenin mutations in primary pigmented nodular adrenocortical disease (PPNAD). Clin Endocrinol (Oxf). 2008;69(3):367–73.

88. Chapman A, Durand J, Ouadi L, Bourdeau I. Identification of genetic alterations of AXIN2 gene in adrenocortical tumors. J Clin Endocrinol Metab. 2011;96(9):E1477–81.

89. Assie G, Libe R, Espiard S, Rizk-Rabin M, Guimier A, Luscap W, et al. ARMC5 mutations in macronodular adrenal hyperplasia with Cushing's syndrome. N Engl J Med. 2013;369(22):2105–14.

90. Espiard S, Drougat L, Libe R, Assie G, Perlemoine K, Guignat L, et al. ARMC5 Mutations in a Large Cohort of Primary Macronodular Adrenal Hyperplasia: Clinical and Functional Consequences. J Clin Endocrinol Metab. 2015;100(6):E926–35.

91. Gagliardi L, Schreiber AW, Hahn CN, Feng J, Cranston T, Boon H, et al. ARMC5 mutations are common in familial bilateral macronodular adrenal hyperplasia. J Clin Endocrinol Metab. 2014;99(9):E1784–92.

92. Alencar GA, Lerario AM, Nishi MY, Mariani BM, Almeida MQ, Tremblay J, et al. ARMC5 mutations are a frequent cause of primary macronodular adrenal Hyperplasia. J Clin Endocrinol Metab. 2014;99(8):E1501–9.

93. Faucz FR, Zilbermint M, Lodish MB, Szarek E, Trivellin G, Sinaii N, et al. Macronodular adrenal hyperplasia due to mutations in an armadillo repeat containing 5 (ARMC5) gene: a clinical and genetic investigation. J Clin Endocrinol Metab. 2014;99(6):E1113–9.

94. Elbelt U, Trovato A, Kloth M, Gentz E, Finke R, Spranger J, et al. Molecular and clinical evidence for an ARMC5 tumor syndrome: concurrent inactivating germline and somatic mutations are associated with both primary macronodular adrenal hyperplasia and meningioma. J Clin Endocrinol Metab. 2015;100(1):E119–28.

95. Correa R, Zilbermint M, Berthon A, Espiard S, Batsis M, Papadakis GZ, et al. The ARMC5 gene shows extensive genetic variance in primary macronodular adrenocortical hyperplasia. Eur J Endocrinol. 2015;173(4):435–40.

96. Suzuki S, Tatsuno I, Oohara E, Nakayama A, Komai E, Shiga A, et al. Germline deletion of armc5 in familial primary macronodular adrenal hyperplasia. Endocr Pract. 2015;21(10):1152–60.

97. Bourdeau IOS, Magne F, Lévesque I, Caceres K, Nolet S, Awadalla P, Tremblay J, Hamet P, Fragoso MC, Lacroix A. ARMC5 mutations in a large French-Canadian family with cortisol-secreting β-adrenergic/vasopressin responsive bilateral macronodular adrenal hyperplasia. Eur J Endocrinol. 2016;174(1):85–96.

98. De Venanzi A, Alencar GA, Bourdeau I, Fragoso MC, Lacroix A. Primary bilateral macronodular adrenal hyperplasia. Curr Opin Endocrinol Diabetes Obes. 2014;21(3):177–84.

99. Zilbermint M, Xekouki P, Faucz FR, Berthon A, Gkourogianni A, Schernthaner-Reiter MH, et al. Primary Aldosteronism and ARMC5 Variants. J Clin Endocrinol Metab. 2015;100(6):E900–9.

100. Fragoso MC, Alencar GA, Lerario AM, Bourdeau I, Almeida MQ, Mendonca BB, et al. Genetics of primary macronodular adrenal hyperplasia. J Endocrinol. 2015;224(1):R31–43.

101. Zhu J, Cui L, Wang W, Hang XY, Xu AX, Yang SX, et al. Whole exome sequencing identifies mutation of EDNRA involved in ACTH-independent macronodular adrenal hyperplasia. Fam Cancer. 2013;12(4):657–67.

102. Forlino A, Vetro A, Garavelli L, Ciccone R, London E, Stratakis CA, et al. PRKACB and Carney complex. N Engl J Med. 2014;370(11):1065–7.

103. Haase M, Willenberg HS, Bornstein SR. Update on the corticomedullary interaction in the adrenal gland. Endocr Dev. 2011;20:28–37.

104. Lefebvre H, Prevost G, Louiset E. Autocrine/paracrine regulatory mechanisms in adrenocortical neoplasms responsible for primary adrenal hypercorticism. Eur J Endocrinol. 2013;169(5):R115–38.

105. Lacroix A, Ndiaye N, Tremblay J, Hamet P. Ectopic and abnormal hormone receptors in adrenal Cushing's syndrome. Endocr Rev. 2001;22(1):75–110.

106. Lacroix A, Baldacchino V, Bourdeau I, Hamet P, Tremblay J. Cushing's syndrome variants secondary to aberrant hormone receptors. Trends Endocrinol Metab. 2004;15(8):375–82.

107. Lacroix A, Bourdeau I, Lampron A, Mazzuco TL, Tremblay J, Hamet P. Aberrant G-protein coupled receptor expression in relation to adrenocortical overfunction. Clin Endocrinol (Oxf). 2010;73(1):1–15.

108. Ghorayeb NE, Bourdeau I, Lacroix A. Multiple aberrant hormone receptors in Cushing's syndrome. Eur J Endocrinol. 2015;173(4):M45–60.

109. Mazzuco TL, Bourdeau I, Lacroix A. Adrenal incidentalomas and subclinical Cushing's syndrome: diagnosis and treatment. Curr Opin Endocrinol Diabetes Obes. 2009;16(3):203–10.

110. Sasano H, Suzuki T, Nagura H. ACTH-independent macronodular adrenocortical hyperplasia: immunohistochemical and in situ hybridization studies of steroidogenic enzymes. Mod Pathol. 1994;7(2):215–9.

111. Wada N, Kubo M, Kijima H, Ishizuka T, Saeki T, Koike T, et al. Adrenocorticotropin-independent bilateral macronodular adrenocortical hyperplasia: immunohistochemical studies of steroidogenic enzymes and post-operative course in two men. Eur J Endocrinol. 1996;134(5):583–7.

112. Antonini SR, Baldacchino V, Tremblay J, Hamet P, Lacroix A. Expression of ACTH receptor pathway genes in glucose-dependent insulinotrophic peptide (GIP)-dependent Cushing's syndrome. Clin Endocrinol (Oxf). 2006;64(1):29–36.

113. Mircescu H, Jilwan J, N'Diaye N, Bourdeau I, Tremblay J, Hamet P, et al. Are ectopic or abnormal membrane hormone receptors frequently present in adrenal Cushing's syndrome? J Clin Endocrinol Metab. 2000;85(10):3531–6.

114. Reznik Y, Lefebvre H, Rohmer V, Charbonnel B, Tabarin A, Rodien P, et al. Aberrant adrenal sensitivity to multiple ligands in unilateral incidentaloma with subclinical autonomous cortisol hypersecretion: a prospective clinical study. Clin Endocrinol (Oxf). 2004;61(3):311–9.

115. Libe R, Coste J, Guignat L, Tissier F, Lefebvre H, Barrande G, et al. Aberrant cortisol regulations in bilateral macronodular adrenal hyperplasia: a frequent finding in a prospective study

of 32 patients with overt or subclinical Cushing's syndrome. Eur J Endocrinol. 2010; 163(1):129–38.

116. Hofland J, Hofland LJ, van Koetsveld PM, Steenbergen J, de Herder WW, van Eijck CH, et al. ACTH-independent macronodular adrenocortical hyperplasia reveals prevalent aberrant in vivo and in vitro responses to hormonal stimuli and coupling of arginine-vasopressin type 1a receptor to 11beta-hydroxylase. Orphanet J Rare Dis. 2013;8:142.

117. Hamet P, Larochelle P, Franks DJ, Cartier P, Bolte E. Cushing syndrome with food-dependent periodic hormonogenesis. Clin Invest Med. 1987;10(6):530–3.

118. Lacroix A, Bolte E, Tremblay J, Dupre J, Poitras P, Fournier H, et al. Gastric inhibitory polypeptide-dependent cortisol hypersecretion—a new cause of Cushing's syndrome. N Engl J Med. 1992;327(14):974–80.

119. Reznik Y, Allali-Zerah V, Chayvialle JA, Leroyer R, Leymarie P, Travert G, et al. Food-dependent Cushing's syndrome mediated by aberrant adrenal sensitivity to gastric inhibitory polypeptide. N Engl J Med. 1992;327(14):981–6.

120. Mazzuco TL, Chabre O, Sturm N, Feige JJ, Thomas M. Ectopic expression of the gastric inhibitory polypeptide receptor gene is a sufficient genetic event to induce benign adrenocortical tumor in a xenotransplantation model. Endocrinology. 2006;147(2):782–90.

121. Albiger NM, Occhi G, Mariniello B, Iacobone M, Favia G, Fassina A, et al. Food-dependent Cushing's syndrome: from molecular characterization to therapeutical results. Eur J Endocrinol. 2007;157(6):771–8.

122. Dall'Asta C, Ballare E, Mantovani G, Ambrosi B, Spada A, Barbetta L, et al. Assessing the presence of abnormal regulation of cortisol secretion by membrane hormone receptors: in vivo and in vitro studies in patients with functioning and non-functioning adrenal adenoma. Horm Metab Res. 2004;36(8):578–83.

123. Bertherat J, Contesse V, Louiset E, Barrande G, Duparc C, Groussin L, et al. In vivo and in vitro screening for illegitimate receptors in adrenocorticotropin-independent macronodular adrenal hyperplasia causing Cushing's syndrome: identification of two cases of gonadotropin/gastric inhibitory polypeptide-dependent hypercortisolism. J Clin Endocrinol Metab. 2005;90(3):1302–10.

124. Preumont V, Mermejo LM, Damoiseaux P, Lacroix A, Maiter D. Transient efficacy of octreotide and pasireotide (SOM230) treatment in GIP-dependent Cushing's syndrome. Horm Metab Res. 2011;43(4):287–91.

125. Karapanou O, Vlassopoulou B, Tzanela M, Stratigou T, Tsatlidis V, Tsirona S, et al. Adrenocorticotropic hormone independent macronodular adrenal hyperplasia due to aberrant receptor expression: is medical treatment always an option? Endocr Pract. 2013;19(3):e77–82.

126. Daidoh H, Morita H, Hanafusa J, Mune T, Murase H, Sato M, et al. In vivo and in vitro effects of AVP and V1a receptor antagonist on Cushing's syndrome due to ACTH-independent bilateral macronodular adrenocortical hyperplasia. Clin Endocrinol (Oxf). 1998;49(3):403–9.

127. Nakamura Y, Son Y, Kohno Y, Shimono D, Kuwamura N, Koshiyama H, et al. Case of adrenocorticotropic hormone-independent macronodular adrenal hyperplasia with possible adrenal hypersensitivity to angiotensin II. Endocrine. 2001;15(1):57–61.

128. Lacroix A, Tremblay J, Rousseau G, Bouvier M, Hamet P. Propranolol therapy for ectopic beta-adrenergic receptors in adrenal Cushing's syndrome. N Engl J Med. 1997;337(20): 1429–34.

129. Mazzuco TL, Thomas M, Martinie M, Cherradi N, Sturm N, Feige JJ, et al. Cellular and molecular abnormalities of a macronodular adrenal hyperplasia causing beta-blocker-sensitive Cushing's syndrome. Arq Bras Endocrinol Metabol. 2007;51(9):1452–62.

130. Mazzuco TL, Chaffanjon P, Martinie M, Sturm N, Chabre O. Adrenal Cushing's syndrome due to bilateral macronodular adrenal hyperplasia: prediction of the efficacy of beta-blockade therapy and interest of unilateral adrenalectomy. Endocr J. 2009;56(7):867–77.

131. Bourdeau I, D'Amour P, Hamet P, Boutin JM, Lacroix A. Aberrant membrane hormone receptors in incidentally discovered bilateral macronodular adrenal hyperplasia with subclinical Cushing's syndrome. J Clin Endocrinol Metab. 2001;86(11):5534–40.

132. Goodarzi MO, Dawson DW, Li X, Lei Z, Shintaku P, Rao CV, et al. Virilization in bilateral macronodular adrenal hyperplasia controlled by luteinizing hormone. J Clin Endocrinol Metab. 2003;88(1):73–7.

133. Feelders RA, Lamberts SW, Hofland LJ, van Koetsveld PM, Verhoef-Post M, Themmen AP, et al. Luteinizing hormone (LH)-responsive Cushing's syndrome: the demonstration of LH receptor messenger ribonucleic acid in hyperplastic adrenal cells, which respond to chorionic gonadotropin and serotonin agonists in vitro. J Clin Endocrinol Metab. 2003;88(1):230–7.

134. Bugalho MJ, Li X, Rao CV, Soares J, Sobrinho LG. Presence of a Gs alpha mutation in an adrenal tumor expressing LH/hCG receptors and clinically associated with Cushing's syndrome. Gynecol Endocrinol. 2000;14(1):50–4.

135. Kero J, Poutanen M, Zhang FP, Rahman N, McNicol AM, Nilson JH, et al. Elevated luteinizing hormone induces expression of its receptor and promotes steroidogenesis in the adrenal cortex. J Clin Invest. 2000;105(5):633–41.

136. Lacroix A, Hamet P, Boutin JM. Leuprolide acetate therapy in luteinizing hormone-dependent Cushing's syndrome. N Engl J Med. 1999;341(21):1577–81.

137. Saner-Amigh K, Mayhew BA, Mantero F, Schiavi F, White PC, Rao CV, et al. Elevated expression of luteinizing hormone receptor in aldosterone-producing adenomas. J Clin Endocrinol Metab. 2006;91(3):1136–42.

138. Teo AE, Garg S, Shaikh LH, Zhou J, Karet Frankl FE, Gurnell M, et al. Pregnancy, Primary Aldosteronism, and Adrenal CTNNB1 Mutations. N Engl J Med. 2015;373(15):1429–36.

139. Vezzosi D, Cartier D, Regnier C, Otal P, Bennet A, Parmentier F, et al. Familial adrenocorticotropin-independent macronodular adrenal hyperplasia with aberrant serotonin and vasopressin adrenal receptors. Eur J Endocrinol. 2007;156(1):21–31.

140. Cartier D, Lihrmann I, Parmentier F, Bastard C, Bertherat J, Caron P, et al. Overexpression of serotonin4 receptors in cisapride-responsive adrenocorticotropin-independent bilateral macronodular adrenal hyperplasia causing Cushing's syndrome. J Clin Endocrinol Metab. 2003;88(1):248–54.

141. Louiset E, Isvi K, Gasc JM, Duparc C, Cauliez B, Laquerriere A, et al. Ectopic expression of serotonin7 receptors in an adrenocortical carcinoma co-secreting renin and cortisol. Endocr Relat Cancer. 2008;15(4):1025–34.

142. Louiset E, Contesse V, Groussin L, Cartier D, Duparc C, Barrande G, et al. Expression of serotonin7 receptor and coupling of ectopic receptors to protein kinase A and ionic currents in adrenocorticotropin-independent macronodular adrenal hyperplasia causing Cushing's syndrome. J Clin Endocrinol Metab. 2006;91(11):4578–86.

143. Contesse V, Reznik Y, Louiset E, Duparc C, Cartier D, Sicard F, et al. Abnormal sensitivity of cortisol-producing adrenocortical adenomas to serotonin: in vivo and in vitro studies. J Clin Endocrinol Metab. 2005;90(5):2843–50.

144. Matsukura S, Kakita T, Sueoka S, Yoshimi H, Hirata Y, Yokota M, et al. Multiple hormone receptors in the adenylate cyclase of human adrenocortical tumors. Cancer Res. 1980;40(10):3768–71.

145. Louiset E, Duparc C, Groussin L, Gobet F, Desailloud R, Barrande G, et al. Abnormal Sensitivity to Glucagon and Related Peptides in Primary Adrenal Cushing's Syndrome. Horm Metab Res. 2014;46(12):876–82.

146. Bram Z, Xekouki P, Louiset E, Keil MF, Avgeropoulos D, Giatzakis C, et al. Does somatostatin have a role in the regulation of cortisol secretion in primary pigmented nodular adrenocortical disease (PPNAD)? A clinical and in vitro investigation. J Clin Endocrinol Metab. 2014;99(5):E891–901.

147. Assie G, Louiset E, Sturm N, Rene-Corail F, Groussin L, Bertherat J, et al. Systematic analysis of G protein-coupled receptor gene expression in adrenocorticotropin-independent macronodular adrenocortical hyperplasia identifies novel targets for pharmacological control of adrenal Cushing's syndrome. J Clin Endocrinol Metab. 2010;95(10):E253–62.

148. Jaiswal N, Sharma RK. Dual regulation of adenylate cyclase and guanylate cyclase: alpha 2-adrenergic signal transduction in adrenocortical carcinoma cells. Arch Biochem Biophys. 1986;249(2):616–9.

149. Pralong FP, Gomez F, Guillou L, Mosimann F, Franscella S, Gaillard RC. Food-dependent Cushing's syndrome: possible involvement of leptin in cortisol hypersecretion. J Clin Endocrinol Metab. 1999;84(10):3817–22.
150. Willenberg HS, Stratakis CA, Marx C, Ehrhart-Bornstein M, Chrousos GP, Bornstein SR. Aberrant interleukin-1 receptors in a cortisol-secreting adrenal adenoma causing Cushing's syndrome. N Engl J Med. 1998;339(1):27–31.
151. Ehrhart-Bornstein M, Haidan A, Alesci S, Bornstein SR. Neurotransmitters and neuropeptides in the differential regulation of steroidogenesis in adrenocortical-chromaffin co-cultures. Endocr Res. 2000;26(4):833–42.
152. Pereira MAA, Araújo RS, Bisi H. Síndrome de Cushing associada à hiperplasia macronodular das adrenais: apresentação de um caso e revisão da literatura. Arq Bras Endocrinol Metabol. 2001;45(6):619–27.
153. Xing Y, Parker CR, Edwards M, Rainey WE. ACTH is a potent regulator of gene expression in human adrenal cells. J Mol Endocrinol. 2010;45(1):59–68.
154. Lefebvre H, Contesse V, Delarue C, Feuilloley M, Hery F, Grise P, et al. Serotonin-induced stimulation of cortisol secretion from human adrenocortical tissue is mediated through activation of a serotonin4 receptor subtype. Neuroscience. 1992;47(4):999–1007.
155. Contesse V, Hamel C, Lefebvre H, Dumuis A, Vaudry H, Delarue C. Activation of 5-hydroxytryptamine4 receptors causes calcium influx in adrenocortical cells: involvement of calcium in 5-hydroxytryptamine-induced steroid secretion. Mol Pharmacol. 1996;49(3): 481–93.
156. Lenglet S, Louiset E, Delarue C, Vaudry H, Contesse V. Activation of 5-HT(7) receptor in rat glomerulosa cells is associated with an increase in adenylyl cyclase activity and calcium influx through T-type calcium channels. Endocrinology. 2002;143(5):1748–60.
157. Raymond JR, Mukhin YV, Gelasco A, Turner J, Collinsworth G, Gettys TW, et al. Multiplicity of mechanisms of serotonin receptor signal transduction. Pharmacol Ther. 2001;92(2-3): 179–212.
158. Louiset E, Stratakis CA, Perraudin V, Griffin KJ, Libe R, Cabrol S, et al. The paradoxical increase in cortisol secretion induced by dexamethasone in primary pigmented nodular adrenocortical disease involves a glucocorticoid receptor-mediated effect of dexamethasone on protein kinase A catalytic subunits. J Clin Endocrinol Metab. 2009;94(7):2406–13.
159. Stratakis CA, Sarlis N, Kirschner LS, Carney JA, Doppman JL, Nieman LK, et al. Paradoxical response to dexamethasone in the diagnosis of primary pigmented nodular adrenocortical disease. Ann Intern Med. 1999;131(8):585–91.
160. Bourdeau I, Lacroix A, Schurch W, Caron P, Antakly T, Stratakis CA. Primary pigmented nodular adrenocortical disease: paradoxical responses of cortisol secretion to dexamethasone occur in vitro and are associated with increased expression of the glucocorticoid receptor. J Clin Endocrinol Metab. 2003;88(8):3931–7.
161. Caticha O, Odell WD, Wilson DE, Dowdell LA, Noth RH, Swislocki AL, et al. Estradiol stimulates cortisol production by adrenal cells in estrogen-dependent primary adrenocortical nodular dysplasia. J Clin Endocrinol Metab. 1993;77(2):494–7.
162. Bram Z, Wils J, Ragazzon B, Risk-Rabin M, Libe R, Young J, et al. β -Estradiol (E2) stimulates cortisol secretion in primary pigmented nodular adrenal disease: an explanation for the increased frequency of Cushing's syndrome in female patients with Carney complex. In: Adrenal tumors: novel causes and mechanisms, oral presentation at the Endocrine Society's 96th Annual Meeting and Expo, June 21–24, 2014, Chicago. p. OR14-2-OR-2.
163. Lefebvre H, Prévost G, Louiset E. Could targeting hormone receptors be an effective strategy in management of adrenal hyperplasia? Int J Endocr Oncol. 2014;1(1):11–4.

Pathogenesis and Treatment of Aggressive Corticotroph Pituitary Tumors

Yang Shen and Anthony P. Heaney

Abstract Although the majority of corticotroph pituitary tumors are microadenomas and amenable to complete surgical resection, a subset exhibits a higher frequency of local invasiveness, tendency to recur, and potential to progress to carcinoma. Specifically, Crooke's cell adenomas, silent corticotroph adenomas, and corticotroph tumors in the setting of Nelson's syndrome are often more aggressive. These tumors may exhibit higher Ki-67 and/or p53 immunostaining though this is not uniform. Emerging molecular markers show promise but have not yet been validated in routine clinical use. Therapy is generally multimodal with surgical debulking to alleviate compressive symptoms, radiation therapy to prevent or delay tumor growth, and medical therapies to manage the many adverse metabolic consequences of hypercortisolism. However, many of these agents do not inhibit tumor growth and recently temozolomide, an alkylating chemotherapy agent, has been demonstrated to offer stabilization and in some instances partial and/or complete regression of these aggressive corticotroph tumors.

Keywords Corticotroph tumors • ACTH-secreting pituitary tumors • Crooke's cell adenoma • Cushing's disease • Pituitary tumors • Pituitary carcinoma • Atypical pituitary tumor • Invasive pituitary tumor • Nelson's syndrome • Temozolomide

Introduction

Pituitary tumors are invariably benign tumors that either are clinically nonfunctioning (~60 % of all cases) or secrete the pituitary hormones prolactin, growth hormone, adrenocorticotrophic hormone, thyroid stimulating hormone, follicle stimulating hormone, or luteinizing hormone, with prolactin-secreting tumors being most common (~30 %). Pituitary carcinoma, defined as metastatic or

Y. Shen, M.D.
Indiana University School of Medicine/IU Health Arnett Clinic, Indianapolis, IN 46202, USA

A.P. Heaney, M.D., Ph.D. (✉)
UCLA School of Medicine, 200 UCLA Medical Plaza, suite 530, Los Angeles, CA 90095, USA
e-mail: aheaney@mednet.ucla.edu

© Springer International Publishing Switzerland 2017
E.B. Geer (ed.), *The Hypothalamic-Pituitary-Adrenal Axis in Health and Disease*, DOI 10.1007/978-3-319-45950-9_5

Table 1 Pituitary corticotroph tumor classification

(a) WHO classification

	Typical	Atypical	Carcinoma
Histologic features	Ki-67 < 3%	Ki-67 ≥ 3%, increased mitoses, extensive p53 staining	Ki-67 ≥ 3%, increased mitoses, extensive p53 staining
Size	Micro/ macroadenoma	Micro/ macroadenoma	Macroadenoma
Invasiveness	+/−	+	+
Metastases or craniospinal dissemination	No	No	Yes

(b) Clinicopathologic grading system proposed by the French Collaborative Study [6]

	Grade 1	Grade 2	Grade 3
Histologic features[a]	(a) – proliferation (b) + Proliferation	(a) – proliferation (b) + Proliferation	+ Proliferation
Size	Micro/ macroadenoma	Micro/ macroadenoma /giant adenoma	Macroadenoma/giant adenoma
Invasiveness[b]	−	+	+
Metastases or craniospinal dissemination	No	No	Yes

[a]Proliferation is defined based on the meeting of at least two of the following three criteria: Mitoses: $n \geq 2$ per 10 HPF; Ki-67 ≥ 3%; p53 > 10 strong positive nuclei per 10 HPF
[b]Invasiveness is defined as histologic or radiological signs of cavernous or sphenoid sinus invasion

craniospinal disseminated tumor at sites not contiguous with the sella, is very rare and seen in only 0.1–0.2% of all cases [1, 2]. However, pituitary tumors often invade surrounding sellar structures such as dura, bone, blood vessels, and nerve sheath [3–5]. These latter tumors are often those that display higher rates of tumor growth or regrowth, exhibit multiple recurrences, are variously labeled "aggressive," and often present a therapeutic challenge. However, whereas there is a WHO definition of an "atypical" pituitary tumor (Table 1), there is no universally accepted definition for an "aggressive" pituitary tumor and this term may mean different things to individual physicians. From an endocrinologist's perspective, in the setting of a corticotroph tumor the rapid appearance of symptoms of hypercortisolism such as rounded face, central obesity, and purple striae over months is evidence of a clinically/biochemically "aggressive" tumor in comparison to a patient who manifests similar symptoms appearing over several years. From the radiology or neurosurgical perspective, a large bulky pituitary corticotroph tumor that exhibits cavernous sinus or bony invasion may be indicators of an "aggressive" phenotype. Pathologists in their analysis interpret histological features such as mitotic rates and expression of proliferative markers to determine the "aggressiveness" of the pituitary tumor. Acknowledging the difficulties in correlating clinical, radiological and histopathological criteria, efforts have been made to develop a broader clinicopathologic classification of pituitary endocrine tumors that may offer more prognostic value, in which greater emphasis is put on proliferation assessment in conjunction with radiological invasiveness [6, 7]. It must be acknowledged that neither classification system has been clinically validated or assessed in prospective fashion. A further term that is sometimes used synonymously with aggressiveness but loosely defined is "recurrence." In discussing tumor recurrence, it is important to consider the time frame as most would agree that tumor recurrence 10 years after a complete resection would not be considered as "aggressive" in comparison to a tumor that recurred 6 months after complete resection. In addition, in literature review it is often difficult to differentiate true recurrence (i.e., growth of tumor after R0 resection) versus growth of residual tumor following subtotal resection. In summary, although no clear consensus on the definitions of an aggressive pituitary corticotroph tumor exists [8], most practitioners would agree that a tumor that exhibits rapid growth (such as presenting with early recurrence after a complete resection) or a tumor displaying hallmarks of high proliferation would be considered aggressive tumors.

Aggressive Corticotroph Tumor Subtypes

Three subtypes of corticotroph pituitary tumors that are known to have more aggressive behaviors have been defined (Table 2). First, the Crooke's cell adenoma (CCA) represents an entity with the distinctive histologic characteristic of extensive keratin deposition (Crooke's hyalinization) within greater than 50 % of the corticotroph tumor cells [12, 13]. The CCA is not to be confused with "Crooke's hyaline change" which was named after pathologist Dr. Arthur Crooke, who first described the phenomenon of keratin deposition in the normal corticotrophs, an appearance found in the setting of excess glucocorticoid either from diseases such as Cushing's syndrome or following exogenous glucocorticoid administration [9–11]. CCAs are rare, with only 80 cases reported [12]. This variant of corticotroph tumor presents with ACTH-dependent hypercortisolism similar to other Cushing's disease cases but is innately aggressive; most present as macroadenomas and exhibit marked cavernous or sphenoid sinus invasion at presentation. Compared to non-Crooke's cell adenomas, CCAs have a higher recurrence rate that approaches 70 % and more frequently progress to pituitary carcinoma [12]. Among the 36 cases that George et al. reported in 2003, 3 patients (5 %) died of the disease (one from multiple local recurrences and two from pituitary carcinoma) versus 0.01 % death rate for all pituitary tumors [13].

The second type of corticotroph tumors that may exhibit aggressive behavior are the so-called "silent" corticotroph adenomas (SCA). SCAs typically exhibit variable immunoreactivity for ACTH and other pro-opiomelanocortin (POMC)-derived peptides. They may secrete ACTH, which may be elevated in the circulation but often the patient does not exhibit clinical signs or biochemical evidence of hypercortisolism.

Table 2 Overview of salient features of corticotroph tumor subtypes

	Aggressive subtypes			Typical corticotroph tumors
	Silent corticotroph adenoma	Crooke's cell adenoma	Nelson's syndrome	
Histological features				
	ACTH immunopositive	ACTH immunopositive	ACTH immunopositive	ACTH immunopositive
		>50 % tumor cell positive for keratin deposition		
Clinical feature of Cushing's				
	Variably present	Present	Present	Present
			History of prior BLA	
Biochemical hypercortisolism				
	Often absent	Present	Present	Present
	Spectrum of increased plasma ACTH			
Radiological features				
Size	Often macroadenoma	Micro- or macroadenoma	Macroadenoma	Often microadenoma
invasiveness	Often present	Often present	Often present	Absent-variable
recurrence/ recurrence rate	Low to medium	High	Recurrence with rapid growth	Low
Clinical course				
Progress to carcinoma	Limited data but increased	Often	Limited data	Unlikely

ACTH; adrenocorticotrophic hormone, BLA; bilateral adrenalectomy

Rarely, SCAs may transform, in the course of the patient's disease, into functional adenomas with patients developing clinical hypercortisolism [14]. SCA as a distinct clinicopathologic entity was first proposed in 1978 when the classification of pituitary tumors was based on the tinctorial properties of the tumor cell cytoplasm, i.e., chromophobic, acidophilic, and basophilic. The case that was reported by Kovacs et al. at that time was a densely granulated basophilic cell adenoma which was immunoactive to ACTH antibodies but the patient was eucortisolemic before and after tumor resection [15]. Later, two morphologic variants of SCAs were defined: Type I are densely granulated basophilic tumors similar to functional corticotroph tumors whereas type II are chromophobic with varying ultrastructural patterns [16]. The morphological difference suggests that there might be variations in clinical phenotype resulting from these SCAs, but due to the small number of cases available, studies have mostly examined the collective features of the SCAs. Initially the mechanism proposed for the "silent" biochemical and clinical features of these tumors invoked impaired ACTH synthesis with enhanced lysosomal degradation of POMC peptides

[15]. More recent studies have demonstrated the incomplete processing of POMC, the precursor peptide of ACTH due to reduced expression of the prohormone convertase (PC1) enzymes PC1-3 [16–18]. Additionally the corticotroph-specific transcription factor TPIT was found to be lower in several SCAs, suggesting altered corticotroph differentiation in at least some of these tumors [17]. Due to the "silent" clinical course of these tumors, many are found as incidentalomas after brain imaging for other reasons or when patients present with symptoms of mass effect [19]. In a large surgical series, most SCAs were macroadenomas with suprasellar extension present in 87–100 % of cases, and compared to nonfunctional adenomas and functional ACTH-secreting tumors, SCAs exhibited a more aggressive clinical course with frequent recurrence [20–23]. Other retrospective reviews from individual institutions reported similar recurrence rates in SCAs as nonfunctioning adenomas [24, 25], but the pace of regrowth tended to be more aggressive [25]. Some patients with SCAs have been noted over time to manifest clinical signs and biochemical evidence of hypercortisolism. Whether this represents a true transformation of the tumor or more likely in the opinion of the authors the tumor as it enlarges attains a threshold of partially active ACTH secretion that can bind the ACTH receptor sufficiently to induce glucocorticoid excess is unclear.

The third setting where corticotroph tumors may behave aggressively is in the setting of Nelson's syndrome. In 1958, Dr. Nelson reported the development of an ACTH-secreting pituitary tumor following bilateral adrenalectomy (BLA) [26] and an early case series in 1979 found that 4 of 12 patients (33 %) treated with bilateral adrenalectomy for Cushing's disease developed pituitary corticotroph tumor growth (Nelson's syndrome). Two out of the 4 patients had spontaneous tumor infarction, one patient died from local tumor invasion despite radiation therapy and another patient had corticotroph tumor regrowth after surgical resection [27]. Nelson's syndrome reminds us of the role of glucocorticoid-mediated negative feedback to control pituitary corticotroph tumor growth whereby removal of cortisol-mediated negative feedback on the pituitary tumor serves as a growth stimulus [28]. A variety of risk factors have been implicated for corticotroph tumor growth after bilateral adrenalectomy including the presence of radiographically visible pituitary tumor remnant, young patient age, duration of Cushing's disease and lack of pituitary radiation prior to BLA. A recent study of 53 patients with Cushing's disease found that short duration of Cushing's before BLA and high plasma ACTH level in the year following BLA were independent predictors for pituitary corticotroph tumor progression, the latter most likely to occur within the first 3 years after BLA [29].

Aggressive Corticotroph Tumors: Role of Histopathological Indicators

As noted in the introduction, the 2004 WHO criteria list Ki-67 ≥ 3 % and extensive p53 immunostaining as indicators of an atypical pituitary tumor or carcinoma. Ki-67 is a well-validated marker expressed during the G1, G2-M, and

S-phase of the cell cycle. Commonly detected by the monoclonal MIB antibody it is reported in the form of Ki-67 labeling index (LI), indicating the number of Ki-67 positive cells in either 4×200 high-powered fields or by less standardized approaches (see later). Pituitary tumors exhibit a very broad range of Ki-67 LI with the vast majority of pituitary adenomas exhibiting Ki-67 LI between 1 and 2 % [30, 31]. In an early study based on 77 cases, Thapar et al. demonstrated significant differences in Ki-67 LI among 37 noninvasive tumors, 33 invasive tumors, and 7 pituitary carcinomas. The authors proposed that a threshold LI of 3 % could be used to distinguish invasive from noninvasive adenomas with 97 % specificity and 73 % sensitivity [5]. This threshold of 3 % was ultimately used in the WHO classification to differentiate atypical pituitary adenomas and carcinomas from typical pituitary tumors. However, prospective studies supporting this cutoff are lacking, and the utility of Ki-67 LI to robustly distinguish benign/typical versus invasive/atypical adenomas is not universally accepted [8, 30, 31]. For example, although one study found that Ki-67 LI was significantly higher in ACTH-secreting tumors versus other functional or nonfunctional tumors [32], that finding was not supported by other studies and despite increased growth in the setting of Nelson's syndrome as previously discussed, no significant association was found between Ki-67 LI and tumor recurrence in patients with Cushing's disease compared to tumors from patients with Nelson's syndrome [33]. Furthermore, the mean Ki-67 LI was relatively low at 0.7–0.8 % in 11 primary Crooke's cell adenomas, although Ki-67 LI was higher at 2.1–6.1 % in the recurrent Crooke's cell tumors. These mostly small single center studies must be interpreted with caution but would appear to highlight limitations of the Ki-67 LI as a stand-alone predictive marker of corticotroph tumor aggressiveness [13].

P53 is a cellular tumor antigen that plays an important role in genomic stability and cell proliferation. In the Thapar study above p53 immunoreactivity was also reported to correlate with pituitary tumor invasiveness and was expressed in 100 % of pituitary carcinoma cases [5]. However, subsequent studies have not observed a clear-cut association between p53 and invasiveness of pituitary tumors [32, 34, 35]. This in large part may be due to the considerable intra and intertumoral variability of p53 tumor expression and we must conclude that the independent role of p53 in predicting pituitary tumor behavior is quite limited.

The situation for these and other immunohistochemical markers is further complicated by the method of analysis for Ki-67 LI which is not standardized. Some pathologists "eyeball" the Ki-67 LI on analyzing variable numbers of tumor sections, more standardized quantitation methods (4×200 high power fields) are labor intensive and computed quantitation analysis is not universally available and may overestimate by counting infiltrating Ki-67 false positive inflammatory cells.

In summary, while the prognostic values of Ki-67 and p53 staining remain controversial, they are presently the most readily available tools to clinicians and it is prudent to monitor corticotroph and other pituitary tumors with Ki-67 LI outside the norm, i.e., >2–3 % more vigilantly.

Emerging Molecular Markers of Corticotroph Tumor Invasion and Aggression

The exact pathogenesis of pituitary tumors including aggressive pituitary cortico-troph is not fully understood but significant advances have been made in the past decade to further our understanding of the transformation of "benign" pituitary tumors to aggressive tumors and pituitary carcinoma. At present, there is no single biomarker that faithfully predicts pituitary tumor behavior [36, 37]. Multiple pathways, including occasional genetic mutations, dysfunctional hormonal and growth factor signaling pathways cooperate to promote pituitary tumor cellular proliferation. Several biomarkers of pituitary tumor aggressiveness have been implicated, though it is important to note the majority are not unique to corticotroph tumors.

For example, invasive pituitary tumors express higher levels of matrix metalloproteinases (MMPs), a class of proteinases that play a key role to break down basement membranes and connective tissues to enable tumor cell access to the extracellular environment. They can also coactivate other family members whereby MMP-2 activates MMP-9 [38]. In a small study, 9 of 10 (90 %) invasive pituitary tumors exhibited functional polymorphisms in the promotor region of the MMP-1 gene resulting in increased MMP-1 transcriptional activity [39, 40]. Additionally, expression of MMP-9 which degrades collagens, elastin, and gelatin was found to be higher in invasive pituitary adenomas [41–44]. In turn MMP-9 can serve to activate protein kinase C (PKC) further contributing to corticotroph tumor aggression [44].

The fibroblast growth factors (FGFs) and their receptors (FGFR) regulate growth, differentiation, migration, and angiogenesis. High levels of FGF mRNA and circulating FGF-2 levels have been reported in aggressive pituitary tumors. Reduction of β-catenin expression resulting in loss of cytoskeletal integrity has been implicated in the process. Also, FGFR4-R388, an FGFR4 allele associated with poor cancer prognosis, was found to be associated with MMP [45].

Vascular endothelial growth factors (VEGFs) and their receptors (VEGFRs) are key signaling proteins essential for tumor angiogenesis. Small case series have reported that pituitary tumors with high VEGF expression have a higher risk of extrasellar growth and recurrence [46]. In support for a role of the VEGF pathway in pituitary tumor aggressiveness, a study of 95 pituitary tumors found that lower expression of an inhibitor of VEGF, called vascular endothelial cell growth inhibitor (VEGI) was associated with suprasellar and sella destruction [47]. A further study reported higher expression of endocan, a proteoglycan involved in neoangiogenesis, in invasive pituitary tumors [48, 49].

Although classic oncogenic mutations such as Ras mutations are uncommon in pituitary tumors [55], a variety of inherited mutations have been implicated in pituitary tumorigenesis. For example, pituitary tumors, including corticotroph tumors, are found in multiple endocrine neoplasia type I (MEN1) and familial pituitary adenoma (FIPA). Some studies have reported that pituitary corticotroph adenomas in those inherited conditions may be larger and more often invasive than sporadic tumors [50, 51]. Both the MEN1 and aryl hydrocarbon receptor interacting protein (AIP) genes are located on chromosome 11q13 [52, 53]. Allelic deletion of 11q13 and an additional 3 loci (13q12–14, 10q, and 1p) and dysregulation of chromosome 11p was found to be more common in aggressive pituitary tumors [48, 54]. Studies in small numbers of invasive versus noninvasive prolactinomas identified ADAMTS6, CRMP1, and DCAMKL3 to be associated with invasion and ASK, CCNB1, AURKB1, CENPE, and PTTG with proliferation [56]. Pituitary tumor transformation gene (PTTG), for example, a member of the securin protein family that regulates sister chromatid separation during mitosis, has been studied extensively and shown to correlate with invasion in several tumor types including corticotroph tumors [57].

Most recently, mutations in ubiquitin-specific protease 8 (USP8), a gene coding a deubiquitinase that inhibits lysosomal degradation of epidermal growth factor receptor (EGFR), were identified in 40 % of corticotroph tumors [58]. Additionally overexpression of the heat shock protein 90 (HSP90) that alters glucocorticoid receptor folding thereby inducing glucocorticoid resistance was demonstrated in corticotroph tumors [59]. However, it is as yet unclear if either USP8 or HSP90 correlates with corticotroph tumor aggressiveness. Potentially, future molecular and histological analysis with established factors such as Ki-67 and p53 could be enhanced with integration of some emerging biomarker candidates such as MMP, PTTG, miRNAs, and chromosome deletion in 11p and 11q. However, the practical application of these biomarkers in routine clinical use as opposed to research studies has not yet been examined.

Role of Surgical Debulking/Resection in Aggressive Corticotroph Tumors

Surgical approaches to either obtain complete near-total resection or significant debulking remain first line therapy in the majority of corticotroph tumors. The wider exposure obtained and the enhanced direct visualization that angulated endoscopes provide may facilitate a more extensive surgical resection of tumors that extend beyond the sella into the cavernous sinuses and other parasellar structures. Occasionally a transcranial approach may be needed in tumors that extend significantly into the suprasellar region. With exceptions, aggressive corticotroph tumors tend to be invasive macroadenomas from presentation, and although it may be possible to achieve a visualized total resection with postoperative imaging showing "no residual tumor," these aggressive corticotroph pituitary tumors tend to recur, typically within 5 years [31]. As noted in prior sections, histopathology assessment may raise the possibility of tumor aggression, alerting the clinician to closely monitor the patient both biochemically (cortisol and ACTH parameters) and by imaging. As in other corticotroph tumors, low (<5 µg/dL) immediate postoperative serum cortisol is a good indicator of immediate remission [60–63]. Thereafter, patients require glucocorticoid replacement for typically 6–12 months. If a patient is able to stop glucocorticoid replacement sooner, this raises concern that they have not ever been fully in remission or have had early recurrence, the latter a potential clinical indicator of an aggressive corticotroph tumor.

Radiation Therapy

Whereas radiation therapy (RT) is not usually effective to induce corticotroph tumor shrinkage it can be helpful to prevent regrowth in subtotally resected corticotroph tumors or to slow growth of an expanding sellar lesion. Radiation can be delivered either as stereotactic radiosurgery (SRS) which involves delivery of high dose radiation typically in a single dose offering good efficacy and enhanced patient convenience, or in small daily dose fractions (fractionated RT) over 5–6 weeks [65]. Fractionated RT is particularly helpful when the tumor approximates radiation sensitive normal tissues that cannot be spared from the RT field. Various forms of radiation therapy exist, including gamma-knife, linear accelerator, cyber-knife, and proton beam therapy that can all be adapted to deliver either SRS or fractionated RT. To date, the greatest experience with SRS has been with gamma knife. Comparing success rates of the various radiation treatments is challenging due to differences in technique, doses administered, duration of follow-up, and definitions of tumor control and biochemical remission [66]. That said, a large retrospective single institution review of proton beam RT showed that actuarial 3-year biochemical remission was achieved in 54 % of 74 patients with persistent Cushing's disease and in 63 % of 8 patients with Nelson's syndrome. Time to biochemical remission was 32 months and 26 months, respectively, and tumor control was achieved in 98 % of the patients with Cushing's disease. The main adverse effect is panhypopituitarism which eventually occurred in 62 % of the 140 patients studied [67]. In another retrospective study involving 96 patients with persistent Cushing's disease treated with gamma knife RT after surgery, 70 % of patients achieved biochemical remission at a median follow-up of 48 months. Median time to remission was 16.6 months and tumor control was achieved in 98 % of patients [68]. As noted in these studies, an additional challenge of RT is delayed biochemical remission, necessitating use of medical therapy until radiation therapy controls the hypercortisolism.

Medical Management

Aggressive corticotroph pituitary tumors similar to any ACTH-secreting tumors may cause complications of hypercortisolism including hyperglycemia, hypertension, venous thromboembolism, and poor wound healing resulting in significant morbidity and mortality. Therefore effective control of hypercortisolism is of paramount importance at all stages in managing these patients, including across potentially definitive therapies such as radiation treatment. Several medical treatments aimed at lowering cortisol levels are currently available [69–71]. An ideal therapy would simultaneously lower ACTH and cortisol levels and offer tumor control with minimal side effects, but no such agent presently exists.

Medical therapies that act at the site of the tumor include the dopamine receptor-2 agonist cabergoline which is generally well tolerated and given its ease of administration can be considered as a medical option for aggressive corticotroph tumors. In patients with Cushing's disease, cabergoline normalized 24-h urinary free cortisol in 40 % of 18 patients and resulted in tumor shrinkage in 4/8 patients treated with doses ranging from 1 to 7 mg/week for 12–24 months [73–75]. However, most would consider D2 agonists weak anti-proliferative agents in corticotroph tumors.

Octreotide, a first generation somatostatin (SMS) analog predominantly targeting the somatostatin receptor subtype-2 (SSTR-2) has been reported to lower ACTH levels and stabilize tumor progression in some patients with Nelson's syndrome [76], but no consistent effect of octreotide is found in patients with Cushing's disease [77, 78]. Pasireotide (SOM 230), a somatostatin receptor ligand with higher binding affinity for SSTR-5, normalized 24-h urinary free cortisol in 20 % of patients with Cushing's disease [79, 80]. Data regarding the action of this agent on corticotroph tumor growth are awaited.

An alternate method to lower serum cortisol is the use of either adrenal steroidogenesis inhibitors such as ketoconazole, metyrapone, and mitotane or the glucocorticoid receptor (GR) antagonist mifepristone.

In one study of 38 patients, 21 of who had not undergone prior pituitary surgery ketoconazole treatment (200–1200 mg/day) normalized 24-h urinary free cortisol in 45 % of patients [81]. A large retrospective multicenter French study similarly reported normalized urinary free cortisol in 49 % of 200 patients [82]. Side effects include nausea, diarrhea (8 %), and skin rash (2 %), and gynecomastia in men (13 %). It is important to point out that ketoconazole like all adrenal- or GR-directed agents will not inhibit tumor growth but nonetheless can be very effective in controlling symptoms of hypercortisolism in combination with other therapies directed at tumor control. Metyrapone is also effective in controlling hypercortisolism. In one study normalization of 24-h urinary free cortisol was reported in 39 of 53 patients (75 %) with Cushing's disease after 1–6 weeks using a mean dose of 2250 mg [83]. Similar response rates were reported in a more recent UK study of 195 patients [84]. As for ketoconazole, gastrointestinal side effects of metyrapone predominate. Hirsutism and acne (70 %) due to androgen accumulation, as well as hypertension and edema (70 %) due to 11-deoxycorticosterone accumulation, can also be seen.

A more recently available method to control symptoms of glucocorticoid excess utilizes mifepristone, a glucocorticoid, androgen, and progesterone receptor antagonist. In a phase III open label study of 50 patients with endogenous Cushing's syndrome who had failed previous therapy, mifepristone led to clinical improvement in hyperglycemia (60 %) and hypertension (38 %) in predefined study subgroups [72]. Given the mechanism of action of the drug to block GR, cortisol cannot be used to guide dose titration and/or monitoring of side effects, and these must be assessed based on clinical symptoms and signs. Serum potassium level should also be monitored due to side effects of hypokalemia.

It must be acknowledged that the majority of these cortisol-lowering therapies have little or no impact on growth of aggressive corticotroph tumors. Indeed, in theory, though not proven in practice, drugs such as adrenal steroid synthesis inhibitors and the GR antagonist mifepristone by removing GR-mediated corticotroph tumor negative feedback could contribute to increased tumor growth. However, much morbidity and mortality in these aggressive corticotroph tumors is due to effects of hypercortisolism, these agents make up a very important component of the treatment regimen. In clinical practice, they are generally used in parallel with other strategies to achieve tumor control such as debulking surgery, radiation therapy, and systemic chemotherapy.

Chemotherapy

No randomized prospective studies of systemic chemotherapy have been conducted for patients with aggressive corticotroph pituitary tumors or indeed other pituitary tumor subtypes. Although aggressive pituitary corticotroph tumors grow and recur, they generally do not exhibit a high proliferative index. Therefore many chemotherapy regimens that offer responses in adenocarcinoma or sarcoma are not effective in patients with pituitary tumors. This aspect is not unique to pituitary tumors and similar observations have been made in other neuroendocrine tumor subtypes of the pancreas and gut. Case reports and small series have demonstrated that temozolomide (TMZ), originally approved for use in refractory glioblastoma multiforme, may offer tumor stabilization in both pituitary carcinoma and aggressive pituitary adenomas [85–87]. TMZ is a lipophilic imadozotetrazine derivative that is converted to a methylating alkylator agent, methyl-triazene-1-yl-imidazole-4-carboxamide (MTIC). MTIC induces DNA damage by base pair mismatch of O^6-methyl-guanine (O^6-meG) with thymidine in the sister chromatid instead of cytosine, resulting in DNA strand breaks and ultimately tumor cell apoptosis [88]. The DNA repair enzyme, O^6-methyl-guanine-DNA methyltransferase (MGMT) restores guanine by direct repair of O^6-meG and studies in gliomas have demonstrated that lower methylated MGMT expression correlates with improved TMZ response [89, 90]. In pituitary tumors, some studies observed a similar correlation between low MGMT expression and good TMZ response [86, 87, 91] although this has not been a ubiquitous finding [92, 93]. Other recent studies have also implicated expression of another DNA mismatch repair protein, MSH6, as a predictor of response to TMZ in atypical pituitary adenomas and carcinomas [94].

There is now reasonable experience of the use of TMZ in treating Crooke's cell adenoma, silent corticotroph tumors, locally aggressive corticotroph tumors, and corticotroph tumors in the setting of Nelson's syndrome, as well as ACTH-secreting pituitary carcinomas refractory to combinations of surgery, radiation therapy, and other medical therapy as discussed previously [91, 92, 95–101]. TMZ has been reported to induce tumor shrinkage and reduction of plasma ACTH levels supporting a direct action of TMZ on corticotroph tumors, and the overall clinical and radiological response rate is ~60 % in aggressive adenomas [8]. TMZ is generally well tolerated, fatigue is common, and hematological toxicity with reduced white blood cell or platelet counts may necessitate dose reduction or occasional drug withdrawal. As for any alkylating agent, TMZ can be associated with a slight increased risk of secondary malignancy (e.g., leukemia or lymphoma). TMZ dosing is based on body surface area (typically 150–200 mg/m^2) and can be given either daily or in cycles (5 days every 28 days). It is unclear at this time whether intermittent dosing or low continuous dosed therapy offers better efficacy or safety profile. Some studies have suggested that treatment-responsive patients can be selected by demonstrating response after three cycles [92]. A modification of the TMZ protocol is the "so-called" CAPTEM regimen in which capecitabine 100 mg PO twice daily is administered on days 1–14 and TMZ 200 mg/m^2 in two divided doses daily on days 10–14 of a 28-day cycle. This combination was initially developed to treat metastatic, well-differentiated neuroendocrine tumors refractory to conventional treatments and was reported to be well tolerated, with thrombocytopenia as the most severe adverse effect (grade 3) [103]. The CAPTEM regimen has also been used in a case series of 4 patients with aggressive ACTH-secreting pituitary tumors refractory to surgery, radiation, and hormonal therapy with reported clinical improvement in all 4 patients and tumor regression in 75 % of this small group of patients [102].

Future Therapeutic Options

As noted, knowledge of the genetic basis of aggressiveness of corticotroph pituitary tumors is expanding. Whole exome sequencing may unravel additional biomarkers of tumor aggression and identify actionable molecular targets in these rare but challenging cases. Potentially, molecular biomarker panels may not only aid in the diagnostic process to identify these tumors earlier but also facilitate personalized therapeutic strategies and portend prognosis [104]. An array of kinase inhibitors, including mTOR and angiogenesis inhibitors, now exist but although some have been shown to reduce cell viability in in vitro cultures of human pituitary tumors other than isolated case reports with variable responses, these agents remain untested in pituitary tumors [105–107]. Additionally, cancer immunomodulation is a rapidly advancing field in oncology and it is intriguing that hypophysitis has emerged as a distinct complication of [108] inhibitors for cytotoxic T-lymphocyte antigen-4 (CTLA-4) and programmed death-1 (PD-1). Additionally, CTLA-4 and PD-1 are expressed in pituitary tissue raising the possibility that these agents may too have a role in treatment of aggressive pituitary tumors.

References

1. Amar AP, Hinton DR, Kriegar MD, Weiss MH. Invasive pituitary adenomas: significance of proliferation parameters. Pituitary. 1999;2:117–22.
2. Heaney AP. Clinical review: pituitary carcinoma: difficult diagnosis and treatment. J Clin Endocrinol Metab. 2011;96:3649–60.

3. Meij BP, Lopes MB, Ellegala DB, Alden TD, Laws ER. The long-term significance of microscopic dural invasion in 354 patients with pituitary adenomas treated with transsphenoidal surgery. J Neurosurg. 2002;92:195–208.
4. Scheithauer BW, Kovacs K, Laws Jr ER, Randall RV. Pathology of invasive pituitary tumors with special reference to functional classification. J Neurosurg. 1986;65:733–44.
5. Thapar K, Kovacs K, Scheithauer BW, Stefaneanu L, Horvath E, Pernicone PJ, Murray D, Laws Jr ER. Proliferative activity and invasiveness among pituitary adenomas and carcinomas: an analysis using the M1B-1 antibody. Neurosurgery. 1996;38:99–106.
6. Trouillas J, Roy P, Sturm N, Dantony E, Cortet-Rudelli C, Viennet G, Bonneville JF, Assaker R, Auger C, Brue T, Cornelius A, Dufour H, Jouanneau E, Francois P, Galland F, Mougel F, Chapuis F, Villeneuve L, Maurage CA, Figarella-Branger D, Raverot G, the members of HYPOPRONOS. A new prognostic clinicopathological classification of pituitary adenomas: a multicentric case-control study of 410 patients with 8 years post-operative follow-up. Acta Neuropathol. 2013;126:123–35.
7. Raverot G, Vasiljevic A, Jouanneau E, Trouillas J. A prognostic clinicopathologic classification of pituitary endocrine tumors. Endocrinol Metab Clin North Am. 2015;44:11–8.
8. Di Ieva A, Rotondo F, Syro L, Cusimano MD, Kovacs K. Aggressive pituitary adenomas—diagnosis and emerging treatments. Nat Rev Endocrinol. 2014;10:423–35.
9. Crooke A. A change in the basophil cells of the pituitary gland common to conditions which exhibit the syndrome attributed to basophil adenoma. J Pathol Bacteriol. 1935;41:339–49.
10. DeCicco FA, Dekker A, Yunis EJ. Fine structure of Crooke's hyaline change in the human pituitary gland. Arch Pathol. 1972;94:65–70.
11. Rotondo F, Cusimano M, Scheithauer BW, Coire C, Horvath E, Kovacs K. Atypical, invasive, recurring Crooke cell adenoma of the pituitary. Hormones. 2012;11:94–100.
12. Di Ieva A, Davidson JM, Syro LV, Rotondo F, Montoya JF, Horvath E, Cusimano MD, Kovacs K. Crooke' cell tumors of the pituitary. Neurosurgery. 2015;76:616–22.
13. George DH, Scheithauer BW, Kovacs K, Horvath E, Young Jr WF, Llyod RV, Meyer FB. Crooke's cell adenoma of the pituitary: an aggressive variant of corticotroph adenoma. Am J Surg Pathol. 2003;27:1330–6.
14. Mete O, Hayhurst C, Alahmadi H, Monsalves E, Gucer H, Gentili F, Ezzat S, Asa SL, Zadeh G. The role of mediators of cell invasiveness, motility, and migration in the pathogenesis of silent corticotroph adenomas. Endocr Pathol. 2013;24:191–8.
15. Kovacs K, Horvath E, Bayley TA, Hassaram ST, Ezrin C. Silent corticotroph cell adenoma with lysosomal accumulation and crinophagy: a distinct clinicopathologic entity. Am J Med. 1978;64:492–9.
16. Tateno T, Izumiyama H, Doi M, Yoshimoto T, Shichiri M, Inoshita N, Hirata Y. Differential gene expression in ACTH-secreting and non-functioning pituitary tumors. Eur J Endocrinol. 2007;6:717–24.
17. Raverot G, Wierinckx A, Jouanneau E, Auger C, Borson-Chazot F, Lachuer J, Trouillas J. Clinical, hormonal and molecular characterization of pituitary ACTH adenomas without (silent corticotroph adenomas) and with Cushing's disease. Eur J Endocrinol. 2010;1:35–43.
18. Ohta S, Nishizawa S, Oki Y, Yokoyama T, Namba H. Significance of absent prohormone convertase 1/3 in inducing clinically silent corticotroph pituitary adenoma of subtype I-immunohistochemical study. Pituitary. 2002;4:221–3.
19. Cooper O, Melmed S. Subclinical hyperfunctioning pituitary adenomas: the silent tumors. Best Pract Res Clin Endocrinol Metab. 2012;26:447–60.
20. Karavitaki N, Ansorge O, Wass JA. Silent corticotroph adenomas. Arq Bras Endocrinol Metab. 2007;51:1314–8.
21. Baldeweg SE, Pollock JR, Powell M, Ahlquist J. A spectrum of behavior in silent corticotroph pituitary adenomas. Br J Neurosurg. 2005;19:38–42.
22. Pawlikowski M, Kunert-Radek J, Radek M. "Silent" corticotropinoma. Neur Endocrinol Lett. 2008;29:347–50.
23. Scheithauer BW, Jaap AJ, Horvath E, Kovacs K, Lloyd RV, Meyer FB, Law Jr ER, Young Jr WF. Clinically silent corticotroph tumors of the pituitary gland. Neurosurg. 2000;47:723–9.

24. Alahmadi H, Lee D, Wilson JR, Hayhurst C, Mete O, Gentili F, Asa S, Zadeh G. Clinical features of silent corticotroph adenomas. Acta Neurochir. 2012;154:1493–8.
25. Bradley KJ, Wass JAH, Turner HE. Non-functioning pituitary adenomas with positive immunoreactivity for ACTH behave more aggressively than ACTH immunonegative tumors but do not recur more frequently. Clin Endocrinol (Oxf). 2003;58:59–64.
26. Nelson DH, Meakin JW, Dealy JW. ACTH-producing tumors of the pituitary gland. N Engl J Med. 1958;259:161–4.
27. Jordan RM, Cook DM, Kendall JW, Kerber CW. Nelson's syndrome and spontaneous pituitary tumor infarction. Arch Intern Med. 1979;139:340–2.
28. Jenkins PJ, Trainer PJ, Plowman PN, Shand WS, Grossman AB, Wass JA, Besser GM. The long-term outcome after adrenalectomy and prophylactic radiotherapy in adrenocorticotrophin-dependent Cushing's syndrome. J Clin Endocrinol Metab. 1995;80:165–71.
29. Assié G, Bahurel H, Coste J, Silvera S, Kujas M, Dugué MA, Karray F, Dousset B, Bertherat J, Legmann P, Bertagna X. Corticotroph tumor progression after adrenalectomy in Cushing's disease: a reappraisal of Nelson's syndrome. J Clin Endocrinol Metab. 2007;92:172–9.
30. Salehi F, Agur A, Scheithauer BW, Kovacs K, Lloyd RV, Cusimano M, Cusimano M. Ki-67 in pituitary neoplasms: a review-Part I. Neurosurgery. 2009;65:429–37.
31. Heaney A. Management of aggressive pituitary adenomas and pituitary carcinomas. J Neurooncol. 2014;117:459–68.
32. Mastronardi L, Guiducci A, Spera C, Puzzilli F, Liberati F, Maira G. Ki-67 labelling index and invasiveness among anterior pituitary adenomas: analysis of 103 cases using the MIB-1 monoclonal antibody. J Clin Pathol. 1999;52:107–11.
33. Scheithauer BW, Gaffey TA, Lloyd RV, Sebo TJ, Kovacs KT, Horvath E, Yapicier O, Young Jr WF, Meyer FB, Kuroki T, Riehle DL, Laws Jr ER. Pathobiology of pituitary adenomas and carcinomas. Neurosurgery. 2006;59:341–53.
34. Horvath E, Kovacs K, Killinger DW, Smyth HS, Platts ME, Singer W. Silent corticotropic adenomas of the human pituitary gland: a histologic, immunocytologic, and ultrastructural study. Am J Pathol. 1980;98:617–38.
35. Kontogeorgos G. Classification and pathology of pituitary tumors. Endocrine. 2005;28:27–35.
36. Kaltsas GA, Grossman AB. Malignant pituitary tumors. Pituitary. 1998;69–81.
37. Gadelha MR, Trivellin G, Hernandez Ramirez LC, Korbonits M. Genetics of pituitary adenomas. Front Horm Res. 2013;41:111–40.
38. Westermarck M, Seth A, Kahari VM. Differential regulation of interstitial collagenase (MMP-1) gene expression by ETS transcription factors. Oncogene. 1997;14:2651–60.
39. Rutter JL, Mitchell TI, Buttice G, Meyers J, Gusella JF, Ozelius LJ, Brinckerhoff CE. A single nucleotide polymorphism in the matrix metalloproteinase-1 promotor creates an Ets binding site and augments transcription. Cancer Res. 1998;58:5321–5.
40. Atlas M, Bayrak OF, Ayan E, Bolukbasi F, Silav G, Coskun KK, Culha M, Sahin F, Sevli S, Elmaci I. The effect of polymorphism sin the promotor region of the MMP-1 gene on the occurrence and invasiveness of hypophyseal adenoma. Acta Neurochir. 2010;152:1611–7.
41. Chakraborti S, Mandal M, Das S, Mandal A, Chakraborti T. Regulation of matrix metalloproteinases: an overview. Mol Cell Biochem. 2003;253:269–85.
42. Kawamoto H, Kawamoto K, Mizoue T, Uozumi T, Arita K, Kurisu K. Matrix metalloproteinase-9 secretion by human pituitary adenomas detected by cell immunoblot analysis. Acta Neurochir. 1996;138:1442–4.
43. Liu W, Matsumoto Y, Okada M, Miyake K, Kunishio K, Kawai N, Tamiya T, Nagao S. Matrix metalloproteinase 2 and 9 expression correlated with cavernous sinus invasion of pituitary adenomas. J Med Invest. 2005;52:151–8.
44. Hussaini IM, Trotter C, Zhao Y, Abdel-Fattah R, Amos S, Xiao A, Agi CU, Redpath GT, Fang Z, Leung GK. Matrix metalloproteinase-9 is differentially expressed in nonfunctioning invasive and noninvasive pituitary adenomas and increases invasion in human pituitary adenoma cell line. Am J Pathol. 2007;170:356–65.
45. Sugiyama N, Varjosalo M, Meller P, Lohi J, Chan KM, Zhou Z, Alitalo K, Taipale J, Keski-Oja J, Lehti K. FGF receptor-4 (FGFR4) polymorphism acts as an activity switch of a membrane type 1 matrix metalloproteinase-FGFR4 complex. Proc Natl Acad Sci U S A. 2010;107:15786–91.

46. Sanchez-Ortiga R, Sanchez-Tejada L, Moreno-Perez O, Riesgo P, Niveiro M, Pico Alfonso AM. Over-expression of vascular endothelial growth factor in pituitary adenomas is associated with extrasellar growth and recurrence. Pituitary. 2013;16:370–7.
47. Jia W, Sander AJ, Jia G, Ni M, Liu X, Lu R, Jiang WG. Vascular endothelial growth inhibitor (VEGI) is an independent indicator for invasion in human pituitary adenomas. Anticancer Res. 2013;33:3815–22.
48. Cornelius A, Cortet-Rudelli C, Assaker R, Kerdraon O, Gevaert MH, Prevot V, Lassalle P, Trouillas J, Delehedde M, Maurage CA. Endothelial expression of endocan is strongly associated with tumor progression in pituitary adenoma. Brain Pathol. 2012;22:757–64.
49. Matano F, Yoshida D, Ishii Y, Tahara S, Teramoto A, Morita A. Endocan, a new invasion and angiogenesis marker of pituitary adenomas. J Neurooncol. 2014;117:485–91.
50. Trouillas J, Labat-Moleur F, Sturm N, Kujas M, Heymann MF, Flgarella-Branger D, Paley M, Mazucca M, Decullier E, Verges B, Chabre O, Calender A, Grouped'etudes des Tumeurs Endocrines. Pituitary tumors and hyperplasia in multiple endocrine neoplasia type 1 syndrome (MEN1): a case-control study in a series of 77 patients versus 2509 non-MEN1 patients. Am J Surg Pathol. 2008;32:534–43.
51. Toledo SP, Lourenco DM, Toledo RA. A differential diagnosis of inherited endocrine tumors and their tumor counterparts. Clinics (San Paulo). 2013;68:1039–56.
52. Tahir A, Chahal HS, Korbonits M. Molecular genetics of the AIP gene in familial pituitary tumorigenesis. Prog Brain Res. 2010;182:229–53.
53. Newey PJ, Thakker RV. Role of multiple endocrine neoplasia type I mutational analysis in clinical practice. Endocr Pract. 2011;17 Suppl 3:8–17.
54. Bates AS, Farrell WE, Bicknell JE, McNicol AM, Talbot JA, Broome JC, Perrett CW, Thakker RV, Clayton RN. Genetic instability in pituitary adenomas reflects aggressive biological activity and has potential value as a prognostic marker. J Clin Endocrinol Metab. 1997;82:818–24.
55. Cai WY, Alexander JM, Hedley-Whyte ET, Scheithauer BW, Jameson JL, Zervas NT, Klibanski A. Ras mutations in human prolactinomas and pituitary carcinomas. J Clin Endocrinol Metab. 1994;78:89–93.
56. Wierinckx A, Auger C, Devauchelle P, Raynaud A, Chevallier P, Jan M, Perrin G, Fevre-Montagne M, Rey C, Figarella-Baranger D, Raverot G, Belin MF, Lachuer J, Trouillas J. A diagnostic maker set for invasion, proliferation and aggressiveness of prolactin pituitary tumors. Endocr Relat Cancer. 2007;14:887–900.
57. Zhang X, Horwitz GA, Heaney AP, Nakashima M, Prezant TP, Bronstein MD, Melmed S. Pituitary tumor transforming gene (PTTG) expression in pituitary adenomas. J Clin Endocrinol Metab. 1999;84:761–7.
58. Reincke M, Sbiera S, Hayakawa A, Theodoropoulou M, Osswald A, Beuschlein F, Meitinger T, Mizuno-Yamasaki E, Kawaguchi K, Saeki Y, Tanaka K, Wieland T, Graf E, Saeger W, Ronchi C, Allolio B, Buchfelder M, Strom TM, Fassnacht M, Koada M. Mutations in the deubiquitinase gene USP8 cause Cushing's disease. Nat Genet. 2015;47:31–8.
59. Riebold M, Kozany C, Freiburger L, Sattler M, Buchfelder M, Hausch F, Stalla GK, Paez-Pereda M. A C-terminal HSP90 inhibitor restores glucocorticoid sensitivity and relieves a mouse allograft model of Cushing disease. Nat Med. 2015;21:276–8.
60. Nieman L, Biller BM, Findling JW, Newell-Price J, Savage MO, Stewart PM, Montori VM. The diagnosis of Cushing's syndrome: an Endocrine Society clinical practice guideline. J Clin Endocrinol Metab. 2008;93:1526–40.
61. Tritos NA, Biller BM. Cushing's disease. Handb Clin Neurol. 2014;124:221–34.
62. Ayala A, Manzano A. Detection of recurrent Cushing's disease: proposal for standardized patient monitoring following transsphenoidal surgery. J Neurooncol. 2014;119:235–42.
63. Roelfsema F, Biermasz NR, Pereira AM. Clinical factors involved in the recurrence of pituitary adenomas after surgical remission: a structured review and meta-analysis. Pituitary. 2012;15:71–83.
64. Zada G, Woodmansee WW, Ramkissoon S, Amadio J, Nose V, Laws Jr ER. Atypical pituitary adenomas: incidence, clinical characteristics, and implications. J Neurosurg. 2011;114:336–44.
65. Tritos NA, Biller BM. Update on radiation therapy in patients with Cushing's disease. Pituitary. 2015;18:263–8.

66. Petit JH, Biller BM, Yock TI, Swearingen B, Coen JJ, Chapman P, Ancukiewicz M, Bussiere M, Klibanski A, Loeffler JS. Proton stereotactic radiotherapy for persistent adrenocorticotropin-producing adenoma. J Clin Endocrinol Metab. 2008;93:393–9.

67. Wattson DA, Tanguturi K, Spiegel DY, Neimieko A, Biller BM, Nachtigall LB, Bussiere MR, Swearingen B, Chapman PH, Loeffler JS, Shih HA. Outcomes of proton therapy for patients with functional pituitary adenomas. Int J Radiat Oncol Biol Phys. 2014;90:532–9.

68. Sheehan JP, Xu Z, Salvetti DJ, Schmitt PJ, Vance ML. Results of gamma knife surgery for Cushing's disease. J Neurosurg. 2013;119:1486–92.

69. Molitch ME. Current approaches to the pharmacological management of Cushing's disease. Mol Cell Endocrinol. 2015;408:185–9.

70. Praw SS, Heaney AP. Medical treatment of Cushing's disease: overview and recent findings. Int J Gen Med. 2009;2:209–17.

71. Fleseriu M. Medical treatment of Cushing disease: new targets, new hope. Endocrinol Metab Clin North Am. 2015;44:51–70.

72. Fleseriu M, Biller BM, Findling JW, Molitch ME, Schteingart DE, Gross C, SEISMIC study investigators. Mifepristone, a glucocorticoid receptor antagonist, produces clinical and metabolic benefits in patients with Cushing's syndrome. J Clin Endocrinol Metab. 2012;97:2039–49.

73. Pivonello R, Ferone D, de Herder WW, Kros JM, De Caro ML, Arvigo M, Annunziato L, Lombardi G, Colao A, Hofland LJ, Lamberts SW. Dopamine receptor expression and function in corticotroph pituitary tumors. J Clin Endocrinol Metab. 2004;89:2452–62.

74. Pivonello R, Faggiano A, DiSalle F, Filippella M, Lombardi G, Colao A. Complete remission of Nelson's syndrome after 1 year treatment with cabergoline. J Endocrinol Invest. 1999;22:860–5.

75. Tritos NA, Schaefer PW, Stein TD. Case records of the Massachusetts General Hospital Case 40-2011. A 52-year-old man with weakness, infections, and enlarged adrenal glands. N Engl J Med. 2011;365:2520–30.

76. Tyrrell JB, Lorenzi M, Gerich JE, Forsham PH. Inhibition by somatostatin of ACTH secretion in Nelson's syndrome. J Clin Endocrinol Metab. 1975;6:1125–7.

77. Strowski MZ, Dashkevicz MP, Parmar RM, Wilkinson H, Kohler M, Schaeffer JM, Blake AD. Somatostatin receptor subtypes 2 and 5 inhibit corticotropin-releasing hormone-stimulated adrenocorticotropin secretion from AtT-20 cells. Neuroendocrinology. 2002;75:339–46.

78. Lamberts SW, Uitterlinden P, Klijn JM. The effect of the long-acting somatostatin analogue SMS 201-995 on ACTH secretion in Nelson's syndrome and Cushing's disease. Acta Endocrinol (Copenh). 1989;120:760–6.

79. Bruns C, Lewis I, Briner U, Meno-tetang G, Weckbecker G. SOM230: a novel somatostatin peptidomimetic with broad somatotropin release inhibiting factor (SRIF) receptor binding and a unique antisecretory profile. Eur J Endocrinol. 2002;5:707–16.

80. Colao A, Petersenn S, Newell-Price J, Findling JW, Gu F, Maldonado M, Schoenherr U, Mills D, Salgodo LR, Biller BM. Pasireotide B2305 Study Group. A 12-month phase 3 study of pasireotide in Cushing's disease. N Engl J Med. 2012;366:914–24.

81. Castinetti F, Morange I, Jaquet P, Conte-Devolx B, Brue T. Ketoconazole revisited: a preoperative or postoperative treatment in Cushing's disease. Eur J Endocrinol. 2008;158:91–9.

82. Castinetti F, Guignat L, Giraud P, Muller M, Kamenicky P, Drui D, Caron P, Luca F, Donadille B, Vantyghem MC, Bihan H, Delemer B, Raverot G, Motte E, Philippon M, Morange I, Conte-Devolx B, Quinquis L, Marinie M, Vezzosi D, Le Bras M, Baudry C, Christin-Maitre S, Goichot B, Chanson P, Young J, Chabre O, Tabarin A, Bertherat J, Brue T. Ketoconazole in Cushing's disease: is it worth a try? J Clin Endocrinol Metab. 2014;99:1623–30.

83. Verhelst JA, Trainer PJ, Howlett TA, Perry L, Rees LH, Grossman AB, Wass JA, Besser GM. Short and long-term responses to metyrapone in the medical management of 91 patients with Cushing's syndrome. Clin Endocrinol (Oxf). 1991;35:169–78.

84. Daniel E, Aylwin S, Mustafa O, Ball S, Munir A, Boelaert K, Chortis V, Cuthbertson DJ, Daousi C, Rajeev SP, Davis J, Cheer K, Drake W, Gunganah K, Grossman A, Gurnell M, Powlson AS, Karavitaki N, Huguet I, Kearney T, Mohit K, Meeran K, Hill N, Rees A, Lansdown AJ, Trainer PJ, Minder AE, Newell-Price J. Effectiveness of metyrapone in treat-

ing Cushing's syndrome: a retrospective multicenter study in 195 patients. J Clin Endocrinol Metab. 2015;100:4146–54.

85. Lim S, Shahinian H, Maya MM, Yong W, Heaney AP. Temozolomide: a novel treatment for pituitary carcinoma. Lancet Oncol. 2006;7:518–20.

86. Hagen C, Schroeder HD, Hansen S, Hagen C, Andersen M. Temozolomide treatment in a pituitary carcinoma and two pituitary macroadenomas resistant to conventional therapy. Eur J Endocrinol. 2009;161:631–7.

87. Ortiz LD, Syro LV, Scheithauer BW, Rotonda F, Uribe H, Fadul CE, Horvath E, Kovacs K. Temozolomide in aggressive pituitary adenomas and carcinomas. Clinics. 2012;67:119–23.

88. Zhang J, Stevens MG, Bradshaw TD. Temozolomide: mechanisms of action, repair and resistance. Curr Mol Pharmacol. 2012;5:102–14.

89. Friedman HS, McLendon RE, Kerby T, Dugan M, Bigner SH, Henry AJ, Ashley DM, Krischer J, Lovell S, Rasheed K, Marchev F, Seman AJ, Cokgor I, Rich J, Stewart E, Colvin OM, Provenzale JM, Bigner DD, Haglund MM, Friedman AH, Modrich PL. DNA mismatch repair and O6-alkylguanine-DNA alkyltransferase analysis and response to Temodal in newly diagnosed malignant glioma. J Clin Oncol. 1998;16:3851–7.

90. Hegi ME, Diserens AC, Gorlia T, Hamou MF, de Tribolet N, Weller M, Kros JM, Hainfellner JA, Mason W, Mariani L, Bromberg JE, Hau P, Mirimanoff RO, Cairncross JG, Janzer RC, Stupp R. MGMT gene silencing and benefit from temozolomide in glioblastoma. N Engl J Med. 2005;352:997–1003.

91. Bengtsson D, Schroder HD, Andersen M, Maiter D, Berinder K, FeldtRasmuseen U, Rasmussen AK, Johannsoon G, Hoybye C, van der Lely AJ, Pegtersson M, Ragnarsson O, Burman P. Long-term outcome and MGMT as a predictive marker in 24 patients with atypical pituitary adenomas and pituitary carcinomas given treatment with temozolomide. J Clin Endocrinol Metab. 2015;100:1689–98.

92. Raverot G, Sturm N, de Fraipont F, Muller M, Salenave S, Caron P, Chabre O, Chanson P, Cortet-Rudelli C, Assaker R, Dufour H, Gaillard S, Francois P, Jouanneau E, Passagia JG, Berneir M, Cornelius A, Figarella-Branger D, Trouillas J, Borson-Chazot F, Brue T. Temozolomide treatment of aggressive pituitary tumors and pituitary carcinomas: a French multicenter experience. J Clin Endocrinol Metab. 2010;95:4592–9.

93. Bush ZM, Longtine JA, Cunningham T, Schiff D, Jane Jr JA, Vance ML, Thorner MO, Laws Jr ER, Lopes MB. Temozolomide treatment for aggressive pituitary tumors: correlation of clinic outcome with O6-methylguanine methyltransferase (MGMT) promotor methylation and expression. J Clin Endocrinol Metab. 2010;95:E280–90.

94. Hirohata T, Asano K, Takano S, Amano K, Isozaki O, Iwai Y, Sakata K, Fukuhara N, Nishioka H, Yamada S, Fujio S, Arita K, Takano K, Tominaga A, Hizuka N, Ikeda H, Osamura RY, Tahara S, Ishii Y, Kawamata T, Shimatsu A, Teramoto A, Matsuno A. DNA mismatch repair protein (MSH6) correlated with the responses of atypical pituitary adenomas and pituitary carcinomas to temozolomide: the national cooperative study by the Japan Society for Hypothalamic and Pituitary Tumors. J Clin Endocrinol Metab. 2013;98:1130–6.

95. Mohammed S, Kovacs K, Mason W, Smyth H, Cusimano MD. Use of temozolomide in aggressive pituitary tumors: case report. Neurosurgery. 2009;64:E773–4.

96. Moyes VJ, Alusi G, Sabin HI, Evanson J, Berney DM, Kovacs K, Monson JP, Plowman PN, Drake WM. Treatment of Nelson's syndrome with temozolomide. Eur J Endocrinol. 2009;160:115–9.

97. Kurowska M, Nowakowski A, Zielinski G, Malicka J, Tarach JS, Maksymowicz M, Denew P. Temozolomide-induced shrinkage of invasive pituitary adenoma in patient with Nelson's syndrome: A case report and review of the literature. Case Rep Endocrinol. 2015;2015:623092. doi:10.1155/2015/623092.

98. Dillard TH, Gultekin SH, Delashaw Jr JB, Yedinak CG, Neuwelt EA, Fleseriu M. Temozolomide for corticotroph pituitary adenomas refractory to standard therapy. Pituitary. 2011;14:80–91.

99. Annamalai AK, Dean AF, Kandasamy N, Kovacs K, Burton H, Halsall DJ, Shaw AS, Antoun NM, Cheow HK, Kirollos RW, Pickard JD, Simpson HL, Jefferies SJ, Burnet NG, Gurnell

M. Temozolomide responsiveness in aggressive corticotroph tumors: a case report and review of the literature. Pituitary. 2012;15:276–87.

100. Moshkin O, Syro LV, Scheithauer BW, Ortiz LD, Fadul CE, Uribe H, Gonzalez R, Cusimano M, Horvath E, Rotondo F, Kovacs K. Aggressive silent corticotroph adenoma progressing to pituitary carcinoma: the role of temozolomide therapy. J Neurooncol. 2015;122:189–96.

101. Curto L, Torre ML, Ferrau F, Pitini V, Altavilla G, Granata F, Longo M, Hofland LJ, Trimarchi F, Cannavo S. Temozolomide-induced shrinkage of a pituitary carcinoma causing Cushing's disease – report of a case and literature review. Scientific World Journal. 2010;10:2132–8.

102. Zacharia BE, Gulati AP, Bruce JN, Carminucci AS, Wardlaw SL, Siegelin M, Remotti H, Lignelli A, Fine RL. High response rates and prolonged survival in patients with corticotroph pituitary tumors and refractory Cushing disease from capecitabine and temozolomide (CAPTEM): a case series. Neurosurgery. 2014;74:E447–55.

103. Fine RL, Gulati AP, Krantz BA, Moss RA, Schreibman S, Tsushima DA, Mowatt KB, Dinnen RD, Mao Y, Stevens PD, Schrope B, Allendorf J, Lee JA, Sherman WH, Chabot JA. Capecitabine and temozolomide (CAPTEM) for metastatic, well-differentiated neuroendocrine cancers: The Pancreas Center at Columbia University experience. Cancer Chemother Pharmacol. 2013;71:663–70.

104. Raverot G, Jouanneau E, Trouillas J. Clinicopathological classification and molecular markers of pituitary tumours for personalized therapeutic strategies. Eur J Endocrinol. 2014;170: R121–32.

105. Monsalves E, Juraschka K, Tateno T, Agnihotri S, Asa SL, Ezzat S, Zadeh G. The PI3K/AKT/mTOR pathway in the pathophysiology and treatment of pituitary adenomas. Endocr Relat Cancer. 2014;21:R331–44.

106. Ortiz LD, Syro LV, Scheithauer BW, Ersen A, Uribe H, Fadul CE, Rotondo F, Horvath E, Kovacs K. Anti-VEGF therapy in pituitary carcinoma. Pituitary. 2012;15:445–9.

107. Jouanneau E, Wierinckx A, Ducray F, Favrel V, Borson-Cchazot F, Honnorat J, Trouillas J, Raverot G. New targeted therapies in pituitary carcinoma resistant to temozolomide. Pituitary. 2012;15:37–43.

108. Corsello SM, Barnabei A, Marchetti P, De Vecchis L, Salvatori R, Torino F. Endocrine side effects induced by immune checkpoint inhibitors. J Clin Endocrinol Metab. 2013;98: 1361–75.

Neoplastic/Pathological and Nonneoplastic/Physiological Hypercortisolism: Cushing Versus Pseudo-Cushing Syndromes

James W. Findling and Hershel Raff

Abstract The diagnosis of endogenous hypercortisolism (Cushing syndrome) is the most challenging problem in clinical endocrinology. Neoplastic (pathological) hypercortisolism is usually due to an ACTH-secreting neoplasm or autonomous cortisol secretion from benign or malignant adrenal neoplasms. Nonneoplastic (physiological) hypercortisolism is common in many medical disorders such as chronic alcoholism, chronic kidney disease, pregnancy, depression/neuropsychiatric disorders, and starvation. The clinical features of hypercortisolism may be apparent in both pathological and physiological hypercortisolism and present a significant diagnostic challenge. A careful history and good examination are usually the most helpful means to identify patients with nonneoplastic/physiological Cushing syndrome. Simple biochemical tests such as late-night salivary cortisol and the overnight 1 mg dexamethasone suppression test have a good negative predictive value and are recommended as first line diagnostic testing in suspected hypercortisolism. Secondary tests such as the DDAVP stimulation test and the dexamethasone-CRH test may be required in some patients to confirm the presence or absence of neoplastic/pathological Cushing syndrome. This review describes the medical disorders and physiological conditions associated with chronic activation of the hypothalamic–pituitary–adrenal axis and provides a rational clinical and biochemical approach to distinguish them from patients with neoplastic/pathological Cushing syndrome.

Keywords Cushing syndrome • Pseudo-Cushing syndrome • Adrenocorticotropic hormone (ACTH) • Cortisol • Alcoholism • Hypothalamic–pituitary–adrenal axis

J.W. Findling, M.D. (✉)
Endocrinology Center and Clinics, Medical College of Wisconsin,
W129 N7055 Northfield Drive, Menomonee Falls, WI 53051, USA
e-mail: james.findling@froedtert.com

H. Raff, Ph.D.
Departments of Medicine, Surgery, and Physiology, Medical College of Wisconsin,
Milwaukee, WI 53215, USA

Endocrine Research Laboratory, Aurora St. Luke's Medical Center, Aurora Research
Institute, 2801 W KK River Pky Suite 245, Milwaukee, WI 53215, USA
e-mail: hraff@mcw.edu

© Springer International Publishing Switzerland 2017
E.B. Geer (ed.), *The Hypothalamic-Pituitary-Adrenal Axis in Health and Disease*, DOI 10.1007/978-3-319-45950-9_6

Introduction and Definitions

Endogenous hypercortisolism due to activation of the hypothalamic–pituitary–adrenal (HPA) axis is an important adaptive response to many types and severities of stress from both external and internal stimuli [1]. This well-appreciated response coordinates an essential increase in the release of energy stores, stimulation of gluconeogenesis, maintenance of blood pressure and tissue perfusion, and attenuation of the inflammatory responses [2]. Chronic sustained or intermittent hypercortisolemic states are recognized in many common physiological situations as well as many medical disorders [3]. Prolonged exposure to cortisol excess often results in a phenotype commonly referred to as Cushing syndrome [4–8]. The term "pseudo-Cushing syndrome" has been used to characterize patients with medical conditions associated with appropriate or inappropriate cortisol excess that do not have a pathological origin from either an adrenocorticotropin (ACTH)-secreting tumor or autonomous cortisol secretion from adrenal nodular disease [9]. Unfortunately, the term "pseudo-Cushing syndrome" has also been applied to patients who may have the common phenotype ascribed to cortisol excess (i.e., the metabolic syndrome), but do not have consistent biochemical evidence of increased activity of the HPA axis. Of course, clinical features of hypercortisolism may be evident in patients with chronic physiological hypercortisolism (for example, in depression, chronic alcoholism, and chronic kidney disease (CKD)) and may be indistinguishable from those with pathological Cushing syndrome [10]. Because of the dynamic range of the HPA axis in these conditions, the biochemical differentiation between physiological and pathological hypercortisolism may be very challenging. Consequently, we think the application of the term "pseudo-Cushing syndrome" is imprecise and vague at best and misleading at worst. We prefer to characterize the Cushing syndromes as either neoplastic endogenous hypercortisolism (pathological) or non-neoplastic (physiological) hypercortisolism (Table 1) with the understanding that sustained cortisol excess in either condition may lead to significant, indistinguishable clinical and metabolic derangements.

The purpose of this chapter is to review the medical disorders and physiological conditions that are known to be associated with chronic activation of the HPA axis and to provide a rational clinical and biochemical approach to help distinguish them from true pathological Cushing syndrome.

HPA Axis Physiology and its Potential Association with Common Disorders

The HPA axis exists primarily to generate a basal, circadian cortisol rhythm and to increase cortisol secretion in response to a wide variety of stimuli collectively termed, rather imprecisely, stress [1, 11, 12]. Attempts have been made to categorize stress into subtypes, such as psychological and physical [1]. These designations

Table 1 Etiologies of chronic hypercortisolism

Neoplastic/pathological hypercortisolism
ACTH-secreting neoplasm
Pituitary (Cushing disease)
Non-pituitary (ectopic ACTH)
Adrenal neoplastic disease
Adrenocortical adenoma
Adrenocortical carcinoma
Bilateral adrenal nodular disease
Primary pigmented micronodular hyperplasia
Primary bilateral macronodular hyperplasia
Nonneoplastic/physiological hypercortisolism
Phenotype similar to neoplastic hypercortisolism
Alcoholism and alcohol withdrawal
Chronic kidney disease stage 5
Depression/neuropsychiatric disease
Glucocorticoid resistance
Uncontrolled diabetes mellitus
Pregnancy
Phenotype not similar to neoplastic hypercortisolism
Starvation/malnutrition—anorexia nervosa
Critical illness
Aging

are particularly relevant to this chapter as we are emphasizing the differences and similarities between chronic stimuli to the HPA axis compared to the truly pathological, endogenous hypercortisolism characteristic of Cushing syndrome and usually due to an endocrine neoplasm.

Figure 1 gives an overview of the general structure of the HPA axis [11, 12]. CNS inputs to the hypothalamus from the circadian rhythm pathways [13] and stress pathways [1, 14–16] elicit an increase in the release of corticotrophin-releasing hormone (CRH) and/or arginine vasopressin into the hypophyseal portal veins which then stimulate the release of preformed ACTH as well as increase the transcription of the proopiomelanocortin (POMC) gene. ACTH is produced from POMC by post-translational processing and released into the systemic circulation. At the adrenal cortex, ACTH binds to and activates the melanocortin 2 (ACTH) receptor leading to an activation of the cAMP-steroidogenic-acute regulatory (StAR) protein cascade which increases steroidogenesis within the adrenal zona fasciculata cell [17, 18]. Cortisol, released into the blood from the adrenal cortex, binds >90 % to plasma proteins (primarily CBG at physiological cortisol concentrations) and circulates throughout the body. At the hypothalamus and anterior pituitary, free (biologically active) cortisol exerts a negative feedback effect via both the glucocorticoid (GR) and mineralocorticoid (MR) receptors [19, 20]. Previous chapters in this book have elaborated in great detail on the mechanisms of action of

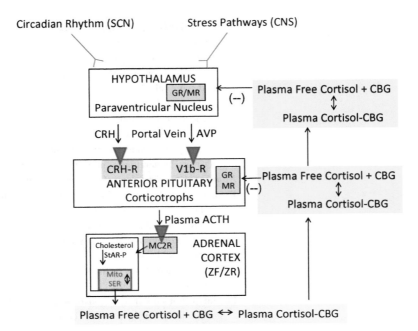

Fig. 1 General organization of the hypothalamic–pituitary–adrenal axis. Suprachiasmatic nucleus (SCN); central nervous system (CNS); corticotrophin-releasing hormone (CRH); arginine vasopressin (AVP); adrenocorticotropic hormone (ACTH); zona fasciculata (ZF); zona reticularis (ZR); melanocortin 2 receptor (MC2R); StAR-P (steroidogenic acute regulatory protein); mitochondria (Mito); smooth endoplasmic reticulum (SER); corticosteroid-binding globulin (CBG, also known as cortisol-binding globulin); Glucocorticoid receptor (GR); mineralocorticoid (MR) receptors. From [11]

circulating cortisol. Suffice it to say here that target tissues for glucocorticoids have an elaborate enzymatic system for protecting the MR from the actions of cortisol at physiological concentrations [21]. This is particularly relevant to this chapter as many of the effects of cortisol excess, such as increased sodium retention and potassium excretion in the kidney, are due to the saturation of these protective mechanisms and the binding of cortisol to the MR [22–24].

A few of the points above deserve elaboration. One of the hallmarks of the diurnal mammal is the circadian rhythm of the HPA axis in which cortisol peaks between 0600 and 0800 h and is at its nadir around midnight or a little later [13, 25]. The HPA axis also exhibits ultradian rhythmicity which may account for some of the "noise" in the assessment of morning cortisol concentrations in plasma [26, 27]. As described later, one of the earliest and most consistent changes in the HPA axis in patients with Cushing syndrome is an increase in the late-night nadir in cortisol [26, 28–32]. As you will see, this is exploited in the diagnosis of Cushing syndrome.

As mentioned above, there are many neuropsychiatric situations in which the normal circadian rhythm of the HPA axis is disrupted [33]. Although chronic increases in HPA activity due to neuropsychiatric disorders are described below, it is also important to point out that patients with these disorders can exhibit an augmented response to stress despite the fact that basal cortisol secretion can be normal [33]. In addition, the termination of the HPA axis response to stimuli in these conditions can also be delayed leading to even more exposure to increased glucocorticoid activity [33]. Therefore, just because a basal cortisol level is within the reference range does not mean the person has not been repetitively exposed to increased cortisol levels during the stresses of everyday life.

Another interesting characteristic of the HPA axis that is becoming of great physiological and psychoneuroendocrinological interest is the increase in cortisol secretion that occurs upon awakening—the cortisol awakening response (CAR)—that is superimposed on the circadian rhythm described above [34–41]. It has been proposed that a change in the CAR is an indication of alterations in arousal, anticipation of the day's events, gender, health status, and the perception of stress. The CAR can easily be assessed by measuring salivary cortisol immediately upon awakening and then at set times thereafter [37]. Of great interest is the possibility that the CAR may reflect changes in neurological function especially in the hippocampus and associated limbic structures [40]. Since endogenous depression is one of the common maladies we will focus on as a state of hypercortisolism, it is interesting that the CAR is increased during episodes of major depression [38].

In summary, it is now clear that exposure to increased cortisol may be quite common in everyday life particularly in people with neuropsychiatric and other medical conditions.

Diagnostic Tests for Endogenous Hypercortisolism

A recent Endocrine Society Guideline has proposed three screening tests for Cushing syndrome that exploit different aspects of the disruption of normal physiology [3, 42]. They are as follows: the failure to achieve the normal nadir in late-night cortisol most commonly performed by the measurement of an increased late-night (typically at bedtime) salivary cortisol; the failure to suppress morning plasma cortisol after an overnight dexamethasone suppression; and an increase in the excretion of free cortisol in the urine. We have extensively reviewed these "first line" tests previously [7, 8, 11, 12, 32, 43] and will only briefly describe them below.

Late-Night Cortisol: It was observed decades ago that patients with severe endogenous Cushing syndrome of any etiology show a disrupted cortisol circadian rhythm [44–47]. More recently, it was demonstrated that patients with milder forms of Cushing syndrome have increased midnight plasma cortisol and that this measurement when done properly can be used to diagnose and rule out Cushing syndrome with accuracy [48, 49]. Considering the challenge of obtaining stress-free late-night blood samples in patients, a major advance came with the development and widespread clinical use of the measurement of late-night salivary cortisol (LNSC) as a surrogate for plasma free cortisol [12, 50, 51]. This test typically has a sensitivity and specificity for endogenous hypercortisolism of >90–95 %, although there are some exceptions and methodological caveats that must always be kept in mind [51, 52].

Low-Dose Dexamethasone Suppression Test: The physiological concept exploited in this test is that ACTH-secreting neoplasms (and obviously ACTH-independent adrenal neoplasms) have attenuated sensitivity to cortisol negative feedback [6, 7, 53]. The failure to fully suppress morning cortisol after an overnight (usually 1 mg) dose of dexamethasone (i.e., cortisol >1.8 μg/dL [50 nmol/L]) has a sensitivity of 95 % in patients with neoplastic Cushing syndrome; however, there are many factors that contribute to false positive results yielding a diagnostic specificity of 85–90 %. Most common causes of misleading results are concomitant use of drugs which accelerate or impair dexamethasone metabolic clearance and the use of estrogen therapy which increases corticosteroid-binding protein (CBG), the major binding protein for cortisol.

Urine Free Cortisol (UFC): The physiological concept is that an increase in the filtered load of free cortisol in the kidney will be reflected in an increase in 24-h UFC excretion. However, this assumption, which was based on old immunoassay data, has recently been questioned [54, 55]. The increasing use of the highly specific measurement of UFC using LC-MS/MS could theoretically resolve some problems related to this test [56]. This may not be true because of the theoretical advantage of measuring cortisol metabolites in the urine [54]. Although measurement of UFC has a sensitivity of only 75 % for the detection of neoplastic Cushing syndrome, marked elevations of urine cortisol (>3–4 times the ULN) are virtually diagnostic of pathological Cushing syndrome [3]. Nonetheless, because of its very poor sensitivity we do not recommend UFC as a first line test in the evaluation of suspected hypercortisolism.

Physiological (Nonneoplastic) Hypercortisolism

There are a variety of circumstances not attributable to an ACTH- or cortisol-secreting neoplasm in which cortisol secretion is chronically increased. This section will discuss some of these rather common situations and will express our opinion that these varied states of mild to severe cortisol excess should not be lumped together, but rather discussed on their own merits. These often subtle situations activate the HPA axis primarily through neural pathways with input to the parvocellular paraventricular nuclei in the hypothalamus (see Fig. 1). In reality as you will see, one mechanism that is a recurring theme in many of these situations is a decrease in sensitivity to glucocorticoid negative feedback. This can lead to mild increases in cortisol levels more equivalent to subclinical Cushing syndrome. However, it is very important to again emphasize that small increases in cortisol can summate over time to considerable glucocorticoid exposure [57].

Alcohol-Induced Hypercortisolism: It has been known for many years that alcohol intake increases cortisol secretion acutely and that actively drinking alcoholics have increased indices of cortisol secretion compared to controls [58, 59]. The mechanism for this increase is thought to be centrally mediated due to increases in CRH and ACTH [60]. In the late 1970s, investigators began to recognize the presence of signs and symptoms of Cushing syndrome in alcoholic patients with resolution of the clinical features and biochemical abnormalities of cortisol excess within 1–2 months after abstinence from alcohol [61, 62].

Experimental studies suggest that the principal stimulus of alcohol-induced cortisol secretion is centrally mediated through hypothalamic CRH [63–65]. Messenger RNA for CRH is increased in the paraventricular nucleus of rats after alcohol administration, and in addition, CRH receptor antagonists abolish the ability of alcohol to stimulate the HPA axis [63, 64]. Alcohol administration does not stimulate the HPA axis in rats after hypophysectomy or suppression of hypothalamic activity with the administration of morphine and pentobarbital [66]. Alcohol-induced increases in vasopressin secretion may also be a factor, since hypothalamic vasopressin of parvo- and magnocellular origin augments the ACTH response to CRH. Removal of endogenous AVP diminishes the alcohol-evoked ACTH secretion in both sham-operated and paraventricular nucleus-lesioned rats [63, 64]. Nonetheless, Cobb et al. have shown that steroid production from isolated perfused rat adrenal glands increases after adding ethanol in the absence of ACTH and that the effect is not enhanced by ACTH administration [67]. However, Elias et al. showed that a large alcohol bolus did not cause an increase in plasma cortisol or ACTH levels and failed to potentiate the effect of exogenous ACTH on cortisol secretion in either alcoholic subjects or normal human subjects [68]. Wand et al. reported increased ACTH levels at 1400 h with concurrently normal cortisol values in 31 actively drinking alcoholics [60]. The normal cortisol in the presence of increased plasma ACTH suggests centrally mediated HPA axis hyperactivity. Withdrawal from alcohol in chronic alcoholics also causes increases in ACTH and cortisol and normalization of HPA axis function may require a few weeks after alcohol cessation [69].

Altered peripheral metabolism of cortisol (particularly in the liver) may contribute to hypercortisolism in these patients. Lamberts et al. reported a patient with alcoholism with clinical Cushing syndrome and suppressed plasma ACTH [70]. They demonstrated a prolonged half-life of cortisol; however, if negative feedback system functioned properly, one would expect cortisol levels to eventually return to normal unless stimulatory input to the hypothalamus was increased. Moreover, some investigators have failed to detect a causal relationship between the impairment of liver function per se and serum cortisol levels [71]. Nonetheless, it is not uncommon for patients with alcohol-induced hypercortisolism to have abnormal liver function studies. Consequently, the presence of persistent liver function abnormalities—particularly if the AST is much greater than the ALT—should raise concern about the possibility of excessive alcohol consumption [72].

Another potential factor in hypercortisolism from excessive alcohol consumption might be interference of the binding of cortisol to plasma proteins with possibly excessive free cortisol concentrations. One study showed a positive correlation between blood alcohol level and the percentage of free plasma cortisol [73]. There was a shift of the fraction of cortisol bound to cortisol-binding globulin to the albumin-bound and unbound fractions. They speculated that there may be an intracellular hypoglucocorticoid state which gives rise to stimulation of the HPA axis in patients with normal cortisol negative feedback control. Since alcohol-induced hypercortisolism is not commonly appreciated in all alcoholic subjects, there may also be genetic influences that have an impact on the effect of alcohol in the HPA axis.

Biochemical studies in patients with alcohol-induced hypercortisolism have shown normal, increased, and occasionally even decreased concentrations of plasma ACTH [60, 70]. Diurnal rhythm is usually absent or attenuated and urinary measurements of corticosteroids are often increased. Overnight dexamethasone suppression testing yields abnormal results in the majority of patients with alcohol-induced hypercortisolism and the ACTH response to CRH is either normal or blunted [70, 74–76]. The dexamethasone-CRH test has been shown to be abnormal in alcohol-induced hypercortisolism and cannot be used to discriminate this from pathological Cushing syndrome [77]. On the other hand, the ACTH response to desmopressin acetate (DDAVP) appears to be absent in alcohol-induced Cushing syndrome (like normal subjects) in contrast to patients with Cushing disease [78]. Finally, using salivary cortisol measurements, it has been shown that alcohol intoxication activates the basal HPA axis but appears to blunt the stress response [79, 80].

In our experience, alcohol-induced hypercortisolism can be difficult to distinguish clinically or biochemically from patients with pathological Cushing syndrome. This is a particular problem if patients are not forthright about the magnitude of their alcohol consumption. The majority of patients have normal or increased plasma ACTH, so the central effects of alcohol in the HPA axis seems to predominate. Nonetheless, we have, like others, observed adrenal nodular disease in patients with alcohol-induced hypercortisolism and this disorder may actually present as an incidental adrenal nodule in some patients [81].

In summary, the etiology of alcohol-induced hypercortisolism is uncertain and, in fact, there may be several interrelated causes that can vary in significance in individual patients. It is clear that it can resolve within a few weeks of alcohol cessation [70, 76], although there may be long-lasting effects on stress responsivity of the HPA axis [80]. Clinical and biochemical features are often indistinguishable from patients with true pathological Cushing syndrome. A high index of suspicion may be necessary to make an accurate diagnosis, and secondary testing with the DDAVP stimulation test would seem to be the best diagnostic option.

Starvation: One of the most important adaptive responses to starvation is activation of the HPA axis to liberate energy stores and stimulate gluconeogenesis in order to maintain plasma glucose in the normal range. Many studies have demonstrated that patients with eating disorders (specifically anorexia nervosa) have activation of the HPA axis with varying degrees of hypercortisolism usually with normal plasma ACTH [46, 82]. Patients with anorexia nervosa have altered HPA dynamics with increases in urinary cortisol excretion and late-night salivary cortisol as well as abnormal dexamethasone suppression [83]. Typically, patients with anorexia nervosa have an attenuated ACTH response to CRH most likely due to the inhibitory feedback of cortisol on the anterior pituitary [46]. The dexamethasone-CRH test may also be abnormal in patients with anorexia [84]. Similar to normal subjects, DDAVP does not stimulate ACTH and, hence, cortisol in patients with anorexia. Furthermore, DDAVP can enhance ACTH and cortisol release after CRH in normal subjects but not in patients with anorexia nervosa [85]. All these findings point to an attenuation of the pituitary corticotroph response to endogenous stimuli.

Anorexia nervosa-induced hypercortisolism correlates with the severity of bone loss in women and has also been shown to be associated with the hypothalamic amenorrhea [86, 87]. In addition, increased bone marrow fat related to cortisol excess has also been reported in patients with anorexia nervosa [83]. It seems likely that other starvation-equivalent disorders may be associated with hypercortisolism and have significant clinical manifestations. For example, patients with prolonged stay in the intensive care unit have been known to have significant myopathy and a catabolic state mediated, in part, by hypercortisolism [88]. Increases in cortisol have also been observed in a study of normal weight women undergoing low-calorie dieting, and increased morning cortisol levels have been demonstrated in women within 6–12 months following bariatric surgery [89]. It seems possible that the chronic wasting and catabolic state seen in many serious chronic conditions (malignancies, cardiac, neurological, or infectious disorders) may be related, in part, to HPA axis activation and endogenous hypercortisolism. It is appreciated that patients with very severe hypercortisolism (usually ectopic ACTH secretion) may present with significant weight loss, edema, and myopathy [90]. In contrast to patients with starvation-induced hypercortisolism, patients with pathological Cushing syndrome almost always have insulin resistance and/or hypertension [5, 7].

In summary, central activation of the HPA axis with endogenous hypercortisolism is a common adaptive response to starvation. These disorders rarely cause any diagnostic confusion from patients with pathological Cushing syndrome. However, it should be pointed out that there are a few reports of anorexia nervosa as the initial clinical feature in patients with pathological Cushing syndrome as a reflection of the remarkably varied neuropsychiatric impact on the brain from cortisol excess [91].

Depression/Neuropsychiatric Disorders: There are a number of neuropsychiatric disorders that can increase or decrease the activity of the HPA axis [33]. It is not possible to go into great detail here. Rather, Table 2 provides a brief summary.

Table 2 Neuropsychiatric disease states that increase HPA axis activity (modified from [33, 92])	
Major depression (melancholic, bipolar, or psychotic):	
	Decreased sensitivity to glucocorticoid negative feedback
	Association with early life stress
Anxiety/panic disorder	
Obsessive-compulsive disorder	
Schizophrenia:	
	Decreased sensitivity to glucocorticoid negative feedback
Autism spectrum disorder:	
	Increased stress response

There is a considerable history of the study of the HPA axis in depression [33]. It is generally accepted that most forms of major depression exhibit increased HPA axis activity. New data suggest that changes in glucocorticoid receptor, and possibly mineralocorticoid receptor sensitivity, lead to resistance to cortisol negative feedback inhibition [92]. Interestingly, successful pharmacotherapy seems to normalize HPA axis activity and the lack of effectiveness of therapy correlates with a persistence of HPA axis hyperactivity [33]. It is well known that patients with major depression (e.g., psychotic depression) often have abnormal low dose dexamethasone suppression as well as elevations of LNSC and UFC. In fact, mental health specialists have utilized not only the low dexamethasone suppression test but also the dexamethasone-CRH test to characterize these disorders and the response to therapy. Since neoplastic/pathological Cushing syndrome is often complicated by significant neuropsychiatric illness, the differentiation from HPA axis activation due to severe forms of depression is challenging. The DDAVP test has not been extensively studied in depressive illness, and variable ACTH and cortisol responses have been reported.

Aging: Healthy aging is associated with an increase in late-night salivary cortisol levels [57]. Again, this increase is not into the pathological range. However, the doubling of late-night salivary cortisol with healthy aging represents a doubling of exposure to bioactive (free) cortisol which is a significant glucocorticoid exposure integrated over time. In fact, this increase correlates with a decrease in bone mineral density—a potential surrogate for glucocorticoid exposure over time [57].

Like depression described above, the mechanism of the increase in HPA axis activity with healthy aging is associated with a decrease in sensitivity to cortisol negative feedback [93–95]. Also of interest is that there appears to be an interaction with aging and the development of depression in terms of the increase in HPA axis activity [96].

Chronic Kidney Disease: CKD and end-stage renal failure have long been known to be associated with dysregulated cortisol excess with abnormal dexamethasone suppression [97–99]. Recently, patients with end-stage renal failure receiving hemodialysis have been shown to have disrupted circadian rhythm and increased late-afternoon to late-night cortisol [25]. The mechanism for the increase in cortisol in CKD5 appears not to be the result of decreased renal clearance of cortisol since plasma ACTH is increased in these patients (Fig. 2). If this was a cortisol clearance defect, ACTH should be suppressed due to negative feedback as occurs with sepsis as described below [100, 101]. It seems clear from these findings that patients with end-stage renal disease have an activation of their HPA axis, presumably of hypothalamic origin. It has also been suggested that this activity correlates with increases in C-reactive protein suggesting that a heightened inflammatory state may be an etiology [25]. The secondary tests outlined below (Dex-CRH or DDAVP) have not been studied in patients with severe CKD.

Other Common Disorders with Subtle Increases in Cortisol: As discussed above, there are many disorders in which cortisol levels are not increased above the reference range but, rather, demonstrate subtle increases that, when integrated over time, can result in significant glucocorticoid exposure. Examples of these are hypertension [102] and type 2 diabetes mellitus (T2DM) [103].

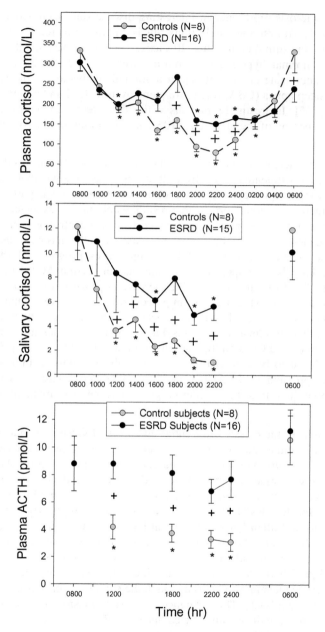

Fig. 2 Circadian rhythm of plasma cortisol (top), salivary cortisol (middle), and plasma ACTH (bottom) in control subjects compared to ESRD subjects. From [25]

Multiple sclerosis represents an interesting disease state with activation of the HPA axis [104]. At first, one might think this is due to the increased inflammatory state and cytokine stimulation of the HPA axis. It seems, rather, that this is due to a disruption of normal hypothalamic control directly due to brain lesions [104]. Another common state one might expect a dramatic activation of the HPA axis is obstructive sleep apnea (OSA) due to frequent arousals [39] and hypopnea leading to hypoxia [105]. Interestingly, patients with OSA seem remarkably resistant to significant activation of the HPA axis [105], although they may have an increased CAR [39].

Pregnancy: The increase in serum-free and salivary cortisol that occurs during pregnancy is well documented [106–108]. It is also well known that serum CBG increases during pregnancy due to the effect of estrogen on hepatic production [106]. The increase in total cortisol that results would not be expected to dramatically increase free (biologically active) cortisol concentrations in the blood. However, the increase in free (bioactive) plasma cortisol that has been demonstrated is due to the increase in plasma ACTH as pregnancy progresses [109]. Several mechanisms have been proposed for the increase in plasma ACTH including the secretion of CRH from the placenta, the increase in progesterone that can act as a glucocorticoid antagonist, a decrease in glucocorticoid negative feedback sensitivity, and production of ACTH from the placenta [106]. It is important to point out that CRH-binding protein also increases during pregnancy which mitigates some of the effect of placental CRH on the maternal pituitary gland [110–112]. In addition, there does appear to be a decrease in corticotroph sensitivity to unbound plasma CRH in late pregnancy [112]. The increase in bioactive (free) cortisol during pregnancy may help to sustain maternal gluconeogenesis to maintain delivery of glucose to the fetus.

Critical Illness: In the classic differential diagnosis of Cushing syndrome, ACTH independence is either due to adrenal autonomy in endogenous disease or due to glucocorticoid therapy in exogenous disease. Any state in which cortisol is increased and ACTH is suppressed can be categorized as ACTH independent. The case of critical illness usually due to sepsis is unusual in this regard and has been a source of confusion for decades [100, 113]. It is acknowledged that early in the development of sepsis and often before the patient has been admitted to the ICU, ACTH does increase (transiently) driving the adrenal gland to increase cortisol production [114]. Therefore, the initial stimulus to the axis seems to be of hypothalamic and pituitary origin. Thereafter, there is a sustained increase in plasma cortisol in the face of decreased plasma ACTH even though the stimulus (i.e., sepsis and hypotension) is still present [101]. This sustained increase in plasma cortisol is thought to be due to a decrease in the metabolic clearance rate of cortisol [100, 101]. Therefore, the final rate of ACTH secretion from the anterior pituitary gland is a balance between stimulatory input from stress pathways and inhibitory cortisol negative feedback. The decrease in CBG concentration often found in septic ICU patients may lead to a state of bioactive cortisol excess [115, 116]. Ironically, the prolonged suppression of ACTH secretion combined with decreased perfusion pressure in patients in the ICU for extended periods of time may eventually result in the risk for the development of adrenal insufficiency [100]. These phenomena do not seem to be unique to severe sepsis as trauma and cirrhotic patients also appear to exhibit some of the same characteristics [117, 118].

Type 2 Diabetes, Insulin Resistance, and the Metabolic Syndrome: One of the ongoing controversies in endocrinology is whether the frequency of pathological Cushing syndrome is more common in patients with T2DM. Although initial reports suggest a higher frequency of Cushing syndrome in patients with T2DM [119, 120], subsequent studies have suggested that only T2DM patients with pathognomonic features of endogenous hypercortisolism should be evaluated for a neoplastic cause [121–123]. It is well known that glucocorticoid therapy causes insulin resistance. A related question is whether subtle disruptions of the HPA axis can contribute to the development of the metabolic syndrome and insulin resistance. Increased late-night salivary cortisol concentrations have been found in poorly controlled patients with T2DM [124]. However, a minimal association of glycemic fluctuations with salivary cortisol excursions has subsequently been described [125]. A fascinating study of a Namibian ethnic group during urbanization found an association of salivary cortisol increases with a disruption of glucose homeostasis [126]. Finally, the concept of "tissue-specific Cushing syndrome" has been suggested in patients with obesity, the metabolic syndrome, and insulin resistance [127, 128]. Specifically, increased adipose expression of 11BHSD1 theoretically generating increased tissue cortisol levels has been proposed [129]. This raises the possibility that medical therapy directed at 11BHSD1 may be useful in the treatment of obesity. In summary, there is moderate evidence that subtle alterations in HPA axis activity can contribute to the development of insulin resistance. In addition, selected patients with clear and specific features of Cushing syndrome and poorly controlled T2DM should be screened for neoplastic causes of endogenous hypercortisolism.

Glucocorticoid Resistance: Glucocorticoid (or primary cortisol) resistance is reviewed in detail elsewhere in this book and therefore will only be discussed in brief here. This is typically a familial receptor-mediated disorder that presents with increased androgen and cortisol production in an otherwise healthy individual [130]. Since the index cases are usually diagnosed in adulthood, the cortisol resistance is partial and accompanied by compensatory increases in circulating pituitary ACTH, and cortisol with excessive secretion of adrenal androgens and adrenal steroid biosynthesis intermediates with salt-retaining activity (e.g., deoxycorticosterone). The clinical manifestations of glucocorticoid resistance include chronic fatigue (possibly due to the result of glucocorticoid deficiency in the central nervous system) and various degrees of hypertension with or without hypokalemic alkalosis and hyperandrogenism [131]. These patients do not have the catabolic features of hypercortisolism such as cutaneous wasting, abdominal striae, myopathy, and low bone density. Nonetheless, the excessive adrenal mineralocorticoid secretion, hypokalemia, and hypertension with hypercortisolemia may be confused with pathological Cushing syndrome. In women, hyperandrogenism can result in hirsutism, menstrual irregularities, and oligomenorrhea with decreased fertility and often mimics the polycystic ovary syndrome [130]. In men, glucocorticoid resistance may lead to infertility and, in children, to precocious puberty [132]. The peripheral tissues are relatively insensitive to cortisol but they maintain sensitivity to androgens and mineralocorticoids. Normal circadian rhythm is maintained in glucocorticoid resistance; however, since the cortisol levels are reset at a higher concentration, late-night salivary cortisol levels will be elevated.

The hypertension of glucocorticoid resistance is volume dependent and associated with low plasma renin activity and sometimes hypokalemic metabolic alkalosis [133]. High circulating concentrations of deoxycorticosterone and cortisol mediate the hypertension. The high cortisol levels overwhelm the intrarenal metabolic clearance of cortisol to cortisone by 11β-hydroxysteroid dehydrogenase type 2 and participate in the generation of hypertension by binding to the mineralocorticoid receptor.

The inheritance patterns are variable, and both autosomal dominant and recessive inheritance have been described [133]. Generally, in the dominant syndromes, the mutant glucocorticoid receptor interferes with the function of the normal receptor causing a so-called dominant-negative effect. In the recessive syndromes, the normal receptor tends to rescue the mutant receptor so that heterozygotes are clinically normal.

Clinical Discrimination of Physiological and Pathological Hypercortisolism

The most important way to separate patients with pathological hypercortisolism from those with a physiological cause is to take a detailed history and perform a good physical examination. Chronic alcoholism, major depressive illness, and use of opioids are often the most difficult historical landmarks to document in patients with hypercortisolism. Many patients with chronic alcohol abuse may underestimate or underreport their alcohol abuse and withhold significant information about such things including binge drinking, alcohol withdrawal syndrome, and arrest for driving while intoxicated [60]. At times, a high index of suspicion is needed to elicit an accurate history and sometimes clues such as persistent increases in liver function tests (particularly when the AST is much greater than the ALT) may provide clues for possible heavy alcohol consumption. Although opioids actually suppress HPA axis function (mediated by hypothalamic CRH suppression), there is an abrupt recovery and actually a hyperactive HPA axis response once the opioid is discontinued or its effect wanes [134]. Consequently, the evaluation of the pituitary–adrenal function in patients taking narcotics may be especially difficult.

As described above, some neuropsychiatric disorders (particularly major depressive illness) may activate the HPA axis and cause dysregulated cortisol hypersecretion. Since neuropsychiatric and neurocognitive dysfunctions are common manifestations of patients with pathological hypercortisolism, the presence of significant melancholia in a patient with hypercortisolism provides special challenges.

Mental health specialists may need to be consulted to help with characterization of neuropsychiatric disorders. Nonetheless, the broad spectrum of neuropsychiatric disorders including obsessive-compulsive disorder, bipolar disorder, schizophrenia, and major depressive illness has all been reported in patients with pathological Cushing syndrome [135].

Poorly controlled diabetes has also been associated with hypercortisolism, but it is not clear what level of hyperglycemia actually activates the HPA axis and may cause diagnostic confusion [124, 136]. Generally, when pathological hypercortisolism causes poorly controlled diabetes mellitus, the clinical and biochemical diagnosis is usually straightforward. However, in many patients it may be necessary to use aggressive hypoglycemic pharmacotherapy to improve glycemic control before inaugurating diagnostic tests for possible Cushing syndrome. Many of the other disorders associated with hypercortisolism such as pregnancy, severe CKD, and chronic intense exercise can usually be easily established with a simple history and routine laboratory tests.

The physical examination may occasionally be helpful; however, the majority of patients where there is diagnostic confusion have relatively mild hypercortisolism, so many of the overt clinical manifestations of Cushing syndrome may be subtle or absent. Moreover, some patients with physiological hypercortisolism (especially alcohol induced) may have overt clinical Cushing syndrome and have some of the more specific physical findings including facial fullness with plethora, violaceous striae, proximal myopathy, and edema [71, 76, 78, 81]. Nonetheless, the majority of patients with true pathological hypercortisolism will have some clear objective clinical finding such as hypertension, diabetes/prediabetes, low bone density with fracture, and hirsutism/oligomenorrhea, as well as some physical evidence of cortisol excess.

Diagnostic Tests: Focus on Physiological vs. Pathological Hypercortisolism

Routine: The physiological explanations for these tests were introduced earlier in this chapter. The presence of consistently normal late-night salivary cortisol concentrations usually excludes the diagnosis of ACTH-dependent Cushing syndrome and no further testing is needed [137]. The only caveat is that some patients with adrenal-dependent Cushing syndrome (mild adrenal-dependent Cushing syndrome) may not have frank increases in late-night salivary cortisol [137]. These patients have an abnormal overnight 1 mg dexamethasone suppression test (post-dexamethasone cortisol >1.8 µg/dL [>50 nmol/L]) [137]. Consequently, the presence of consistently normal late-night salivary cortisol concentrations and normal suppression of cortisol after low dose dexamethasone suppression virtually excludes pathological hypercortisolism and no further testing is usually needed [137]. Although both of these tests have an excellent negative predictive value, their specificity is not 100 % and there are many causes of false positive tests. If the index of suspicion for pathological Cushing syndrome is low, then repeat testing complemented by UFC measurements over time may be the best strategy. Cyclical or intermittent Cushing syndrome is another phenomenon that may be associated with discordant testing and further confusion [138]. If diagnostic uncertainty prevails and the patient is restless and not willing to wait, then second line tests may be helpful to distinguish pathological and physiological hypercortisolism.

Imaging: Imaging studies should never be used to distinguish pathological and physiological hypercortisolism. The presence of small, faint, and sometimes imaginary abnormalities on magnetic resonance imaging (MRI) of the pituitary in 10–20 % of normal subjects [139] will only cause further diagnostic confusion and patient angst, so imaging of the pituitary should only be considered when a diagnosis of pathological ACTH-dependent hypercortisolism is established. Although bilateral inferior petrosal sinus ACTH sampling (IPSS) with CRH stimulation is a useful invasive diagnostic study in the differential diagnosis of ACTH-dependent Cushing syndrome, IPSS cannot distinguish states of pathological hypercortisolism from those of physiological origin [140]. Obviously, computed tomography (CT) of the abdomen with the finding of an incidental adrenal nodule is often the prelude to the consideration of hypercortisolism [81]. Nonetheless, adrenal imaging should not be performed as an index of adrenal function. For example, some patients with very large bilateral macronodular hyperplasia may have normal cortisol secretion while normal-sized adrenal glands are observed in patients with severe Cushing disease [141, 142].

Secondary Tests

DDAVP Stimulation: It is well known that corticotroph adenomas can harbor specific vasopressin receptors (V1b) and that DDAVP can elicit an ACTH response in patients with Cushing disease [10, 143, 144] In contrast, normal subjects and patients with physiological states of hypercortisolism appear to have a limited or attenuated response to DDAVP [10, 143, 144]. The significant ACTH-releasing activity of DDAVP in Cushing disease may be due to the high density of vasopressin-sensitive receptors on ACTH-producing tumor cells as well as the increased number of corticotrophs in the adenoma [145, 146]. Studies have shown good sensitivity, specificity, positive predictive value, and negative predictive value for the DDAVP stimulation test in the differential diagnosis between physiological and pathological hypercortisolism [10, 143, 144]. Serial blood samples for ACTH and cortisol measurements are secured from an indwelling venous catheter at baseline and at 15, 30, 45, and 60 min after 10 µg of DDAVP is administered intravenously. The test should be performed in the morning. The majority of studies have shown that patients with pathological hypercortisolism (ACTH dependent) will have an incremental increase in plasma ACTH of 24–30 pg/mL or a peak plasma ACTH response >60–75 pg/mL. However, many studies did not provide data from normal subjects [147].

A study by Moro et al. used a 27 pg/mL peak increase in plasma ACTH as a cutoff for Cushing disease and correctly identified 18 of 20 patients with mild Cushing disease from 29 of 30 individuals with physiological causes of hypercortisolism yielding a diagnostic accuracy of 94 % [148]. More recently, Rollin et al. studied a total of 68 patients with proven Cushing disease and compared them with 56 patients with suspected ACTH-dependent Cushing syndrome [143]. According to a receiver–operator curve analysis, an ACTH peak of 72 pg/mL following DDAVP administration provided a specificity of 95 % and sensitivity of 91 % in the correct diagnosis of Cushing disease yielding a positive predictive value of 95 %. An absolute ACTH increment more than 37 pg/mL above baseline was only observed in 2 of 56 patients without Cushing disease. Neither of these studies measured late-night salivary cortisol and the investigators acknowledged that more simplified testing may have provided a correct differential diagnosis without secondary DDAVP testing.

In patients with chronic alcoholism, there appears to be an absent ACTH and cortisol response to DDAVP but only a few patients have been carefully studied [78]. It appears that patients with depression usually have a blunted ACTH/cortisol response to DDAVP but variable results have been reported. A possible limitation to the test is the variability in ACTH assays in reference laboratories across the world. Normative data for most ACTH assays after DDAVP stimulation test are lacking and the dynamic range of the ACTH and cortisol responses in normal subjects (non-obese and obese) is unclear. In addition, some patients with ectopic ACTH-secreting tumors and hypercortisolism may not have an ACTH response to DDAVP providing further potential confusion when using this test [143].

It has also been recently shown that a positive ACTH response to DDAVP (before or after dexamethasone suppression) may actually be the earliest diagnostic indicator of recurrent Cushing disease preceding elevations of both urinary cortisol and late-night salivary cortisol [149]. Despite its limitations, the DDAVP stimulation test is simple and relatively inexpensive. We think it is currently the best secondary test to consider in patients with ACTH-dependent hypercortisolism in order to establish the presence or absence of a pathological cause.

Dexamethasone-CRH Test: Introduced in 1993, the dexamethasone-CRH test has been promoted as a means of distinguishing patients with true pathological hypercortisolism due to Cushing disease from those with hypercortisolism from a physiological cause [9]. Although some protocols have varied from the initial published approach, usually dexamethasone (0.5 mg) is given orally for eight doses prior to the morning administration of CRH (1 µg/kg or 100 µg) and cortisol measurements are obtained at baseline, 15, and 30 min. This initial report found that a serum cortisol concentration exceeding 1.4 µg/dL (39 nmol/L) was considered predictive of true Cushing disease with 100 % specificity. Recently, Alwani et al. reported 73 patients with clinical features of hypercortisolism and insufficient suppression of cortisol after 1 mg dexamethasone and/or an increased secretion of urine cortisol [10]. Fifty-three of these patients were eventually found to have true Cushing disease and 20 patients were classified as pseudo-Cushing syndrome. Using receiver operator curve analysis, an optimal cutoff value for serum cortisol concentration of

3.1 µg/dL (87 nmol/L) at 15 min had the best sensitivity (94%) and specificity (100%). This study also used a late-night salivary cortisol level >9.5 nmol/L as predictive of Cushing disease in 94% of patients with a negative predictive value of 100%. They also measured a midnight to morning ratio of serum cortisol and found that a ratio >0.67 was highly suggestive of Cushing disease with a positive predictive value of 100%. Moreover, a midnight serum cortisol concentration >8.8 µg/dL (243 nmol/L) had a positive value of 98% in predicting true Cushing disease. Defined assessment of midnight serum cortisol levels and the dexamethasone-CRH test was performed in 53 patients (35 Cushing disease and 18 pseudo-Cushing syndrome) and discordant results were found in four patients. Because of the small sample size of patients with Cushing syndrome in this study over a 12-year period, it was impossible to demonstrate any benefit of combining results of two second line tests to discriminate Cushing disease from those with pseudo-Cushing syndrome.

Since the introduction of the dexamethasone-CRH test, a lower diagnostic performance has been described with a positive predictive value of 80–86% and a negative predictive value of 92–100% using different threshold values with a 15-min post-CRH cortisol concentration (1.6–4.0 µg/dL [44–110 nmol/L]) [10, 140]. The reliability of the dexamethasone-CRH test may also be limited by differences in CRH preparations (ovine or human) as well as variation in cortisol and ACTH assays. Many commonly prescribed medications that alter dexamethasone metabolism will significantly decrease the specificity of the Dex-CRH test from 96 to 70% using a 15-min post-CRH cortisol cutoff of 1.4 µg/dL (39 nmol/L) [150]. The test is also quite cumbersome. Although it can be executed on an outpatient basis, reliability of patients taking dexamethasone every 6 h 2 days prior to the test is always a concern.

It should also be noted that the dexamethasone-CRH test is commonly employed by psychiatrists in the diagnosis of patients with depression [151]. The protocols employed by mental health specialists are less challenging for the patient (a single dose of dexamethasone usually administered the night before the test) but the fact that patients with depressive disorders tend to have augmented cortisol responses to CRH after dexamethasone creates significant clinical concern about the predicted value of a positive test in patients who are depressed and have evidence of biochemical hypercortisolism [151].

Diagnostic Pearls and Summary

The diagnosis of Cushing syndrome (particularly when it is mild) is the most challenging problem in clinical endocrinology. The differentiation between true pathological Cushing syndrome and states of physiological hypercortisolism is a common diagnostic conundrum. By definition, the degree of hypercortisolism is mild since patients with prodigious cortisol excess usually do not pose a diagnostic challenge.

Although many experts have claimed that patients with mild Cushing disease will eventually declare themselves over time, in our experience, this is not always true. Many endocrine disorders remain clinically mild for many years and sometimes decades before overt clinical manifestations are apparent. The same is likely to be so for mild pathological hypercortisolism. For example, patients with mild adrenal-dependent dysregulated cortisol excess probably have only a few physical (or biochemical) changes over many years of follow-up evaluation. The mild hypercortisolism accompanying the many common conditions we have reviewed (CKD5, depression, alcohol abuse and withdrawal, starvation) may have important clinical implications. More research is needed to characterize the mechanism and magnitude of the impact of cortisol excess in these disorders in order to consider therapeutic intervention.

The definitive diagnosis of hypercortisolism should not be established until the endocrinologist is satisfied with the presence of clinical findings as well as biochemical studies that show consistent and sustained abnormalities. When there is a history of heavy alcohol use, narcotic use, severe depressive illness, or poorly controlled diabetes mellitus, these issues should be addressed before a diagnosis of true pathological Cushing syndrome is established especially if the degree of cortisol excess is mild. If biochemical abnormalities are consistent and there is still some clinical doubt about the presence or absence of true Cushing syndrome, a second line test should be considered.

Current evidence suggests that the DDAVP stimulation test is the most useful due to its simplicity and its very good diagnostic performance. There are several caveats: this test has not been thoroughly evaluated in obese patients with the metabolic syndrome and patients with poorly controlled diabetes. The ACTH/cortisol response to DDAVP is also not well characterized in depressive illness. As previously mentioned, one study showed 15 % of patients with simple obesity may have a positive response [148]. Variability in ACTH assays and different diagnostic criteria also compromise interpretation of this test. Nonetheless, a peak ACTH response to DDAVP >70 pg/mL or incremental response >27 pg/mL seem to provide a high positive predictive test for Cushing disease. In experienced hands and with properly established reference ranges, the dexamethasone-CRH test may provide some additional diagnostic utility and some investigators have shown that these tests have similar diagnostic performance [144].

When the diagnostic biochemical studies are discordant and do not correlate with clinical findings, there should be suspicion that the patient does not have pathological Cushing syndrome. A patient can often become frustrated by the lack of certainty and it is reasonable to offer the patient another opinion from an experienced endocrinologist. As we have stated previously [5], if you have never missed the diagnosis of Cushing syndrome or have not been humbled by trying to establish its cause, you should refer your patients with suspected hypercortisolism to someone who has.

References

1. Jacobson L. Hypothalamic-pituitary-adrenocortical axis regulation. Endocrinol Metab Clin North Am. 2005;34(2):271–92. vii.
2. Stahn C, Buttgereit F. Genomic and nongenomic effects of glucocorticoids. Nat Clin Pract Rheumatol. 2008;4(10):525–33.
3. Nieman LK, Biller BM, Findling JW, Newell-Price J, Savage MO, Stewart PM, et al. The diagnosis of Cushing's syndrome: an Endocrine Society Clinical Practice Guideline. J Clin Endocrinol Metab. 2008;93(5):1526–40.
4. Findling JW, Raff H. Newer diagnostic techniques and problems in Cushing's disease. Endocrinol Metab Clin North Am. 1999;28(1):191–210.
5. Findling JW, Raff H. Diagnosis and differential diagnosis of Cushing's syndrome. Endocrinol Metab Clin North Am. 2001;30(3):729–47.
6. Raff H, Findling JW. A physiologic approach to diagnosis of the Cushing syndrome. Ann Intern Med. 2003;138(12):980–91.
7. Findling JW, Raff H. Screening and diagnosis of Cushing's syndrome. Endocrinol Metab Clin North Am. 2005;34(2):385–402.
8. Findling JW, Raff H. Cushing's Syndrome: important issues in diagnosis and management. J Clin Endocrinol Metab. 2006;91(10):3746–53.
9. Yanovski JA, Cutler Jr GB, Chrousos GP, Nieman LK. Corticotropin-releasing hormone stimulation following low-dose dexamethasone administration. A new test to distinguish Cushing's syndrome from pseudo-Cushing's states. JAMA. 1993;269(17):2232–8.
10. Alwani RA, Schmit Jongbloed LW, de Jong FH, van der Lely AJ, de Herder WW, Feelders RA. Differentiating between Cushing's disease and pseudo-Cushing's syndrome: comparison of four tests. Eur J Endocrinol. 2014;170(4):477–86.
11. Raff H, Sharma ST, Nieman LK. Physiological basis for the etiology, diagnosis, and treatment of adrenal disorders: Cushing's syndrome, adrenal insufficiency, and congenital adrenal hyperplasia. Compr Physiol. 2014;4(2):739–69.
12. Raff H, Carroll T. Cushing's syndrome: from physiological principles to diagnosis and clinical care. J Physiol. 2015;593(3):493–506.
13. Spiga F, Walker JJ, Terry JR, Lightman SL. HPA axis-rhythms. Compr Physiol. 2014;4(3):1273–98.
14. McEwen BS, Sapolsky RM. Stress and cognitive function. Curr Opin Neurobiol. 1995;5(2):205–16.
15. Sage D, Maurel D, Bosler O. Involvement of the suprachiasmatic nucleus in diurnal ACTH and corticosterone responsiveness to stress. Am J Physiol Endocrinol Metab. 2001;280(2):E260–9.
16. Gold PW. The organization of the stress system and its dysregulation in depressive illness. Mol Psychiatry. 2015;20(1):32–47.
17. Gallo-Payet N, Payet MD. Mechanism of action of ACTH: beyond cAMP. Microsc Res Tech. 2003;61(3):275–87.
18. Gallo-Payet N, Battista MC. Steroidogenesis-adrenal cell signal transduction. Compr Physiol. 2014;4(3):889–964.
19. Keller-Wood ME, Dallman MF. Corticosteroid inhibition of ACTH secretion. Endocr Rev. 1984;5(1):1–24.
20. Keller-Wood M. Hypothalamic-pituitary-adrenal axis-feedback control. Compr Physiol. 2015;5(3):1161–82.
21. Gomez-Sanchez E, Gomez-Sanchez CE. The multifaceted mineralocorticoid receptor. Compr Physiol. 2014;4(3):965–94.
22. Edwards CR, Stewart PM, Burt D, Brett L, McIntyre MA, Sutanto WS, et al. Localisation of 11 beta-hydroxysteroid dehydrogenase—tissue specific protector of the mineralocorticoid receptor. Lancet. 1988;2(8618):986–9.

23. Stewart PM, Whorwood CB, Walker BR. Steroid hormones and hypertension: the cortisol-cortisone shuttle. Steroids. 1993;58(12):614–20.
24. Tomlinson JW, Stewart PM. Cortisol metabolism and the role of 11beta-hydroxysteroid dehydrogenase. Best Pract Res Clin Endocrinol Metab. 2001;15(1):61–78.
25. Raff H, Trivedi H. Circadian rhythm of salivary cortisol, plasma cortisol, and plasma ACTH in end-stage renal disease. Endocr Connect. 2013;2(1):23–31.
26. Raff H, Raff JL, Findling JW. Late-night salivary cortisol as a screening test for Cushing's syndrome. J Clin Endocrinol Metab. 1998;83(8):2681–6.
27. Lightman SL, Conway-Campbell BL. The crucial role of pulsatile activity of the HPA axis for continuous dynamic equilibration. Nat Rev Neurosci. 2010;11(10):710–8.
28. Raff H. Salivary cortisol: a useful measurement in the diagnosis of Cushing's syndrome and the evaluation of the hypothalamic-pituitary adrenal axis. Endocrinologist. 2000;10:9–17.
29. Raff H. The role of salivary cortisol determinations in the diagnosis of Cushing's syndrome. Curr Opin Endocrinol Diabetes. 2004;11:271–5.
30. Raff H. Utility of salivary cortisol measurements in Cushing's syndrome and adrenal insufficiency. J Clin Endocrinol Metab. 2009;94(10):3647–55.
31. Raff H. Update on late-night salivary cortisol for the diagnosis of Cushing's syndrome: methodological considerations. Endocrine. 2013;44(2):346–9.
32. Raff H. Cushing Syndrome: Update on Testing. Endocrinol Metab Clin North Am. 2015;44(1):43–50.
33. Jacobson L. Hypothalamic-pituitary-adrenocortical axis: neuropsychiatric aspects. Compr Physiol. 2014;4(2):715–38.
34. Chida Y, Steptoe A. Cortisol awakening response and psychosocial factors: a systematic review and meta-analysis. Biol Psychol. 2009;80(3):265–78.
35. Clow A, Hucklebridge F, Stalder T, Evans P, Thorn L. The cortisol awakening response: more than a measure of HPA axis function. Neurosci Biobehav Rev. 2010;35(1):97–103.
36. Clow A, Hucklebridge F, Thorn L. The cortisol awakening response in context. Int Rev Neurobiol. 2010;93:153–75.
37. Clow A, Thorn L, Evans P, Hucklebridge F. The awakening cortisol response: methodological issues and significance. Stress. 2004;7(1):29–37.
38. Dedovic K, Ngiam J. The cortisol awakening response and major depression: examining the evidence. Neuropsychiatr Dis Treat. 2015;11:1181–9.
39. Elder GJ, Wetherell MA, Barclay NL, Ellis JG. The cortisol awakening response—applications and implications for sleep medicine. Sleep Med Rev. 2014;18(3):215–24.
40. Fries E, Dettenborn L, Kirschbaum C. The cortisol awakening response (CAR): facts and future directions. Int J Psychophysiol. 2009;72(1):67–73.
41. Roa SL, Elias PC, Castro M, Moreira AC. The cortisol awakening response is blunted in patients with active Cushing's disease. Eur J Endocrinol. 2013;168(5):657–64.
42. Guignat L, Bertherat J. The diagnosis of Cushing's syndrome: an Endocrine Society Clinical Practice Guideline: commentary from a European perspective. Eur J Endocrinol. 2010;163(1):9–13.
43. Raff H. Cushing's syndrome: diagnosis and surveillance using salivary cortisol. Pituitary. 2012;15(1):64–70.
44. Krieger DT, Allen W, Rizzo F, Krieger HP. Characterization of the normal temporal pattern of plasma corticosteroid levels. J Clin Endocrinol Metab. 1971;32(2):266–84.
45. Fehm HL, Voigt KH. Pathophysiology of Cushing's disease. Pathobiol Annu. 1979;9:225–55.
46. Glass AR, Zavadil III AP, Halberg F, Cornelissen G, Schaaf M. Circadian rhythm of serum cortisol in Cushing's disease. J Clin Endocrinol Metab. 1984;59(1):161–5.
47. Laudat MH, Cerdas S, Fournier C, Guiban D, Guilhaume B, Luton JP. Salivary cortisol measurement: a practical approach to assess pituitary-adrenal function. J Clin Endocrinol Metab. 1988;66(2):343–8.
48. Newell-Price J, Trainer P, Perry L, Wass J, Grossman A, Besser M. A single sleeping midnight cortisol has 100% sensitivity for the diagnosis of Cushing's syndrome. Clin Endocrinol (Oxf). 1995;43(5):545–50.

49. Papanicolaou DA, Yanovski JA, Cutler Jr GB, Chrousos GP, Nieman LK. A single midnight serum cortisol measurement distinguishes Cushing's syndrome from pseudo-Cushing states. J Clin Endocrinol Metab. 1998;83(4):1163–7.
50. Carroll T, Raff H, Findling JW. Late-night salivary cortisol measurement in the diagnosis of Cushing's syndrome. Nat Clin Pract Endocrinol Metab. 2008;4(6):344–50.
51. Carroll T, Raff H, Findling JW. Late-night salivary cortisol for the diagnosis of Cushing syndrome: a meta-analysis. Endocr Pract. 2009;15(4):335–42.
52. Bukan AP, Dere HB, Jadhav SS, Kasaliwal RR, Budyal SR, Shivane VK, et al. The performance and reproducibility of late-night salivary cortisol estimation by enzyme immunoassay for screening Cushing disease. Endocr Pract. 2015;21(2):158–64.
53. Findling JW, Raff H, Aron DC. The low-dose dexamethasone suppression test: a reevaluation in patients with Cushing's syndrome. J Clin Endocrinol Metab. 2004;89(3):1222–6.
54. Raff H, Auchus RJ, Findling JW, Nieman LK. Urine free cortisol in the diagnosis of Cushing's syndrome: is it worth doing, and if so, how? J Clin Endocrinol Metab. 2015;100(2):395–7.
55. Alexandraki KI, Grossman AB. Is urinary free cortisol of value in the diagnosis of Cushing's syndrome? Curr Opin Endocrinol Diabetes Obes. 2011;18(4):259–63.
56. Ceccato F, Barbot M, Zilio M, Frigo AC, Albiger N, Camozzi V, et al. Screening tests for Cushing's syndrome: urinary free cortisol role measured by LC-MS/MS. J Clin Endocrinol Metab. 2015;100(10):3856–61.
57. Raff H, Raff JL, Duthie EH, Wilson CR, Sasse EA, Rudman I, et al. Elevated salivary cortisol in the evening in healthy elderly men and women: correlation with bone mineral density. J Gerontol A Biol Sci Med Sci. 1999;54(9):M479–83.
58. Inder WJ, Joyce PR, Wells JE, Evans MJ, Ellis MJ, Mattioli L, et al. The acute effects of oral ethanol on the hypothalamic-pituitary-adrenal axis in normal human subjects. Clin Endocrinol (Oxf). 1995;42(1):65–71.
59. Waltman C, Blevins Jr LS, Boyd G, Wand GS. The effects of mild ethanol intoxication on the hypothalamic-pituitary-adrenal axis in nonalcoholic men. J Clin Endocrinol Metab. 1993;77(2):518–22.
60. Wand GS, Dobs AS. Alterations in the hypothalamic-pituitary-adrenal axis in actively drinking alcoholics. J Clin Endocrinol Metab. 1991;72(6):1290–5.
61. Rees LH, Besser GM, Jeffcoate WJ, Goldie DJ, Marks V. Alcohol-induced pseudo-Cushing's syndrome. Lancet. 1977;1(8014):726–8.
62. Smalls AG, Kloppenborg PW, Njo KT, Knoben JM, Ruland CM. Alcohol-induced Cushingoid syndrome. Br Med J. 1976;2(6047):1298.
63. Rivier C, Bruhn T, Vale W. Effect of ethanol on the hypothalamic-pituitary-adrenal axis in the rat: role of corticotropin-releasing factor (CRF). J Pharmacol Exp Ther. 1984;229(1):127–31.
64. Rivier C, Imaki T, Vale W. Prolonged exposure to alcohol: effect on CRF mRNA levels, and CRF- and stress-induced ACTH secretion in the rat. Brain Res. 1990;520(1-2):1–5.
65. Ogilvie KM, Lee S, Rivier C. Role of arginine vasopressin and corticotropin-releasing factor in mediating alcohol-induced adrenocorticotropin and vasopressin secretion in male rats bearing lesions of the paraventricular nuclei. Brain Res. 1997;744(1):83–95.
66. Ellis FW. Effect of ethanol on plasma corticosterone levels. J Pharmacol Exp Ther. 1966;153(1):121–7.
67. Cobb CF, Van Thiel DH, Gavaler JS, Lester R. Effects of ethanol and acetaldehyde on the rat adrenal. Metabolism. 1981;30(6):537–43.
68. Elias AN, Meshkinpour H, Valenta LJ, Grossman MK. Pseudo-Cushing's syndrome: the role of alcohol. J Clin Gastroenterol. 1982;4(2):137–9.
69. Besemer F, Pereira AM, Smit JW. Alcohol-induced Cushing syndrome. Hypercortisolism caused by alcohol abuse. Neth J Med. 2011;69(7):318–23.
70. Lamberts SW, de Jong FH, Birkenhager JC. Biochemical characteristics of alcohol-induced pseudo-Cushing's syndrome [proceedings]. J Endocrinol. 1979;80(2):62P–3.
71. Smals AG, Njo KT, Knoben JM, Ruland CM, Kloppenborg PW. Alcohol-induced Cushingoid syndrome. J R Coll Physicians Lond. 1977;12(1):36–41.

72. Nyblom H, Berggren U, Balldin J, Olsson R. High AST/ALT ratio may indicate advanced alcoholic liver disease rather than heavy drinking. Alcohol Alcohol. 2004;39(4):336–9.

73. Hiramatsu R, Nisula BC. Effect of alcohol on the interaction of cortisol with plasma proteins, glucocorticoid receptors and erythrocytes. J Steroid Biochem. 1989;33(1):65–70.

74. Proto G, Barberi M, Bertolissi F. Pseudo-Cushing's syndrome: an example of alcohol-induced central disorder in corticotropin-releasing factor-ACTH release? Drug Alcohol Depend. 1985;16(2):111–5.

75. Dackis CA, Stuckey RF, Gold MS, Pottash AL. Dexamethasone suppression test testing of depressed alcoholics. Alcohol Clin Exp Res. 1986;10(1):59–60.

76. Lamberts SW, Klijn JG, de Jong FH, Birkenhager JC. Hormone secretion in alcohol-induced pseudo-Cushing's syndrome. Differential diagnosis with Cushing disease. JAMA. 1979;242(15):1640–3.

77. Hundt W, Zimmermann U, Pottig M, Spring K, Holsboer F. The combined dexamethasone-suppression/CRH-stimulation test in alcoholics during and after acute withdrawal. Alcohol Clin Exp Res. 2001;25(5):687–91.

78. Coiro V, Volpi R, Capretti L, Caffarri G, Chiodera P. Desmopressin and hexarelin tests in alcohol-induced pseudo-Cushing's syndrome. J Intern Med. 2000;247(6):667–73.

79. Adinoff B, Ruether K, Krebaum S, Iranmanesh A, Williams MJ. Increased salivary cortisol concentrations during chronic alcohol intoxication in a naturalistic clinical sample of men. Alcohol Clin Exp Res. 2003;27(9):1420–7.

80. Lovallo WR, Dickensheets SL, Myers DA, Thomas TL, Nixon SJ. Blunted stress cortisol response in abstinent alcoholic and polysubstance-abusing men. Alcohol Clin Exp Res. 2000;24(5):651–8.

81. Kapcala LP. Alcohol-induced pseudo-Cushing's syndrome mimicking Cushing's disease in a patient with an adrenal mass. Am J Med. 1987;82(4):849–56.

82. Miller KK. Endocrine dysregulation in anorexia nervosa update. J Clin Endocrinol Metab. 2011;96(10):2939–49.

83. Bredella MA, Fazeli PK, Miller KK, Misra M, Torriani M, Thomas BJ, et al. Increased bone marrow fat in anorexia nervosa. J Clin Endocrinol Metab. 2009;94(6):2129–36.

84. Duclos M, Corcuff JB, Roger P, Tabarin A. The dexamethasone-suppressed corticotrophin-releasing hormone stimulation test in anorexia nervosa. Clin Endocrinol (Oxf). 1999; 51(6):725–31.

85. Foppiani L, Sessarego P, Valenti S, Falivene MR, Cuttica CM, Giusti DM. Lack of effect of desmopressin on ACTH and cortisol responses to ovine corticotropin-releasing hormone in anorexia nervosa. Eur J Clin Invest. 1996;26(10):879–83.

86. Lawson EA, Donoho D, Miller KK, Misra M, Meenaghan E, Lydecker J, et al. Hypercortisolemia is associated with severity of bone loss and depression in hypothalamic amenorrhea and anorexia nervosa. J Clin Endocrinol Metab. 2009;94(12):4710–6.

87. Misra M, Miller KK, Almazan C, Ramaswamy K, Lapcharoensap W, Worley M, et al. Alterations in cortisol secretory dynamics in adolescent girls with anorexia nervosa and effects on bone metabolism. J Clin Endocrinol Metab. 2004;89(10):4972–80.

88. Van den Berghe G. Novel insights into the neuroendocrinology of critical illness. Eur J Endocrinol. 2000;143(1):1–13.

89. Valentine AR, Raff H, Liu H, Ballesteros M, Rose JM, Jossart GH, et al. Salivary cortisol increases after bariatric surgery in women. Horm Metab Res. 2011;43(8):587–90.

90. Findling JW, Raff H. Ectopic ACTH. In: Mazzaferri EL, Samaan NA, editors. Endocrine tumors. Cambridge: Blackwell Scientific; 1993. p. 554–66.

91. Black MM, Hall R, Kay DW, Kilborn JR. Anorexia nervosa in Cushing's syndrome. J Clin Endocrinol Metab. 1965;25:1030–4.

92. Pariante CM, Lightman SL. The HPA axis in major depression: classical theories and new developments. Trends Neurosci. 2008;31(9):464–8.

93. Wilkinson CW, Peskind ER, Raskind MA. Decreased hypothalamic-pituitary-adrenal axis sensitivity to cortisol feedback inhibition in human aging. Neuroendocrinology. 1997; 65(1):79–90.

94. Wilkinson CW, Petrie EC, Murray SR, Colasurdo EA, Raskind MA, Peskind ER. Human glucocorticoid feedback inhibition is reduced in older individuals: evening study. J Clin Endocrinol Metab. 2001;86(2):545–50.

95. Gupta D, Morley JE. Hypothalamic-pituitary-adrenal (HPA) axis and aging. Compr Physiol. 2014;4(4):1495–510.

96. Belvederi MM, Pariante C, Mondelli V, Masotti M, Atti AR, Mellacqua Z, et al. HPA axis and aging in depression: systematic review and meta-analysis. Psychoneuroendocrinology. 2014;41:46–62.

97. Letizia C, Mazzaferro S, De CA, Cerci S, Morabito S, Cinotti GA, et al. Effects of haemodialysis session on plasma beta-endorphin, ACTH and cortisol in patients with end-stage renal disease. Scand J Urol Nephrol. 1996;30(5):399–402.

98. N'Gankam V, Uehlinger D, Dick B, Frey BM, Frey FJ. Increased cortisol metabolites and reduced activity of 11beta-hydroxysteroid dehydrogenase in patients on hemodialysis. Kidney Int. 2002;61(5):1859–66.

99. Wallace EZ, Rosman P, Toshav N, Sacerdote A, Balthazar A. Pituitary-adrenocortical function in chronic renal failure: studies of episodic secretion of cortisol and dexamethasone suppressibility. J Clin Endocrinol Metab. 1980;50(1):46–51.

100. Boonen E, Van den Berghe G. Endocrine responses to critical illness: novel insights and therapeutic implications. J Clin Endocrinol Metab. 2014;99(5):1569–82.

101. Boonen E, Vervenne H, Meersseman P, Andrew R, Mortier L, Declercq PE, et al. Reduced cortisol metabolism during critical illness. N Engl J Med. 2013;368(16):1477–88.

102. Kidambi S, Kotchen JM, Grim CE, Raff H, Mao J, Singh RJ, et al. Association of adrenal steroids with hypertension and the metabolic syndrome in blacks. Hypertension. 2007;49(3):704–11.

103. Hackett RA, Steptoe A, Kumari M. Association of diurnal patterns in salivary cortisol with type 2 diabetes in the Whitehall II study. J Clin Endocrinol Metab. 2014;99(12):4625–31.

104. Huitinga I, Erkut ZA, van Beurden D, Swaab DF. The hypothalamo-pituitary-adrenal axis in multiple sclerosis. Ann N Y Acad Sci. 2003;992:118–28.

105. Raff H, Ettema SL, Eastwood DC, Woodson BT. Salivary cortisol in obstructive sleep apnea: the effect of CPAP. Endocrine. 2011;40(1):137–9.

106. Lindsay JR, Nieman LK. The hypothalamic-pituitary-adrenal axis in pregnancy: challenges in disease detection and treatment. Endocr Rev. 2005;26(6):775–99.

107. Lopes LM, Francisco RP, Galletta MA, Bronstein MD. Determination of nighttime salivary cortisol during pregnancy: comparison with values in non-pregnancy and Cushing's disease. Pituitary. 2016;19(1):30–8.

108. Jung C, Ho JT, Torpy DJ, Rogers A, Doogue M, Lewis JG, et al. A longitudinal study of plasma and urinary cortisol in pregnancy and postpartum. J Clin Endocrinol Metab. 2011;96(5):1533–40.

109. Carr BR, Parker Jr CR, Madden JD, MacDonald PC, Porter JC. Maternal plasma adrenocorticotropin and cortisol relationships throughout human pregnancy. Am J Obstet Gynecol. 1981;139(4):416–22.

110. Suda T, Iwashita M, Tozawa F, Ushiyama T, Tomori N, Sumitomo T, et al. Characterization of corticotropin-releasing hormone binding protein in human plasma by chemical crosslinking and its binding during pregnancy. J Clin Endocrinol Metab. 1988;67(6):1278–83.

111. Sasaki A, Shinkawa O, Yoshinaga K. Placental corticotropin-releasing hormone may be a stimulator of maternal pituitary adrenocorticotropic hormone secretion in humans. J Clin Invest. 1989;84(6):1997–2001.

112. Thomson M. The physiological roles of placental corticotropin releasing hormone in pregnancy and childbirth. J Physiol Biochem. 2013;69(3):559–73.

113. Findling JW, Waters VO, Raff H. The dissociation of renin and aldosterone during critical illness. J Clin Endocrinol Metab. 1987;64(3):592–5.

114. Vermes I, Beishuizen A. The hypothalamic-pituitary-adrenal response to critical illness. Best Pract Res Clin Endocrinol Metab. 2001;15(4):495–511.

115. Beishuizen A, Thijs LG, Vermes I. Patterns of corticosteroid-binding globulin and the free cortisol index during septic shock and multitrauma. Intensive Care Med. 2001;27(10):1584–91.

116. Nenke MA, Rankin W, Chapman MJ, Stevens NE, Diener KR, Hayball JD, et al. Depletion of high-affinity corticosteroid-binding globulin corresponds to illness severity in sepsis and septic shock; clinical implications. Clin Endocrinol (Oxf). 2015;82(6):801–7.

117. Bartanusz V, Corneille MG, Sordo S, Gildea M, Michalek JE, Nair PV, et al. Diurnal salivary cortisol measurement in the neurosurgical-surgical intensive care unit in critically ill acute trauma patients. J Clin Neurosci. 2014;21(12):2150–4.

118. Thevenot T, Dorin R, Monnet E, Qualls CR, Sapin R, Grandclement E, et al. High serum levels of free cortisol indicate severity of cirrhosis in hemodynamically stable patients. J Gastroenterol Hepatol. 2012;27(10):1596–601.

119. Catargi B, Rigalleau V, Poussin A, Ronci-Chaix N, Bex V, Vergnot V, et al. Occult Cushing's syndrome in type-2 diabetes. J Clin Endocrinol Metab. 2003;88(12):5808–13.

120. Leibowitz G, Tsur A, Chayen SD, Salameh M, Raz I, Cerasi E, et al. Pre-clinical Cushing's syndrome: an unexpected frequent cause of poor glycaemic control in obese diabetic patients. Clin Endocrinol (Oxf). 1996;44(6):717–22.

121. Krarup T, Krarup T, Hagen C. Do patients with type 2 diabetes mellitus have an increased prevalence of Cushing's syndrome? Diabetes Metab Res Rev. 2012;28(3):219–27.

122. Terzolo M, Reimondo G, Chiodini I, Castello R, Giordano R, Ciccarelli E, et al. Screening of Cushing's syndrome in outpatients with type 2 diabetes: results of a prospective multicentric study in Italy. J Clin Endocrinol Metab. 2012;97(10):3467–75.

123. Newsome S, Chen K, Hoang J, Wilson JD, Potter JM, Hickman PE. Cushing's syndrome in a clinic population with diabetes. Intern Med J. 2008;38(3):178–82.

124. Liu H, Bravata DM, Cabaccan J, Raff H, Ryzen E. Elevated late-night salivary cortisol levels in elderly male type 2 diabetic veterans. Clin Endocrinol (Oxf). 2005;63(6):642–9.

125. Bellastella G, Maiorino MI, De BA, Vietri MT, Mosca C, Scappaticcio L, et al. Serum but not salivary cortisol levels are influenced by daily glycemic oscillations in type 2 diabetes. Endocrine. 2016;53(1):220–6.

126. Kann PH, Munzel M, Hadji P, Daniel H, Flache S, Nyarango P, et al. Alterations of cortisol homeostasis may link changes of the sociocultural environment to an increased diabetes and metabolic risk in developing countries: a prospective diagnostic study performed in cooperation with the Ovahimba people of the Kunene region/northwestern Namibia. J Clin Endocrinol Metab. 2015;100(3):E482–6.

127. Constantinopoulos P, Michalaki M, Kottorou A, Habeos I, Psyrogiannis A, Kalfarentzos F, et al. Cortisol in tissue and systemic level as a contributing factor to the development of metabolic syndrome in severely obese patients. Eur J Endocrinol. 2015;172(1):69–78.

128. Anagnostis P, Athyros VG, Tziomalos K, Karagiannis A, Mikhailidis DP. Clinical review: The pathogenetic role of cortisol in the metabolic syndrome: a hypothesis. J Clin Endocrinol Metab. 2009;94(8):2692–701.

129. Paterson JM, Morton NM, Fievet C, Kenyon CJ, Holmes MC, Staels B, et al. Metabolic syndrome without obesity: Hepatic overexpression of 11beta-hydroxysteroid dehydrogenase type 1 in transgenic mice. Proc Natl Acad Sci U S A. 2004;101(18):7088–93.

130. Malchoff CD, Malchoff DM. Glucocorticoid resistance and hypersensitivity. Endocrinol Metab Clin North Am. 2005;34(2):315–26. viii.

131. Bronnegard M, Werner S, Gustafsson JA. Primary cortisol resistance associated with a thermolabile glucocorticoid receptor in a patient with fatigue as the only symptom. J Clin Invest. 1986;78(5):1270–8.

132. Malchoff CD, Javier EC, Malchoff DM, Martin T, Rogol A, Brandon D, et al. Primary cortisol resistance presenting as isosexual precocity. J Clin Endocrinol Metab. 1990; 70(2):503–7.

133. Charmandari E, Kino T, Ichijo T, Chrousos GP. Generalized glucocorticoid resistance: clinical aspects, molecular mechanisms, and implications of a rare genetic disorder. J Clin Endocrinol Metab. 2008;93(5):1563–72.

134. Oltmanns KM, Fehm HL, Peters A. Chronic fentanyl application induces adrenocortical insufficiency. J Intern Med. 2005;257(5):478–80.
135. Pivonello R, Simeoli C, De Martino MC, Cozzolino A, De LM, Iacuaniello D, et al. Neuropsychiatric disorders in Cushing's syndrome. Front Neurosci. 2015;9:129.
136. Oltmanns KM, Dodt B, Schultes B, Raspe HH, Schweiger U, Born J, et al. Cortisol correlates with metabolic disturbances in a population study of type 2 diabetic patients. Eur J Endocrinol. 2006;154(2):325–31.
137. Findling JW. Evolution, global warming, smart phones, and late-night salivary cortisol. Endocr Pract. 2015;21(2):205–7.
138. Atkinson AB, Kennedy AL, Carson DJ, Hadden DR, Weaver JA, Sheridan B. Five cases of cyclical Cushing's syndrome. Br Med J (Clin Res Ed). 1985;291(6507):1453–7.
139. Hall WA, Luciano MG, Doppman JL, Patronas NJ, Oldfield EH. Pituitary magnetic resonance imaging in normal human volunteers: occult adenomas in the general population. Ann Intern Med. 1994;120(10):817–20.
140. Yanovski JA, Cutler Jr GB, Doppman JL, Miller DL, Chrousos GP, Oldfield EH, et al. The limited ability of inferior petrosal sinus sampling with corticotropin-releasing hormone to distinguish Cushing's disease from pseudo-Cushing states or normal physiology. J Clin Endocrinol Metab. 1993;77(2):503–9.
141. Drougat L, Espiard S, Bertherat J. Genetics of primary bilateral macronodular adrenal hyperplasia: a model for early diagnosis of Cushing's syndrome? Eur J Endocrinol. 2015;173(4):M121–31.
142. Pojunas KW, Daniels DL, Williams AL, Thorsen MK, Haughton VM. Pituitary and adrenal CT of Cushing syndrome. AJR Am J Roentgenol. 1986;146(6):1235–8.
143. Rollin G, Costenaro F, Gerchman F, Rodrigues TC, Czepielewski MA. Evaluation of the DDAVP Test in the Diagnosis of Cushing's Disease. Clin Endocrinol (Oxf). 2015;82(6):793–800.
144. Tirabassi G, Faloia E, Papa R, Furlani G, Boscaro M, Arnaldi G. Use of the desmopressin test in the differential diagnosis of pseudo-Cushing state from Cushing's disease. J Clin Endocrinol Metab. 2010;95(3):1115–22.
145. Arnaldi G, de Keyzer Y, Gasc JM, Clauser E, Bertagna X. Vasopressin receptors modulate the pharmacological phenotypes of Cushing's syndrome. Endocr Res. 1998;24(3-4):807–16.
146. Tsagarakis S, Tsigos C, Vasiliou V, Tsiotra P, Kaskarelis J, Sotiropoulou C, et al. The desmopressin and combined CRH-desmopressin tests in the differential diagnosis of ACTH-dependent Cushing's syndrome: constraints imposed by the expression of V2 vasopressin receptors in tumors with ectopic ACTH secretion. J Clin Endocrinol Metab. 2002;87(4):1646–53.
147. Pecori Giraldi F, Pivonello R, Ambrogio AG, De Martino MC, De Martin M, Scacchi M, Colao A, Toja PM, Lombardi G, Cavagnini F. The dexamethasone-suppressed corticotropin-releasing hormone stimulation test and the desmopressin test to distinguish Cushing's syndrome from pseudo-Cushing's states. Clin Endocrinol (Oxf). 2007;66(2):251–7.
148. Moro M, Putignano P, Losa M, Invitti C, Maraschini C, Cavagnini F. The desmopressin test in the differential diagnosis between Cushing's disease and pseudo-Cushing states. J Clin Endocrinol Metab. 2000;85(10):3569–74.
149. Bou KR, Baudry C, Guignat L, Carrasco C, Guibourdenche J, Gaillard S, et al. Sequential hormonal changes in 21 patients with recurrent Cushing's disease after successful pituitary surgery. Eur J Endocrinol. 2011;165(5):729–37.
150. Valassi E, Swearingen B, Lee H, Nachtigall LB, Donoho DA, Klibanksi A, Biller BMK. Concomitant medication use can confound interpretation of the combined dexamethasone-corticotropin releasing hormone test in Cushing's syndrome. J Clin Endocrinol Metab. 2009;94(12):4851–9.
151. Ising M, Kunzel HE, Binder EB, Nickel T, Modell S, Holsboer F. The combined dexamethasone/CRH test as a potential surrogate marker in depression. Prog Neuropsychopharmacol Biol Psychiatry. 2005;29(6):1085–93.

Imaging Strategies for Localization of ACTH-Secreting Tumors

Lynnette K. Nieman and Ahmed M. Gharib

Abstract The causes of ACTH-dependent Cushing's syndrome include corticotrope tumors that secrete ACTH (Cushing's disease) and tumors outside the pituitary gland that secrete ACTH "ectopically" (Ectopic ACTH secretion). Since pituitary tumors are much more common, imaging usually begins with a pituitary MRI. The specific protocol used for the study influences the ability to identify a tumor, but even the best protocols do not identify more than 80 % of these tumors.

Ectopic ACTH-secreting tumors occur most commonly in the thorax but may be found in the neck, abdomen, or pelvis. Structural imaging with CT (and MRI as an adjunctive modality) is the mainstay but is complemented by function imaging, usually with somatostatin analogs. Since many tumors are occult at initial presentation, imaging is repeated at intervals until a tumor is identified and (hopefully) resected.

Keywords Cushing's disease • Ectopic ACTH • Cortisol • ACTH • Inferior petrosal sinus sampling

Introduction

Once the diagnosis of ACTH-dependent Cushing's syndrome is made, biochemical testing is used to discriminate between ectopic and pituitary tumoral production of ACTH. A corticotrope tumor (Cushing's disease) is the most common cause of ACTH-dependent Cushing's syndrome. If inferior petrosal sinus sampling is planned, a pituitary MRI should be obtained beforehand to exclude a 6 mm or larger pituitary mass that might obviate the need for the invasive sampling procedure. If sampling will not be done, or if there is a very high clinical suspicion of ectopic

L.K. Nieman, M.D. (✉)
Diabetes, Endocrinology and Obesity Branch, National Institute of Diabetes and Digestive and Kidney Diseases, Bethesda, MD, USA
e-mail: NiemanL@nih.gov

A.M. Gharib, M.D.
Biomedical and Metabolic Imaging Branch, National Institute of Diabetes and Digestive and Kidney Diseases, Bethesda, MD, USA
e-mail: agharib@mail.nih.gov

© Springer International Publishing Switzerland 2017
E.B. Geer (ed.), *The Hypothalamic-Pituitary-Adrenal Axis in Health and Disease*, DOI 10.1007/978-3-319-45950-9_7

tumoral production, the pituitary MRI might be deferred until other data are collected. However, most clinicians obtain a pituitary MRI as an initial step regardless of the planned evaluation. Details about a pituitary MRI are below.

When an ectopic ACTH-secreting tumor is suspected, biochemical testing may suggest the type of tumor (e.g., elevated calcitonin), but imaging must be performed for localization. Available imaging techniques are either functional or structural; the latter yielding good information about anatomy and more limited information about function and vice versa. Because of these differences, anatomical imaging with CT is the mainstay for tumor identification, and MRI and the functional imaging techniques provide very useful ancillary information. The available modalities are discussed below.

Imaging Studies for Localization of a Corticotrope Tumor (Cushing's Disease)

A dedicated pituitary MRI examination is the gold standard for identification of a pituitary lesion. Spin echo MRI protocols were the first to gain widespread popularity and continue to be the most commonly used pulse sequence in the evaluation of the pituitary gland [1]. The spin echo sequence is made up of two radiofrequency pulses — one pulse that excites the spins in the tissue (repetition time, TR) and a subsequent 180° pulse that refocuses a resultant "echo" (echo time, TE). T1-weighted images use a short TR (500–700 ms) and TE (15–25 ms). As a result, tissues that relax more quickly (such as fat) present as bright signal. Tumors have longer T1 relaxation times and show as a dark signal [2]. Based on this, T1-weighted spin echo MRI has been recommended for the routine evaluation of pituitary adenomas [3–5].

MRI performed by the standard T1-weighted spin echo technique only detects up to 60 % of corticotrope tumors, perhaps because they tend to be microadenomas with signal and enhancing characteristics similar to normal pituitary tissue [6]. A number of parameters influence the final T1-weighted spin echo MR image, particularly the length of the TR and TE intervals [7]. As shown in the Table 1, other variables that affect sensitivity include magnetic field strength (greater sensitivity with higher magnetic field) and field of view (FOV) or spatial resolution, which optimally focuses on the pituitary gland (12×12 cm) rather than the entire brain. Additionally, thin interleaved slice images of 3 mm or less improve resolution [2,7]. The use of a T1 contrast agent also enhances detection of pituitary adenomas, which take up contrast more slowly than surrounding normal tissue [7].

Dynamic spin echo techniques (dMRI) take advantage of the differential uptake of contrast by tumors vs. normal pituitary tissue. By obtaining multiple images immediately after contrast injection, a "dynamic" MRI is obtained. These require rapid imaging techniques called spoiled gradient recalled echo (GRE) in order to capture the proper enhancement phase of the tumor (discussed below). Based on relatively small studies, it seems that dMRI has better sensitivity than conventional SE technique, but may identify more false positive lesions, suggesting an important loss of specificity [8].

Table 1 Studies comparing types of MRI sequence, magnet strength, or parameters of MR protocols

Reference	Type of MRI sequence	Magnet strength (T)	Matrix size	TR/TE (ms)	FOV (cm)	Slice thickness (mm)	Sens[b]	FP[c]
Kasaliwal et al. [9]	Dynamic contrast spin echo (DC-SE)	1.5	256×138	n/a	21×21	2, interleaved	*16/24 (67%)*	0
Kasaliwal et al. [9]	3D-spoiled gradient echo	1.5	256×205	n/a	16×16	1, no gap	*21/24 (88%)*	0
Tabarin et al. [8]	DC-SE	1.0	210×256	575/15	30×30	3, no gap	*11/14 (79%)*	*3*
Tabarin et al. [8]	T1 SE	1.0	256×256	450/14	20×20	3	*8/14 (57%)*	*0*
Patronas et al. [6]	SPGR	1.5	160×256	9.6/2.3	12 or 18	1, no gap	*40/50 (80%)*	*2-4/50*
Patronas et al. [6]	T1 SE	1.5	192×256Corticotrope tumors:MRI sequence, magnet strength/MR protocols	400/9	12	3, interleaved	*25/50 (50%)*	*1-2/50*
Chowdhury et al. [51]	T1 SE at NIH	1.5		400/10.3±0.5	12×12	3	*18/18*	
Chowdhury et al. [51]	T1 SE not at NIH	<1.5 n=5, 1.5 n=11		492±19/17.2±1.2	17±0.6×18±0.7	3	*2/18*	
De Rotte et al. [52]	T1 SE[a], T2 SE	7	n/a	3952/37	25×25	n/a	8/9	n/a
De Rotte et al. [52]	T1 SE, T2 SE dMRI	1.5	n/a	n/a	n/a	n/a	5/9	n/a

The italic numerals indicate important differences between the comparator groups

[a] 3D T1-weighted magnetization-prepared inversion recovery (MPIR) SE

[b] Sens = Sensitivity

[c] FP = False positive result

Besides the spin echo technique, other MRI protocols have been used for the detection of corticotropinomas. For example, spoiled gradient recalled (SPGR) acquisition in the steady state improved the tumor detection rate compared to T1 spin echo imaging (80 % vs. 49 %), at the expense of a higher false positive rate (2 % vs. 4 %) [6]. Another study comparing dMRI with spoiled gradient echo (SGE) sequences found the SGE protocol to have better sensitivity [9].

Limited numbers of patients have been studied at both 1.5 and 3 T magnet strength [10,11]; both studies suggest improved sensitivity with the higher magnet strength.

Factors Affecting Interpretation of Pituitary Lesions on MRI

In a study of 100 healthy volunteers, 10 % had a pituitary lesion on T1 SE MRI imaging, with a 3–6 mm diameter [12]. In a study of 201 patients with Cushing's disease who had surgical confirmation of the location of the tumor, 14 % had a false positive lesion on MRI [13]. Similarly, in a study of 66 patients with ectopic ACTH secretion, 17 (26 %) had an abnormal pituitary MRI, 13 of whom had previous unsuccessful pituitary exploration [14]. In another study, 6 of 26 patients with ectopic ACTH secretion had a lesion on pituitary MRI, but only one had a diameter >6 mm (96 % specificity for 6 mm criterion) [15]. Taken together, these data indicate that a lesion on pituitary MRI does not necessarily correspond to a corticotrope adenoma. Such a lesion does provide a location to target during transsphenoidal surgery, however.

Non-pituitary (Ectopic) Location of Corticotrope Tumors

When reviewing imaging studies to identify corticotrope tumors, it is important to recognize that these may occur rarely in a non-pituitary location along the developmental path of Rathke's pouch: in the nasal cavity [16], the sphenoid sinus [17], and clivus. They may also occur in locations proximal to, but outside of the anterior pituitary gland, including the infundibulum [18], parasellar location [19], posterior pituitary [20], and cavernous sinus. The imaging results for these areas should be reviewed in patients in whom biochemical data suggest Cushing's disease but the pituitary MRI is negative and in those with unsuccessful transsphenoidal exploration.

Positron Emission Tomography Approaches to Localization of Corticotrope Tumors

A few studies have evaluated the use of ^{11}C-methionine or ^{18}F-FDG positron emission tomography (PET) for the localization of pituitary adenomas. The essential amino acid methionine is taken up into tissues that have increased protein synthesis.

Physiologic uptake is present in normal pituitary. In one study, 7 of 10 patients with Cushing's disease had asymmetric uptake in the pituitary gland at the site of a lesion seen by SPGR MRI. These were all confirmed to be ACTH-secreting tumors after surgical resection [21]. Another study compared the ability of ^{18}F-FDG to image metabolically active tissue with the sensitivity of T1 SE or SPGR MRI. ^{18}F-FDG PET localized tumor in 4 patients, all of whom had a less than 180 % increase in ACTH after CRH stimulation. ^{18}F-FDG PET also detected two adenomas not identified by T1 SE, but did not improve the sensitivity of SPGR MRI [22].

Imaging Studies for Localization of an Ectopic ACTH-Producing Tumor

Having assigned a diagnosis of presumed ectopic ACTH secretion based on biochemical testing, the next challenge is to locate a possible tumor. Although biochemical tumor markers are not uniformly helpful, they may suggest what to image first. For example, elevated calcitonin or plasma free metanephrines may point to the thyroid or adrenal gland; on the other hand, chromogranin A is not specific, and urinary 5-HIAAA is often not abnormal in patients with foregut carcinoids, perhaps because these often do not express the enzyme aromatic L-amino-acid decarboxylase needed for serotonin synthesis.

Although the initial description of the ectopic ACTH syndrome highlighted overt and metastatic tumors, slow growing, often occult tumors represent the majority of cases in 2016. As a result, imaging identification and surgical removal of the tumor are critical to successful treatment [23]. Despite the use of anatomical imaging techniques like computed tomography (CT) and magnetic resonance imaging (MRI), up to 50 % of ectopic ACTH-secreting tumors are not found on initial imaging [24].

Anatomic Imaging

If no biochemical marker suggests an anatomic source, given that about 50 % of these tumors arise in the chest (Table 2), computed tomography (CT) of the thorax, using thin slice thickness (1–2 mm), is a cost-effective initial imaging strategy. If a clear-cut lesion is identified, then additional imaging may not be needed. However, in many series, tumors remain occult, or occur elsewhere, and additional imaging with different modalities over time is needed [25].

Additional imaging includes CT imaging of the neck, abdomen, and pelvis, as well as MRI of these areas and the chest. Neuroendocrine tumors may be "bright" on T2 sequences that utilize fat-suppression techniques, making these sequences an important part of an MRI protocol [26]. The use of "triple phase" CT imaging may improve detection of intestinal and pancreatic tumors and hepatic metastases. This involves imaging before injection of iodinated contrast, followed by three phases after contrast injection at a rapid rate (2–3 mL/s). These phases include a late

Table 2 Types of non-corticotrope tumors reported to secrete ACTH

Type of tumor-producing ACTH	Number				
	Reference (n)				
	Salgado et al. [53]	Aniszewski et al. [54]	Ilias et al. [14]	Isidori et al. [55]	Ejaz et al. [34]
	n=25	n=106	n=73	n=40	n=43
Pulmonary c'oid	10	28	35	12	9
Pancreatic c'oid	3	17	1	3	
Medullary thyroid Ca		9	2	3	5
Thymic carcinoids	4	5	5	2	3
Pheochromocytoma	5	3	5	1	
Gastrinoma			6		
Non-specific NET		7	13	2	3
Small cell lung Ca		12	3	7	9
Other tumors[a]	1	9	3	5	6
Occult	2	17	17	5	

C'oid = carcinoid; Ca = Cancer

[a]Olfactory esthesioneuroblastoma, mesothelioma, glomus tumor, other carcinoid tumors (hepatic, appendix, tumorlets, disseminated GI carcinoid), tumors of the esophagus, stomach, pancreas, larynx, trachea, salivary gland, Leydig cell, breast, ovary, cervix, kidney, gallbladder, prostate, hepatocellular carcinoma, melanoma, leukemia, lymphoma, ostomyeloma [56]

arterial phase of enhancement, at 20–45 s after the start of the injection, followed by a third imaging at 60–70 s after the start of injection, for the portal venous phase [27]. A delayed phase scan may also be obtained at 3 min to better characterize liver lesions if present.

MRI and CT provide the best anatomic/structural resolution of tumors, and are complementary, having about 90 % combined sensitivity [14,24].

Functional ("Molecular") Imaging

Functional imaging, also called "molecular imaging," reduces false positive results because it relies on the specific properties of tumor cells, not just their anatomic characteristics. However, tumors lacking the relevant somatostatin receptors, increased metabolic rate (FDG-PET), or amine precursor uptake (F-DOPA) have false negative results [28].

In the United States, somatostatin receptor scintigraphy is commercially available using [^{111}In-DTPA-D-Phe]-pentetreotide (Octreoscan™, OCT) at a 6 mCi dose. The ability of OCT to identify the tumors depends on multiple factors, including the dose of the radiopharmaceutical, the type and degree of somatostatin receptor expression, and tumor size [28–30]. Relatively small case series report that OCT detects 4/12 [31], 6/6 [32], 10/18 [33], 12/20 [34], and 5/16 tumors [35]. A larger series of 39 patients found a sensitivity of 41 %, but with a false positive rate of 27 % [36]. A systematic review of the literature found an overall OCT detection rate of 48.9 % (84/172) [24].

More recently, ^{68}Ga-labeled somatostatin analogs (DOTATATE, DOTATOC, and DOTANOC, collectively referred to as SSTR-PET/CT) have been studied, primarily in European centers. These PET radiopharmaceuticals have high affinity for the somatostatin receptor subtype 2 (SSTR2) and deliver a lower total body radiation dose than octreotide. Thus, somatostatin receptor imaging with ^{68}Ga-labeled somatostatin analogs should not only have higher sensitivity for tumor detection because of the advantages of PET imaging over gamma scintigraphy, but it also has improved radiation exposure compared to OCT.

Initial studies that included primarily gastrointestinal–pancreatic neuroendocrine tumors suggested that ^{68}Ga-DOTA-conjugated peptides have high sensitivity, about 95 %, for the identification of tumor, with high specificity, around 90 % [37,38]. More recently, a few studies evaluated pulmonary neuroendocrine tumors. Kayani et al. demonstrated positive uptake in all 11 typical and 2 of 5 atypical tumors [39]. Another group also reported very high sensitivity (19/20 patients) [40]. However, neither of these studies included patients with ACTH-secreting tumors, and nearly all tumors were more than 1 cm in diameter and easily detected by conventional imaging. The tumor diameter is 1 cm or less in many patients with ACTH-secreting pulmonary neuroendocrine tumors.

In a recent study of 12 patients with ectopic ACTH secretion, imaging identified 13 tumors in 11 patients. Twelve of these lesions were identified by contrast-enhanced CT (sensitivity 92.3 %), which also detected five false positive lesions. ^{68}Ga-DOTANOC PET/CT identified 9/13 lesions (sensitivity 69.2 %), ranging in size from 7 to 5 cm, with no false positive lesions [41]. A systematic review of the literature found an overall detection rate of SSTR-PET/CT of 81.8 % (18/22) [24].

[^{18}F]-Fluorodeoxyglucose (FDG)-PET has been used for years for tumor localization (22), reflecting the increased glycolytic metabolic rate of lung, bone, and colorectal cancers compared to normal tissue [42]. A systematic review of the literature found that FDG-PET detected 51.7 % of tumors (46/89) [24]. However, in general, FDG PET does not detect (or suggest) any tumors that are not identified by CT and/or MRI. In ectopic ACTH syndrome, FDG-PET is most likely to detect metabolically active tumors or adrenal pheochromocytomas [43, 44].

Neuroendocrine tumors such as foregut carcinoids have been classified as APUDomas based on demonstration of **a**mine **p**recursor **u**ptake and **d**ecarboxylation [45]. In particular, tryptophan is taken up and hydroxylated to 5-hydroxy-tryptophan (5-HTP). Carcinoid tumors that express the enzyme aromatic amino acid decarboxylase (usually the mid-gut carcinoids) can decarboxylate 5-HTP to serotonin (5-hydroxytryptamine or 5-HT). Sundin and colleagues demonstrated that these tumors take up and retain [^{11}C]-5-HTP, allowing visualization via PET [46]. Similarly, the tumors take up and decarboxylate L-3,4-dihydroxyphenylalanine (DOPA) [47]. The activity of L-DOPA decarboxylase is increased in these tumors [48]. A systematic review of the literature found that F-DOPA-PET had a sensitivity of 57.1 % (12/21) [24].

Possible Future Directions

A 3 T MRI scanner increases the strength of magnetic field compared to conventional 1.5 T MRI, allowing for a stronger signal and therefore improved signal-to-noise ratio. Free breathing techniques (such as diaphragm navigator) are used to avoid breath holding, which may be difficult for patients who are volume overloaded. The combination of higher signal and decreased motion artifacts may improve resolution (approached that of CT) to allow for better delineation of small lesions [57, 58]. However, to date, no study compares the diagnostic accuracy of the 1.5 vs. 3 T scanners in this patient population.

11C-5-hydroxy-tryptophan positron emission tomography also takes advantage of the APUD system, but has been studied in very few patients with ectopic ACTH secretion [49].

Three-dimensional reconstruction and the ability to co-register anatomic and functional imaging will likely lead to improved locations and detection rates.

Recommendations Regarding Imaging of Ectopic ACTH Secreting Tumors

Nearly 20 years later, de Herder et al.'s analysis [50] that no single imaging technique has optimal accuracy is still accurate. If biochemical markers are not helpful, a reasonable approach is to perform thin slice CT of the thorax, followed by MRI and somatostatin imaging, preferably using a ^{68}Ga-SSTR tracer if the CT scan is negative. One might then progress to full body imaging by CT and MRI. Using two different types of imaging (anatomic and functional) should help reduce the rate of overall false positive lesions, assuming that they would not be concordant in both studies.

It is important to recognize the critical input of our radiology and nuclear medicine colleagues, both in terms of details of the imaging techniques and in identifying often very small lesions [36]. Nearly all studies show that tumors were best detected by correlating different imaging modalities. Knowledge of the fact that these tumors are often quite small and occur in locations that are unusual (e.g., epicardiac fat) or difficult to visualize or interpret (retrocardiac or pancreatic) may assist in their identification.

If tumors are not identified at initial evaluation, we recommend that the patient be referred to a highly specialized center to obtain additional imaging and interpretation by an experienced team of radiologists.

Further investigations in patients with different tumor types and amounts of tumor burden are necessary to confirm and extend previous findings and determine the best imaging studies and /or their combinations for the detection of ectopic ACTH-producing tumors.

References

1. Atlas SW. Magnetic resonance imaging of the brain and spine. 3rd ed. Baltimore: Lippincott Williams & Wilkins; 2002.
2. Rajan SS. MRI: a conceptual overview. New York: Springer; 1998.
3. Doppman JL, Frank JA, Dwyer AJ, Oldfield EH, Miller DL, Nieman LK, Chrousos GP, Cutler Jr GB, Loriaux DL. Gadolinium DTPA enhanced MR imaging of ACTH-secreting microadenomas of the pituitary gland. J Comput Assist Tomogr. 1988;12:728–35.
4. Dwyer AJ, Frank JA, Doppman JL, Oldfield EH, Hickey AM, Cutler GB, Loriaux DL, Schiable TF. Pituitary adenomas in patients with Cushing disease: initial experience with Gd-DTPA-enhanced MR imaging. Radiology. 1987;163:421–6.
5. Peck WW, Dillon WP, Norman D, Newton TH, Wilson CB. High resolution MR imaging of pituitary microadenomas at 1.5 T: experience with Cushing disease. AJR Am J Roentgenol. 1989;152:145–51.
6. Patronas N, Bulakbasi N, Stratakis C, Lafferty A, Oldfield EH, Doppman J, Nieman LK. Spoiled gradient recalled acquisition in the steady state technique is superior to conventional postcontrast spin echo technique for magnetic resonance imaging detection of adrenocorticotropin-secreting pituitary tumors. J Clin Endocrinol Metab. 2003;88:1665–9.
7. Stadnik T, Stevenaert A, Beckers A, Luypaert R, Buisseret T, Osteaux M. Pituitary microadenomas: diagnosis with two-and three-dimensional MR imaging at 1.5 T before and after injection of gadolinium. Radiology. 1990;176:419–28.
8. Tabarin A, Laurent F, Catargi B, Olivier-Puel F, Lescene R, Berge J, Galli FS, Drouillard J, Roger P, Guerin J. Comparative evaluation of conventional and dynamic magnetic resonance imaging of the pituitary gland for the diagnosis of Cushing's disease. Clin Endocrinol (Oxf). 1998;49:293–300.
9. Kasaliwal R, Sankhe SS, Lila AR, Budyal SR, Jagtap VS, Sarathi V, Kakade H, Bandgar T, Menon PS, Shah NS. Volume interpolated 3D-spoiled gradient echo sequence is better than dynamic contrast spin echo sequence for MRI detection of corticotropin secreting pituitary microadenomas. Clin Endocrinol (Oxf). 2013;78:825–30.
10. Erickson D, Erickson B, Watson R, Patton A, Atkinson J, Meyer F, Nippoldt T, Carpenter P, Natt N, Vella A, Thapa P. 3 Tesla magnetic resonance imaging with and without corticotropin releasing hormone stimulation for the detection of microadenomas in Cushing's syndrome. Clin Endocrinol (Oxf). 2010;72:793–9.
11. Stobo DB, Lindsay RS, Connell JM, Dunn L, Forbes KP. Initial experience of 3 Tesla versus conventional field strength magnetic resonance imaging of small functioning pituitary tumours. Clin Endocrinol (Oxf). 2011;75:673–7.
12. Hall WA, Luciano MG, Doppman JL, Patronas NJ, Oldfield EH. Pituitary magnetic resonance imaging in normal human volunteers: occult adenomas in the general population. Ann Intern Med. 1994;120:817–20.
13. Wind JJ, Lonser RR, Nieman LK, DeVroom HL, Chang R, Oldfield EH. The lateralization accuracy of inferior petrosal sinus sampling in 501 patients with Cushing's disease. J Clin Endocrinol Metab. 2013;98:2285–93.
14. Ilias I, Torpy DJ, Pacak K, Mullen N, Wesley RA, Nieman LK. Cushing's syndrome due to ectopic corticotropin secretion: twenty years' experience at the National Institutes of Health. J Clin Endocrinol Metab. 2005;90:4955–62.
15. Yogi-Morren D, Habra MA, Faiman C, Bena J, Hatipoglu B, Kennedy L, Weil RJ, Hamrahian AH. Pituitary MRI findings in patients with pituitary and ectopic ACTH-dependent Cushing syndrome: does a 6-mm pituitary tumor size Cut-Off value exclude ectopic ACTH syndrome? Endocr Pract. 2015;21:1098–103.
16. Gurazada K, Ihuoma A, Galloway M, Dorward N, Wilhelm T, Khoo B, Bouloux PM. Nasally located ectopic ACTH-secreting pituitary adenoma (EAPA) causing Nelson's syndrome: diagnostic challenges. Pituitary. 2014;17:423–9.

17. Flitsch J, Schmid SM, Bernreuther C, Winterberg B, Ritter MM, Lehnert H, Burkhardt T. A pitfall in diagnosing Cushing's disease: ectopic ACTH-producing pituitary adenoma in the sphenoid sinus. Pituitary. 2015;18:279–82.

18. Mason RB, Nieman LK, Doppman JL, Oldfield EH. Selective excision of adenomas originating in or extending into the pituitary stalk with preservation of pituitary function. J Neurosurg. 1997;87:343–51.

19. Pluta RM, Nieman L, Doppman JL, et al. Extrapituitary parasellar microadenoma in Cushing's disease. J Clin Endocrinol Metab. 1999;84:2912–23.

20. Weil RJ, Vortmeyer AO, Nieman LK, Devroom HL, Wanebo J, Oldfield EH. Surgical remission of pituitary adenomas confined to the neurohypophysis in Cushing's disease. J Clin Endocrinol Metab. 2006;91:2656–64.

21. Koulouri O, Steuwe A, Gillett D, Hoole AC, Powlson AS, Donnelly NA, Burnet NG, Antoun NM, Cheow H, Mannion RJ, Pickard JD, Gurnell M. A role for 11C-methionine PET imaging in ACTH-dependent Cushing's syndrome. Eur J Endocrinol. 2015;173:M107–20.

22. Chittiboina P, Montgomery BK, Millo C, Herscovitch P, Lonser RR. High-resolution(18) F-fluorodeoxyglucose positron emission tomography and magnetic resonance imaging for pituitary adenoma detection in Cushing disease. J Neurosurg. 2015;122:791–7.

23. Nieman LK, Biller BM, Findling JW, Murad MH, Newell-Price J, Savage MO, Tabarin A, Endocrine Society. Treatment of Cushing's syndrome: an endocrine society clinical practice guideline. J Clin Endocrinol Metab. 2015;100:2807–31.

24. Isidori AM, Sbardella E, Zatelli MC, Boschetti M, Vitale G, Colao A, Pivonello R, ABC Study Group. Conventional and nuclear medicine imaging in ectopic Cushing's syndrome: a systematic review. J Clin Endocrinol Metab. 2015;100:3231–44.

25. Alexandraki KI, Grossman AB. The ectopic ACTH syndrome. Rev Endocr Metab Disord. 2010;11:117–26.

26. Sookur PA, Sahdev A, Rockall AG, Isidori AM, Monson JP, Grossman AB, Reznek RH. Imaging in covert ectopic ACTH secretion: a CT pictorial review. Eur Radiol. 2009; 19:1069–78.

27. Tamm EP, Bhosale P, Lee JH, Rohren EM. State-of-the-art imaging of pancreatic neuroendocrine tumors. Surg Oncol Clin N Am. 2016;25:375–400.

28. Papotti M, Croce S, Bello M, Bongiovanni M, Allia E, Schindler M, Bussolati G. Expression of somatostatin receptor types 2, 3 and 5 in biopsies and surgical specimens of human lung tumours. Correlation with preoperative octreotide scintigraphy. Virchows Arch. 2001;439:787–97.

29. Kwekkeboom DJ, Krenning EP, Bakker WH, Oei HY, Kooij PP, Lamberts SW. Somatostatin analogue scintigraphy in carcinoid tumours. Eur J Nucl Med. 1993;20:283–92.

30. McCarthy KE, Espenan GD, Cronin M, Anthony LB, Woltering EA. Visualization of somatostatin receptor-positive lesions with 111-In-pentetreotide is dose dependent. Dig Dis Sci. 1998;43:1876.

31. Tabarin A, Valli N, Chanson P, Bachelot Y, Rohmer V, Bex-Bachellerie V, Catargi B, Roger P, Laurent F. Usefulness of somatostatin receptor scintigraphy in patients with occult ectopic adrenocorticotropin syndrome. J Clin Endocrinol Metab. 1999;84:1193–202.

32. Tsagarakis S, Christoforaki M, Giannopoulou H, Rondogianni F, Housianakou I, Malagari C, Rontogianni D, Bellenis I, Thalassinos N. A reappraisal of the utility of somatostatin receptor scintigraphy in patients with ectopic adrenocorticotropin Cushing's syndrome. J Clin Endocrinol Metab. 2003;88:4754–8.

33. Torpy DJ, Chen CC, Mullen N, Doppman JL, Carrasquillo JA, Chrousos GP, Nieman LK. Lack of utility of (111)In-pentetreotide scintigraphy in localizing ectopic ACTH producing tumors: follow-up of 18 patients. J Clin Endocrinol Metab. 1999;84:1186–92.

34. Ejaz S, Vassilopoulou-Sellin R, Busaidy NL, Hu MI, Waguespack SG, Jimenez C, Ying AK, Cabanillas M, Abbara M, Habra MA. Cushing syndrome secondary to ectopic adrenocorticotropic hormone secretion: the University of Texas MD Anderson Cancer Center Experience. Cancer. 2011;117:4381–9.

35. Özkan ZG, Kuyumcu S, Balköse D, Ozkan B, Aksakal N, Yılmaz E, Sanlı Y, Türkmen C, Aral F, Adalet I. The value of somatostatin receptor imaging with In-111 Octreotide and/or Ga-68

DOTATATE in localizing ectopic ACTH producing tumors. Mol Imaging Radionucl Ther. 2013;22:49–55.

36. Zemskova MS, Gundabolu B, Sinaii N, Chen CC, Carrasquillo JA, Whatley M, Chowdhury I, Gharib AM, Nieman LK. Utility of various functional and anatomic imaging modalities for detection of ectopic adrenocorticotropin-secreting tumors. J Clin Endocrinol Metab. 2010;95:1207–19.

37. Gabriel M, Decristoforo C, Kendler D, Dobrozemsky G, Heute D, Uprimny C, Kovacs P, Von Guggenberg E, Bale R, Virgolini IJ. 68Ga-DOTA-Tyr3-Octreotide PET in neuroendocrine tumors: comparison with somatostatin receptor scintigraphy and CT. J Nucl Med. 2007;48:508–18.

38. Koukouraki S, Strauss LG, Georgoulias V, Eisenhut M, Haberkorn U, Dimitrakopoulou-Strauss A. Comparison of the pharmacokinetics of 68Ga-DOTATOC and [18F] FDG in patients with metastatic neuroendocrine tumours scheduled for 90Y-DOTATOC therapy. Eur J Nucl Med Mol Imaging. 2006;33:1115–22.

39. Kayani I, Conry BG, Groves AM, Win T, Dickson J, Caplin M, Bomanji JB. A comparison of 68Ga-DOTATATE and 18F-FDG PET/CT in pulmonary neuroendocrine tumors. J Nucl Med. 2009;50:1927–32.

40. Jindal T, Kumar A, Venkitaraman B, Dutta R, Kumar R. Role of (68)Ga-DOTATOC PET/CT in the evaluation of primary pulmonary carcinoids. Korean J Intern Med. 2010;25:386–91.

41. Goroshi M, Jadhav S, Lila AR, Kasaliwal R, Khare S, Yerawar C, Hira P, Phadke U, Shah H, Lele V, Malhotra G, Bandgar TR, Shah N. Comparison of 68Ga-DOTANOC PET/CT and CECT in localisation of tumors in EAS. Endocr Connect. 2016;5(2):83–91.

42. Conti PS, Lilien DL, Hawley K, Keppler J, Grafton ST, Bading JR. PET and [18F]-FDG in oncology: a clinical update. Nucl Med Biol. 1996;23:717–35.

43. Kumar J, Spring M, Carroll PV, Barrington SF, Powrie JK. 18Flurodeoxyglucose positron emission tomography in the localization of ectopic ACTH-secreting neuroendocrine tumours. Clin Endocrinol (Oxf). 2006;64:371–4.

44. Timmers HJ, Kozupa A, Chen CC, Carrasquillo JA, Ling A, Eisenhofer G, Adams KT, Solis D, Lenders JW, Pacak K. Superiority of fluorodeoxyglucose positron emission tomography to other functional imaging techniques in the evaluation of metastatic SDHB associated pheochromocytoma and paraganglioma. J Clin Oncol. 2007;25:2262–9.

45. Pearse AG. The cytochemistry and ultrastructure of polypeptide hormone-producing cells of the APUD series and the embryologic, physiologic and pathologic implications of the concept. J Histochem Cytochem. 1969;17:303–13.

46. Sundin A, Eriksson B, Bergstrom M, Långström B, Oberg K, Orlefors H. PET in the diagnosis of neuroendocrine tumors. Ann N Y Acad Sci. 2004;1014:246–57.

47. Bergstrom M, Eriksson B, Oberg K, Sundin A, Ahlström H, Lindner KJ, Bjurling P, Långström B. In vivo demonstration of enzyme activity in endocrine pancreatic tumors: Decarboxylation of carbon-11-DOPA to carbon-11-dopamine. J Nucl Med. 1996;37:32–7.

48. Gazdar AF, Helman LJ, Israel MA, Russell EK, Linnoila RI, Mulshine JL, Schuller HM, Park JG. Expression of neuroendocrine cell markers L-dopa decarboxylase, chromogranin A, and dense core granules in human tumors of endocrine and nonendocrine origin. Cancer Res. 1988;48:4078–82.

49. Nikolaou A, Thomas D, Kampanellou C, Alexandraki K, Andersson LG, Sundin A, Kaltsas G. The value of 11C-5-hydroxy-tryptophan positron emission tomography in neuroendocrine tumor diagnosis and management: experience from one center. J Endocrinol Invest. 2010;33:794–9.

50. de Herder WW, Lamberts SW. Tumor localization: the ectopic ACTH syndrome. J Clin Endocrinol Metab. 1999;84:1184–5.

51. Chowdhury IN, Sinaii N, Oldfield EH, Patronas N, Nieman LK. A change in pituitary magnetic resonance imaging protocol detects ACTH-secreting tumours in patients with previously negative results. Clin Endocrinol (Oxf). 2010;72:502–6.

52. de Rotte AA, Groenewegen A, Rutgers DR, Witkamp T, Zelissen PM, Meijer FJ, van Lindert EJ, Hermus A, Luijten PR, Hendrikse J. High resolution pituitary gland MRI at 7.0 Tesla: a clinical evaluation in Cushing's disease. Eur Radiol. 2016;26:271–7.

53. Salgado LR, Fragoso MC, Knoepfelmacher M, Machado MC, Domenice S, Pereira MA, de Mendonça BB. Ectopic ACTH syndrome: our experience with 25 cases. Eur J Endocrinol. 2006;155:725–33.
54. Aniszewski JP, Young Jr WF, Thompson GB, Grant CS, van Heerden JA. Cushing syndrome due to ectopic adrenocorticotropic hormone secretion. World J Surg. 2001;25:934–40.
55. Isidori AM, Kaltsas GA, Pozza C, Frajese V, Newell-Price J, Reznek RH, et al. The ectopic adrenocorticotrophin syndrome: clinical features, diagnosis, management and long-term follow-up. J Clin Endocrinol Metab. 2006;91:371–7.
56. Wajchenberg BL, Mendonca BB, Liberman B, Pereira MA, Carneiro PC, Wakamatsu A, Kirschner MA. Ectopic adrenocorticotropic hormone syndrome. Endocr Rev. 1994;15:752–87.
57. Bhansali A, Walia R, Rana SS, Dutta P, Radotra BD, Khandelwal N, Bhadada SK. Ectopic Cushing's syndrome: experience from a tertiary care centre. Indian J Med Res. 2009;129:33–41.
58. Pinker K, Ba-Ssalamah A, Wolfsberger S, Mlynarik V, Knosp E, Trattnig S. The value of high-field MRI (3T) in the assessment of sellar lesions. Eur J Radiol. 2005;54:327–34.

Surgical Treatment of Cushing's Disease

Hekmat Zarzour, Margaret Pain, Joshua Bederson, and Kalmon D. Post

Abstract Cushing's disease, as noted in this book, has very serious consequences for those affected. Accurate endocrine diagnosis is crucial as often the adenomas causing the ACTH excess are not large enough to be visualized on imaging studies. While a pituitary adenoma is causative in over 85 % of patients, this often needs confirmation with petrosal sinus sampling and measurements of circulating ACTH. Surgery with the intent of complete removal of the adenoma is usually the first-line of treatment. This is almost always done via a transsphenoidal approach with either microscopic or endoscopic techniques. In this chapter, we will discuss the imaging and surgical techniques for these microadenomas, as well as the more common reasons for failure of accurate diagnosis and treatment.

Keywords Cushing's disease • Transsphenoidal surgery • Pseudocapsule • Endocrinopathy • Inferior petrosal sinus sampling • Microsurgery • Endoscopic surgery

Introduction

At present, surgical resection is considered the gold standard in the treatment of Cushing's disease (CD). While chemotherapeutic and radiotherapeutic treatments have been developed, they are generally not first-line therapies. The most common surgical approach (the transsphenoidal adenomectomy) [1] is minimally invasive and well tolerated by most patients. It avoids exposure of the brain to the extracranial compartment with a low rate of postoperative complications. Second, as most Cushing's tumors are low grade, complete resection provides the opportunity of an immediate and lasting cure. Finally, surgery can be a repeated treatment in the case of persistent or recurrent disease.

H. Zarzour, M.D. • M. Pain, M.D. • J. Bederson, M.D. • K.D. Post, M.D. (✉)
Department of Neurosurgery, Mount Sinai Pituitary Care and Research Center,
One Gustave Levy Place, New York, NY 10029, USA
e-mail: kalmon.post@mountsinai.org

© Springer International Publishing Switzerland 2017
E.B. Geer (ed.), *The Hypothalamic-Pituitary-Adrenal Axis in Health
and Disease*, DOI 10.1007/978-3-319-45950-9_8

149

There are several factors that can be modified to improve success rates through surgery. Higher remission rates are observed at high volume surgical centers and are typically quoted to be between 65 and 98 % [2]. Recurrence of disease is common however, and can range from 2 to 35 % in long-term follow-up [2]. Perhaps, the most critical aspect of successful surgery is achieving a complete resection through meticulous dissection of the adenoma pseudocapsule.

Surgical Indications

As ACTH-secreting adenomas tend to be microadenomas, the primary goal of most surgical interventions is relief of the underlying endocrinopathy. Less frequently, surgery is performed to reduce mass effect of the tumor on surrounding structures or to confirm a diagnosis of Cushing's disease. ACTH-secreting tumors are rarely macroadenomas. Most series report 5–9 % of tumors to be greater than 1 cm in maximal diameter [3–5]. These lesions are remarkable for the fact that most lesions become symptomatic due to endocrine imbalance, while still remaining small in size. The systemic effects of hypercortisolism are observed in more than half of the patients affected and include centripetal obesity, hypertension, hypercholesterolemia, hirsutism, and psychological difficulties, less often diabetes mellitus, osteoporosis, "moon" facies, myopathy, menstrual irregularities, atherosclerosis, headache, and dermatologic abnormalities [6].

Preoperative Evaluation

All patients with clinical evidence of Cushing's disease should receive a full workup to determine the source of the hypercortisolemia. This evaluation will be discussed in another chapter. Referral to a neurosurgeon is only necessary if the source of the endocrinopathy is suspected to be the pituitary gland.

Surgical planning is greatly facilitated by acquiring radiographic evidence of a tumor. MRI is frequently used to locate the tumor within the sella turcica. In comparison to other pituitary tumors, ACTH-secreting adenomas tend to be smaller in size and often located along the midline [7]. Standard MRI may detect larger tumors, but when it fails to do so, higher field strength or different views can be done to maximize the sensitivity of the study. 3 T MRI was significantly more sensitive ($p < 0.016$) for detection of pituitary microadenomas than 1.5 T MRI [8]. However, no difference was reported between 3 T and the 3 T o-CRH examinations [8]. Spoiled gradient recalled acquisition in the steady state has higher sensitivity (80 %, confidence interval: 68–91 %; vs. 49 %, confidence interval: 34–63 %), with lower false positive rate (2 % vs. 4 %) compared with standard T1-weighted spin echo [9] in detection of ACTH-secreting pituitary tumors. However, care in interpreting MRI still must be exercised because incorrect lateralization can occur [10, 11]. In addition,

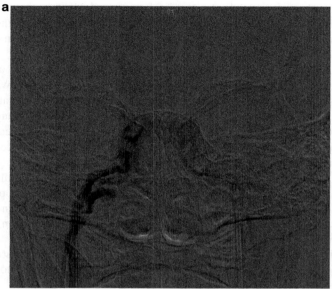

	0 Minutes	2 Minutes	5 Minutes	10 Minutes	15 Minutes
Left Petrosal Sinus	369 pg/mL	n/a	n/a	1106 pg/mL	1138 pg/mL
Right Petrosal Sinus	41 pg/mL	87 pg/mL	100 pg/mL	99 pg/mL	145 pg/mL
Peripheral Source	29 pg/mL	31 pg/mL	53 pg/mL	69 pg/mL	72 pg/mL

Fig. 1 (**a**) Digitally subtracted cerebral venogram for inferior petrosal sinus sampling. Catheters are placed in the bilateral inferior petrosal sinus. Contrast dye has been injected and both inferior petrosal sinus and cavernous sinus are opacified. (**b**) Results of inferior petrosal sinus sampling, values are concentration of ACTH (pg/mL). Results demonstrate likely left-sided source of ACTH secretion

ACTH-secreting microadenomas are not detectable in 40–50 % of patients [12]. In such cases, inferior petrosal sinus sampling can be helpful, and the indication for transsphenoidal surgery of the pituitary gland is based only on biochemical data indicating the origin of hypercortisolism to be the sella [13] (see Fig. 1).

In addition to tumor detection, MRI provides valuable information for the surgeon to assess surrounding structural anatomy. As tumor size increases, this anatomy is more likely to become distorted. In particular, compression or invasion of the cavernous sinus occurs with larger tumors. Sol et al. [14] retrospectively studied 63 patients who underwent transsphenoidal surgery for pituitary adenoma and compared the preoperative MRI with intraoperative findings for cavernous sinus invasion. If T1 sequence with contrast did not show periarterial enhancement, invasion was highly probable (positive predictive value, 86%; $P<0.001$); in the same study, no enhancement of the medial wall of the cavernous sinus on T2 sequence and the lesion crossing lateral inner carotid line revealed invasion in 87.5% and 85%, respectively [14].

Although CT is rarely used for adenoma localization, it can aid in operative planning and navigation during surgery. While historically, orientation within the posterior nasopharynx and sphenoid sinus could be complicated if the patient had abnormal anatomy, this is rarely the case today because of advances in imaging and intraoperative navigation. Modern navigation platforms can fuse preoperative high-resolution CT and MRI to utilize bony anatomy for enhanced accuracy in surgical planning and intraoperative navigation.

Improper localization of the midline of the anterior sella wall can result in high risk of injury to the patient. For example, a lateral, rather than a midline, opening of the anterior wall of the sella exposes the cavernous sinus, cranial nerves, and carotid artery to possible injury [15]. In addition to a higher risk for complication, the risk that the surgeon may not be able to access the tumor is also increased.

If endocrine studies are diagnostic for Cushing's disease and imaging studies are negative for any definitive pathology, further tests should be performed to confirm the diagnosis of Cushing's disease before surgery.

IPSS can be used to corroborate a pituitary source for ACTH hypersecretion [13]. IPSS is indicated for excluding extrasellar ectopic ACTH secretion and to suggest the laterality of the tumor within the sella turcica. IPSS can also be helpful when cortico-trophin-releasing hormone (CRH) and 8 mg dexamethasone stress test results are equivocal. Access to both inferior petrosal sinuses is achieved with endovascular catheters directed at each side via the femoral veins. Baseline sampling of ACTH is performed and then compared with the local concentration produced in each sinus by CRH stimulation. While the study can help to suggest the gross laterality of the tumor, the results can be compromised by several factors. Improper catheterization of the inferior petrosal sinus, alternate flow of the sinus into the cavernous sinus, and anomalous venous drainage can all lead to false lateralization. Generally, the study has high sensitivity and specificity for identifying the cause of hypercortisolism (80–100% sensitivity, greater than 95% specificity) [16]. This test might help guide the surgeon intraoperatively in cases where no distinct tumor is found during operative exploration. However, IPSS correctly predicted the side of the pituitary gland that contained the tumor only in 69%, whereas the tumor was located contralaterally in 31% [17]. IPSS is an invasive procedure and carries certain risks. Among the more common complications noted are tinnitus and otalgia (1–2%) and groin swelling and hematoma (2–3%) [16]. Rarely, more serious complications have been reported, including, but not limited to nerve palsy, subarachnoid hemorrhage, and brainstem infarction [18]. IPSS is also indicated in postoperative cases when no tumor was found within the sella but the patient continues to demonstrate hypercortisolism and IPSS had not been done preoperatively. Positive IPSS in these cases might be sugges-tive of a pituitary adenoma with an abnormal location such as the cavernous sinus, posterior gland, or pituitary stalk, which could have been missed during surgery. It lends support for re-exploration [1, 19, 20].

Preoperative Challenges for Cushing's Disease

Undetectable adenomas on preoperative MRI and invasive adenomas are two of the main challenges in Cushing's disease. Remission rates for microadenomas that are detected on preoperative MRI are high [11, 12]. It is well known that remission rates are lower for patients with negative MRIs [15, 17]. Finding the adenoma in these cases is not easy. Adenoma invasion presents another surgical challenge for Cushing's disease [21]. When the cavernous sinus and surrounding structures are invaded by the tumor, total adenoma resection is almost impossible and dangerous to achieve [21]. Remission rates are lower for invasive tumors compared with those where a complete resection can be achieved [22–24]. The likelihood of invasion increases with tumor size, so while larger tumors may be more easily identified, a total resection can still be difficult. Adenomas associated with dural invasion tend to be larger (2–37 mm) comparing to noninvasive tumors (2.5–12 mm) [25]. Unfortunately, dural invasion is not well characterized by preoperative imaging and tends to underestimate the prevalence of invasive tumors (22 % of cases) [25]. This is compared with an estimated 34 % of patients who had histologically confirmed dural invasion [25]. If the invasion is limited to the dural medial wall and does not penetrate the cavernous sinus, complete resection is probably achievable with a high rate of remission [25–27]. However, once the medial wall of the cavernous sinus is breached, surgical remission is unlikely and additional treatment is frequently required [27]. Although with endoscopic techniques, medial cavernous sinus tumors may be seen and resected [28].

Endoscopic vs. Microscopic vs. Transcranial Approaches

The radiographic location of the tumor, size of the tumor, the presence of invasion and/or compression of the surrounding structures, and surgeon experience dictate the choice of approach. In most cases, the adenoma is intrasellar or not visible. In such cases, a transsphenoidal approach is the preferred surgical approach. As the tumor grows in size or has supradiaphragmatic extension, the technical difficulties of a transsphenoidal approach increase, although this approach is still often preferred. When there is significant tumor above the sella and a total resection is technically difficult, a debulking procedure can be performed. In the postoperative months, the remaining tumor frequently descends into the sella, and a second surgery can be performed at that time to complete the resection. In rare cases, if suprasellar tumors are eccentric intracranially and not completely accessible transsphenoidally, they can be accessed transcranially through a pterional or subfrontal approach. As stated previously, the major location of extrasellar extension of pituitary adenomas is the cavernous sinus. Tumor in this location is generally not amenable to safe resection by either surgical approach, and adjuvant therapy is usually required [25, 27]. But, as noted above, endoscopic approaches may enhance the ability to resect Knosp grade 2 and 3 tumors [29].

Both the surgical microscopic and endoscopic techniques are commonly used to access the sella through a transsphenoidal approach. Microscopic surgery has been considered to be the standard of care for many years, and in experienced hands it is associated with minimal morbidity and mortality. Jankowski et al. [30] introduced the endoscope to pituitary surgery in 1992. With advancements in optics and operator experience in endoscopy, this method is becoming increasingly popular. Endoscopy offers two main advantages in surgery of the sella: enhanced visualization of the entire surgical field and ability to extend the standard opening of the skull base. Most endoscopes project a two-dimensional image that may hamper depth perception for some surgeons who are accustomed to the operating microscope. Bimanual surgery and the ability to control surgical bleeding are thought to be relatively more difficult with purely endoscopic techniques, but these limitations are decreasing as experience accrues and endoscopic technology improves.

Gao et al. in 2014 [31] performed a systematic review comparing the results of endoscopic to microscopic surgery. Their search of all articles published after 1992 included a total of 15 studies and 1014 patients. They found a higher rate of gross total resection and lower rate of septal perforation in the endoscopy group but no significant difference in the rate of complication or length of surgery. Additionally, the review reported a significantly shorter hospital stay for endoscopy patients but the reasons were not clear. They concluded that the endoscopic transsphenoidal approach is safer and more effective than microscopic surgery. Higgans et al. [32] retrospectively analyzed 19 subjects who underwent endoscopic excision and 29 subjects who underwent microscopic excision. They analyzed demographics information, tumor characteristics, operative details, length of hospital stay, intraoperative and postoperative complications, level of postoperative pain, recurrence rate, use of computed tomography (CT) image guidance, and length of follow-up. They concluded that the two techniques have similar intraoperative characteristics and immediate complication rates. Alahmadi et al. operated on 42 patients (15 macroadenomas and 27 microadenomas) using both techniques and concluded that there was no significant difference in remission rates between the two techniques ($p=0.757$).

Surgical Techniques

Specific details of the procedure and operating room setup are not discussed here, as many details are dependent on surgeon preferences. Whether the sphenoid is approached through a sub-labial incision, trans-nasal microscopy, or endoscopy is largely based on operator preference. We carry out our microscopic transsphenoidal approach from the right nostril generally with the patient in the supine position, head tilted to left and slightly turned to the right. We always prepare the belly for a possible fat graft. Image guidance is used on all cases. Image-based surgical navigation or C-arm fluoroscopy is based on surgeon preference and specific patient anatomy. In reoperations, image-based intraoperative navigation is the method of choice. Care should be taken to keep the nasal septal mucosal incision in soft tissue about 2 cm from the external nostril. Following the incision, the remainder of the approach can be carried out by blunt submucosal dissection. Fluoroscopy or intra-operative navigation tools confirm the trajectory to the sphenoid sinus and sella.

Few studies describe and report outcomes of pituitary surgery that focus on the method for adenoma removal once the sella has been opened, but this is the critical portion of the operation [33–35]. The anterior pituitary gland has its own thin capsule that separates it from the surrounding dura, sella, and cavernous sinus. The gland con-tains a collagen matrix that gives it a firm texture and allows it to be distinguished from the adenoma (which tends to have a soft consistency) and posterior pituitary gland. As an adenoma grows in size, it causes compression on the normal pituitary tissue and displaces it to form an interface to the normal gland. This compression forms a smooth wrapping around the adenoma and is termed «pseudocapsule» [35]. Careful dissection within the pseudocapsule, using it as surgical plane, is the key for total and successful resection. Pseudocapsules can be found in tumors as small as 2–3 mm in diameter but tend to be absent in tumors less than 1 mm because the compression caused is insuf-ficient at smaller sizes [35]. Appreciation of the pseudocapsule is important to ensure gross total resection as well as to diagnose dural invasion.

Using an endoscope or microscope, broad exposure of the pituitary gland is required to allow visualization of the entire anterior lobe (see Fig. 2). Exposure of the anterior sella wall is complete when the faint blue edge of the cavernous sinus can be visualized on either side of the field. Various incisions of the dura are used but we prefer an «H» opening. Cruciate or box incisions are also used. In this area of the dura, there are often large venous channels that can lead to rapid bleeding at this stage of the surgery. A variety of surgical techniques and tools can be used to slow or stop the bleeding but both the surgeon and anesthesiologist should be aware of the potential for significant bleeding.

After dural incision, the surface of the anterior gland is carefully inspected for areas of irregularity or discoloration. Some authors have reported the use of a micro-Doppler for visualization with varying degrees of success [36]. In our opin-ion, visualization of the pseudocapsule is the most consistent finding to locate the tumor. Once the possible location of the adenoma is identified, the pituitary capsule is incised sharply and then the pseudocapsule is dissected. We try to avoid piece-meal resection, when possible. Special care is taken not to lose any of the speci-mens in the suction.

Sectioning of the gland is performed if no adenoma can be detected after gross inspection of the anterior and lateral surfaces. Incisions are made horizontally or vertically at 2 mm intervals until a tumor is uncovered or the posterior gland is iden-tified. If the adenoma is uncovered, then dissection of the pseudocapsule is per-formed with attempted gross total resection. If the pseudocapsule or surrounding dura is breached, careful inspection is performed to ensure that no areas of dural invasion are missed.

Parasellar ectopic ACTH-producing tumors have been reported [20]. They may be suspected if no adenoma is found after all abovementioned steps have been per-formed. In this case, careful inspection of the gland back to the neurohypophysis is recommended. If no tumor can be identified, a partial or total hypophysectomy can

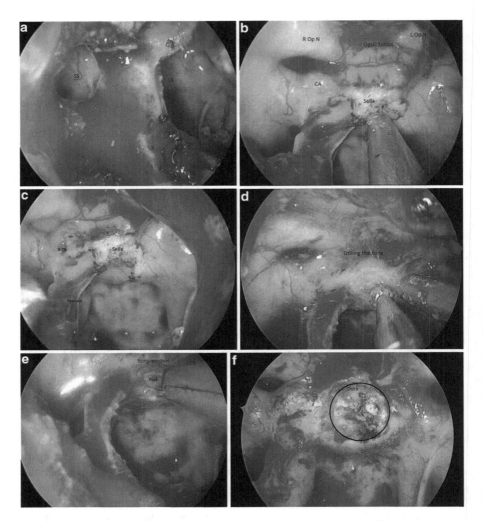

Fig. 2 Endoscopic transsphenoidal approach for pituitary adenomectomy. (**a**) 0° endoscope view, after elevation of the posterior septal mucosa and removal of the vomer. Bilateral sphenoid ostia are visualized opening to the sphenoid sinus (SS). (**b**) Anterior wall of the sphenoid sinus and piece of the sphenoid septum has been removed between the sphenoid ostia, endoscope advanced into the sphenoid sinus. Mucosa has been removed from the sella, and suction is directed toward the sella. Bony prominence of the right and left optic nerve (R Op N, L Op N), bilateral carotid artery (CA) labeled. (**c**) Septum directed toward the right carotid artery (R CA) has been removed. (**d**) Diamond burr drill used to remove the bone over the sella turcica. (**e**) Kerrison rongeur used to remove the remaining bone. Dura outside pituitary exposed. (**f**) Final bony opening with dura exposed. (**g**) Dural has been opened with bayoneted scalpel. The tumor is being debulked with a curette and suction. (**h**) Tumor has been debulked. Suprasellar arachnoid has descended into the surgical field. No residual tumor has been identified

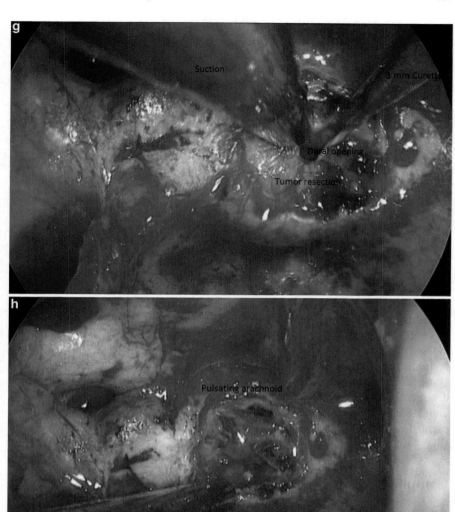

Fig. 2 (continued)

be considered. Because panhypopituitarism develops after total resection, a partial resection is preferred and can be directed based on the results of the preoperative IPSS. With this method, a high rate of remission (92 %, 24 of 26 patients) can be achieved through partial hypophysectomy [26].

Jagannathan et al. [37] determined the success of using the pseudocapsule as a surgical capsule through a retrospective review of 261 patients. Tumor was identified radiographically in only 135 patients (52 %). However, through meticulous exploration of the sella and identification of the pseudocapsule, the group was able to attain remission in 252 cases (97 %). In the remaining 9 patients, remission was achieved for 4 with repeated surgery. Further evidence for the efficacy of this method of dissection was found in the rate at which patients became hypocortisolemic after surgery. Using the pseudocapsule as a guide in dissection, patients became hypocortisolemic 19.4 h after surgery, which was more rapid than other methods of dissection and suggested a more complete resection [38]. This further suggests that identification of the pseudocapsule is critical for achieving a gross total resection.

Depending on the clinical situation, more aggressive or more conservative resection may be indicated. For the seriously debilitated or elderly patient, transsphenoidal surgery may be attempted first but if no adenoma is found, it may be appropriate to perform complete hypophysectomy to minimize the need for repeated surgery.

After completing the resection of the adenoma, we routinely inspect for possible CSF leak with a Valsalva maneuver. Any evidence of communication of cerebrospinal fluid with the sella mandates intrasellar packing in addition to obliteration of the sphenoid by fat taken from the abdomen. Closure of the surgical site is accomplished by placing a piece of fat within the sella, followed by a piece of the vomer taken during the approach to repair the broached anterior sellar wall. If no bone is available, we use a biodegradable substitute. We generally do not use spinal drains. To enable these materials to coalesce, nasal tampons are placed and maintained for the first two days after surgery.

In the perioperative period, antibiotics are administered, but no glucocorticoids. Our goal is to test serum cortisol and ACTH levels the next morning to determine the success of the surgery. We find reports of frozen pathologic specimens to be unreliable indicators of complete resection and prefer to use the entire specimen for permanent sectioning.

Complications

The overall rate of complication in transsphenoidal surgery for Cushing's disease appears to be relatively low. Some of the problems caused by chronic hypercortisolism put the patient at a higher risk for surgery than individuals undergoing transsphenoidal surgery for other functioning microadenomas [5]. Patil et al. [39] analyzed the nationwide database of patients who underwent transsphenoidal resection of a pituitary tumor for Cushing's disease between 1993 and 2002. They analyzed length of stay, rates of inpatient complications, death, and adverse outcomes. Of the 3525 cases studied, the in-hospital mortality rate was 0.7 % and total complication rate was 42.1 % (DI — 15 %, fluid and electrolyte abnormality — 12.5 %, postoperative neurological symptom — 5.6 %, postoperative bleeding — 2.6 %, pulmonary sign — 1.7 %, CSF leak — 1.4 %, diplopia or ptosis or CNIII, IV, VI deficit — 0.7 %, cardiac symptom — 0.7 %, DVT + PE — 0.7 %, iatrogenic panhypopituitarism — 0.5 %, and infection — 0.4 %). Advanced age and multiple preoperative comorbidities were identified as important risk factors. Prevedello et al. reported 93 % rate of panhypopituitarism following total hypophysectomy and 20 % rate of Nelson's syndrome after bilateral adrenalectomy [40]. Looking specifically at pediatric patients undergoing surgery for Cushing's disease, Lonser et al. reported rare complications in their cohort of 200 patients. These complications included: DI (5 %), seizure due to sodium abnormality (1.5 %), maxillary fracture with transient diplopia (1.5 %), and delayed pseudotumor cerebri (2 %). Cerebral vasculitis was noticed in one patient after they sustained a postoperative thalamic infarction.

Management of Recurrent and Persistent Disease

Postoperative remission is defined as normalization or insufficiency of circadian cortisol secretion. If a gross total resection is achieved, restoration of the normal hypothalamic-pituitary-adrenal axis function can take months to years (mean 20.8, range 0.5–84 months) [41]. The majority of patients experience hypocortisolism after surgery and require replacement therapy until normal axis activity can be reestablished. Positive prognostic factors for initial postoperative remission include gross identification of the tumor during surgery, immunohistochemical identification of ACTH-producing adenoma tissue, and the existence of a noninvasive adenoma [26].

A morning serum cortisol of less than 1 mg/dL after surgery had a positive predictive value for lasting remission of 96 % [26], while morning cortisol value of 2 μg/dL or less has 93 % of sustained remission of CD for at least 5 years [42]. While eucortisolism can indicate remission, patients with this finding require close follow-up as they frequently have recurrences. Persistent hypercortisolism on day one usually indicates a significant amount of residual functioning tumor. In these cases, collaboration between the neurosurgeon, pathologist, radiologist, and endocrinologist is needed to determine the course of further management.

When tumor is identified in pathological specimens but the patient remains hypercortisolemic, there is high likelihood for invasion of the cavernous sinus or surrounding dura. In some cases, repeated imaging can shed light on the location of the residual tumor and in such cases, repeated surgery is advised. If an ACTH-positive adenoma was found during the first surgical procedure, we usually advise re-exploration with a more vigorous resection of surrounding tissue. The question of how soon after the initial surgery should a second procedure be done is often raised. We choose to give at least several weeks or months of follow-up with endocrine data, as some patients will drift into normal or low values over a longer period of time. Radiation therapy is also an option in these cases.

If pathologic specimens fail to demonstrate any tumor, then one must suspect another cause of Cushing's syndrome or atypical/ectopic location, or that the tumor was missed during surgery. If IPSS had not been initially performed, it should be performed at this time. If presumed microadenoma was removed with negative pathologic examination, there may be a distinct adenoma in the remaining portion of the gland or ectopic source of ACTH. Many surgeons will re-explore the gland. Most surgeons will suggest early repeat surgical intervention to investigate the portion that was not inspected during first surgery [15, 43–46]. Ram et al. reviewed 13 % of 222 patients with persistent hypercortisolism, with early reoperation in most of these patients, and they were able to induce remission at the second operation in 70 % of these patients, indicating the need of an aggressive resection in an attempt to induce remission [43]. Friedman et al. reported a higher remission rate if adenomas were identified during surgical re-exploration; if an adenoma was not identified then partial or total hypophysectomy was performed with 42 % remission rate and 50 % hypopituitarism [15].

All patients should be retested at regular intervals postoperatively. A significant percentage of patients will relapse after initial remission, with rates ranging from 2 to 35 % at long-term follow-up [2]. The decision-making process for further therapy is similar to that described above with the probability of regrowth of the adenoma. Stereotactic radiotherapy or radiosurgery is the chief modality of adjuvant therapy used to achieve remission in cases of Cushing's disease not responding to surgical therapy alone [10]. There are other new medications that can also be tried as discussed in another chapter.

Long-Term Outcomes

In 11 retrospective studies on Cushing's including 1167 patients analyzed, early remission ranged from 65 to 98 %; however, disease relapse occurred with rates ranging from 2 to 35 % at long-term follow-up [2]. Factors associated with failure to achieve remission include the presence of residual tumor, failed identification of the tumor, invasion of the cavernous sinus, and ectopic source of ACTH production [47]. Patil et al. [46] reported 36 patients who underwent repeat TS surgery for recurrent Cushing's disease. The median time to recurrence after initial successful TS surgery was 36 months (range, 4 months–16 years). Remission after repeat TS surgery was observed in 22 (61 %) of the 36 patients. Two of the 22 patients presented with a second recurrence at 6 and 11 months. In the remaining 36 patients, stereotactic radiosurgery, adrenalectomy, and ketoconazole were used with remission achieved in 30 (83.3 %).

Recurrence rates tend to be higher in patients with postoperative eucortisolism compared to hypocortisolism and with longer follow-up. Postoperative hypocortisolism without recovery of the HPA axis is a good indicator of remission, but does not indicate a permanent cure. On average, CD recurrence occurs within 0.5–5 years of successful surgery, but it has happened as late as 30 years after initial surgery [41, 48, 49]. Patterns of recurrence suggest that most recurrence is local. For 43 patients in whom an adenoma was identified in the initial surgery, the recurrence was found at the same site, but with dural invasion that was not recognized on preoperative MRI [47]. Dimopoulou et al. [48] reported the outcome of 120 patients, of which 36 patients had revision with mean follow-up time of 79 months. The remission rates for patients were 71 % and 42 % for initial surgery and revision, respectively. Patients with early hypocortisolism were 0.7 times less likely to have disease recurrence compared to those with postoperative eucortisolism. Castinetti et al. reported the outcome of 40 patients with Cushing's disease treated with gamma knife with a mean follow-up of 54.7 months. Median margin dose was 29.5 Gy. Seventeen patients (42.5 %) were in remission after a mean of 22 months (range 12–48 months), with lower target volume in the remission group vs. those with persistent disease [50].

Bilateral adrenalectomy is considered if ACTH-dependent Cushing's syndrome is refractory to other treatment modalities including surgery, radiosurgery, and medical therapies. Bilateral adrenalectomy is relatively safe (median surgical morbidity 15 %; median surgical mortality 3 %) with excellent outcome [51]. Long-term complications include the development of adrenal crisis and Nelson's syndrome [51].

Conclusions

The significant morbidity caused by hypercortisolism merits aggressive treatment of the underlying cause. Successful diagnosis and surgical treatment can provide immediate remission while maintaining pituitary function. Transsphenoidal surgery is the initial and most effective treatment for Cushing's disease, but is not possible for all patients and recurrences are noted. Re-exploration is recommended in such cases. Using the surgical pseudocapsule to guide microsurgical resection is crucial. The pseudocapsule allows an exact and total tumor resection enabling a higher remission rate with minimal complications [37]. Pathological confirmation is preferable, as the rate of relapse is higher with lack of histological confirmation. Without histopathologically confirmed tumor, close monitoring is recommended so that early intervention can be performed, if needed [23]. Endoscopic adenoma excision is a reasonable alternative to the traditional method of microscopic sellar mass excision, and it is preferred in invasive cases. Bilateral adrenalectomy is the last treatment option and is frequently considered after radiosurgery and medical therapy have been exhausted.

References

1. Weil RJ, Vortmeyer AO, Nieman LK, DeVroom HL, Wanebo J, Oldfield EH. Surgical remission of pituitary adenomas confined to the neurohypophysis in Cushing's disease. J Clin Endocrinol Metab. 2006;91(7):2656–64.
2. Dallapiazza RF, Oldfield EH, Jane Jr JA. Surgical management of Cushing's disease. Pituitary. 2015;18(2):211–6.
3. Chandler WF, Schteingart DE, Lloyd RV, McKeever PE, Ibarra-Perez G. Surgical treatment of Cushing's disease. J Neurosurg. 1987;66(2):204–12.
4. Guilhaume B, Bertagna X, Thomsen M, Bricaire C, Vila-Porcile E, Olivier L, et al. Transsphenoidal pituitary surgery for the treatment of Cushing's disease: results in 64 patients and long term follow-up studies. J Clin Endocrinol Metab. 1988;66(5):1056–64.
5. Fahlbusch R, Buchfelder M, Muller OA. Transsphenoidal surgery for Cushing's disease. J R Soc Med. 1986;79(5):262–9.

6. Mampalam TJ, Tyrrell JB, Wilson CB. Transsphenoidal microsurgery for Cushing disease. A report of 216 cases. Ann Intern Med. 1988;109(6):487–93.

7. Bonneville JF, Bonneville F, Cattin F. Magnetic resonance imaging of pituitary adenomas. Eur Radiol. 2005;15(3):543–8.

8. Erickson D, Erickson B, Watson R, Patton A, Atkinson J, Meyer F, et al. 3 Tesla magnetic resonance imaging with and without corticotropin releasing hormone stimulation for the detection of microadenomas in Cushing's syndrome. Clin Endocrinol (Oxf). 2010;72(6):793–9.

9. Patronas N, Bulakbasi N, Stratakis CA, Lafferty A, Oldfield EH, Doppman J, et al. Spoiled gradient recalled acquisition in the steady state technique is superior to conventional postcontrast spin echo technique for magnetic resonance imaging detection of adrenocorticotropin-secreting pituitary tumors. J Clin Endocrinol Metab. 2003;88(4):1565–9.

10. Knappe UJ, Ludecke DK. Transnasal microsurgery in children and adolescents with Cushing's disease. Neurosurgery. 1996;39(3):484–92. discussion 92-3.

11. Peck WW, Dillon WP, Norman D, Newton TH, Wilson CB. High-resolution MR imaging of pituitary microadenomas at 1.5 T: experience with Cushing disease. AJR Am J Roentgenol. 1989;152(1):145–51.

12. Tabarin A, Laurent F, Catargi B, Olivier-Puel F, Lescene R, Berge J, et al. Comparative evaluation of conventional and dynamic magnetic resonance imaging of the pituitary gland for the diagnosis of Cushing's disease. Clin Endocrinol (Oxf). 1998;49(3):293–300.

13. Jehle S, Walsh JE, Freda PU, Post KD. Selective use of bilateral inferior petrosal sinus sampling in patients with adrenocorticotropin-dependent Cushing's syndrome prior to transsphenoidal surgery. J Clin Endocrinol Metab. 2008;93(12):4624–32.

14. Sol YL, Lee SK, Choi HS, Lee YH, Kim J, Kim SH. Evaluation of MRI criteria for cavernous sinus invasion in pituitary macroadenoma. J Neuroimaging. 2014;24(5):498–503.

15. Friedman RB, Oldfield EH, Nieman LK, Chrousos GP, Doppman JL, Cutler Jr GB, et al. Repeat transsphenoidal surgery for Cushing's disease. J Neurosurg. 1989;71(4):520–7.

16. Giraldi FP, Cavallo LM, Tortora F, Pivonello R, Colao A, Cappabianca P, et al. The role of inferior petrosal sinus sampling in ACTH-dependent Cushing's syndrome: review and joint opinion statement by members of the Italian Society for Endocrinology, Italian Society for Neurosurgery, and Italian Society for Neuroradiology. Neurosurg Focus. 2015;38(2):5.

17. Wind JJ, Lonser RR, Nieman LK, DeVroom HL, Chang R, Oldfield EH. The lateralization accuracy of inferior petrosal sinus sampling in 501 patients with Cushing's disease. J Clin Endocrinol Metab. 2013;98(6):2285–93.

18. Gandhi CD, Meyer SA, Patel AB, Johnson DM, Post KD. Neurologic complications of inferior petrosal sinus sampling. AJNR Am J Neuroradiol. 2008;29(4):760–5.

19. Oldfield EH, Vance ML. A cryptic cause of Cushing's disease. J Clin Endocrinol Metab. 2013;98(12):4593–4.

20. Pluta RM, Nieman L, Doppman JL, Watson JC, Tresser N, Katz DA, et al. Extrapituitary parasellar microadenoma in Cushing's disease. J Clin Endocrinol Metab. 1999;84(8):2912–23.

21. Nishioka H, Fukuhara N, Horiguchi K, Yamada S. Aggressive transsphenoidal resection of tumors invading the cavernous sinus in patients with acromegaly: predictive factors, strategies, and outcomes. J Neurosurg. 2014;121(3):505–10.

22. Kakade HR, Kasaliwal R, Khadilkar KS, Jadhav S, Bukan A, Khare S, et al. Clinical, biochemical and imaging characteristics of Cushing's macroadenomas and their long-term treatment outcome. Clin Endocrinol (Oxf). 2014;81(3):336–42.

23. Pouratian N, Prevedello DM, Jagannathan J, Lopes MB, Vance ML, Laws ER. Outcomes and management of patients with Cushing's disease without pathological confirmation of tumor resection after transsphenoidal surgery. J Clin Endocrinol Metab. 2007;92(9):3383–8.

24. Zielinski G, Podgorski JK, Koziarski A, Warczynska A, Zgliczynski W, Makowska A. [Surgical treatment of invasive pituitary adenomas (somatotropinoma or corticotropinoma)]. Neurol Neurochir Pol. 2003;37(6):1239–55.

25. Lonser RR, Ksendzovsky A, Wind JJ, Vortmeyer AO, Oldfield EH. Prospective evaluation of the characteristics and incidence of adenoma-associated dural invasion in Cushing disease. J Neurosurg. 2012;116(2):272–9.

26. Lonser RR, Wind JJ, Nieman LK, Weil RJ, DeVroom HL, Oldfield EH. Outcome of surgical treatment of 200 children with Cushing's disease. J Clin Endocrinol Metab. 2013;98(3):892–901.
27. Oldfield EH. Editorial: management of invasion by pituitary adenomas. J Neurosurg. 2014;121(3):501–3.
28. McLaughlin N, Eisenberg AA, Cohan P, Chaloner CB, Kelly DF. Value of endoscopy for maximizing tumor removal in endonasal transsphenoidal pituitary adenoma surgery. J Neurosurg. 2013;118(3):613–20.
29. Juraschka K, Khan OH, Godoy BL, Monsalves E, Kilian A, Krischek B, et al. Endoscopic endonasal transsphenoidal approach to large and giant pituitary adenomas: institutional experience and predictors of extent of resection. J Neurosurg. 2014;121(1):75–83.
30. Jankowski R, Auque J, Simon C, Marchal JC, Hepner H, Wayoff M. Endoscopic pituitary tumor surgery. Laryngoscope. 1992;102(2):198–202.
31. Gao Y, Zhong C, Wang Y, Xu S, Guo Y, Dai C, et al. Endoscopic versus microscopic transsphenoidal pituitary adenoma surgery: a meta-analysis. World J Surg Oncol. 2014;12:94.
32. Higgins TS, Courtemanche C, Karakla D, Strasnick B, Singh RV, Koen JL, et al. Analysis of transnasal endoscopic versus transseptal microscopic approach for excision of pituitary tumors. Am J Rhinol. 2008;22(6):649–52.
33. Kawamata T, Kubo O, Hori T. Surgical removal of growth hormone-secreting pituitary adenomas with intensive microsurgical pseudocapsule resection results in complete remission of acromegaly. Neurosurg Rev. 2005;28(3):201–8.
34. Oldfield EH. Surgical management of Cushing's disease: a personal perspective. Clin Neurosurg. 2011;58:13–26.
35. Oldfield EH, Vortmeyer AO. Development of a histological pseudocapsule and its use as a surgical capsule in the excision of pituitary tumors. J Neurosurg. 2006;104(1):7–19.
36. Watson JC, Shawker TH, Nieman LK, DeVroom HL, Doppman JL, Oldfield EH. Localization of pituitary adenomas by using intraoperative ultrasound in patients with Cushing's disease and no demonstrable pituitary tumor on magnetic resonance imaging. J Neurosurg. 1998;89(6):927–32.
37. Jagannathan J, Smith R, DeVroom HL, Vortmeyer AO, Stratakis CA, Nieman LK, et al. Outcome of using the histological pseudocapsule as a surgical capsule in Cushing disease Clinical article. J Neurosurg. 2009;111(3):531–9.
38. Monteith SJ, Starke RM, Jane Jr JA, Oldfield EH. Use of the histological pseudocapsule in surgery for Cushing disease: rapid postoperative cortisol decline predicting complete tumor resection. J Neurosurg. 2012;116(4):721–7.
39. Patil CG, Lad SP, Harsh GR, Laws Jr ER, Boakye M. National trends, complications, and outcomes following transsphenoidal surgery for Cushing's disease from 1993 to 2002. Neurosurg Focus. 2007;23(3), E7.
40. Prevedello DM, Pouratian N, Sherman J, Jane Jr JA, Vance ML, Lopes MB, et al. Management of Cushing's disease: outcome in patients with microadenoma detected on pituitary magnetic resonance imaging. J Neurosurg. 2008;109(4):751–9.
41. Aranda G, Ensenat J, Mora M, Puig-Domingo M, Martinez de Osaba MJ, Casals G, et al. Long-term remission and recurrence rate in a cohort of Cushing's disease: the need for long-term follow-up. Pituitary. 2015;18(1):142–9.
42. Chen JCT, Amar AP, Choi S, Singer P, Couldwell WT, Weiss MH. Transsphenoidal microsurgical treatment of Cushing disease: postoperative assessment of surgical efficacy by application of an overnight low-dose dexamethasone suppression test. J Neurosurg. 2003;98(5):967–73.
43. Ram Z, Nieman LK, Cutler Jr GB, Chrousos GP, Doppman JL, Oldfield EH. Early repeat surgery for persistent Cushing's disease. J Neurosurg. 1994;80(1):37–45.
44. Valderrabano P, Aller J, Garcia-Valdecasas L, Garcia-Uria J, Martin L, Palacios N, et al. Results of repeated transsphenoidal surgery in Cushing's disease. Long-term follow-up. Endocrinol Nutr. 2014;61(4):176–83.
45. Bertagna X, Guignat L. Approach to the Cushing's disease patient with persistent/recurrent hypercortisolism after pituitary surgery. J Clin Endocrinol Metab. 2013;98(4):1307–18.

46. Patil CG, Veeravagu A, Prevedello DM, Katznelson L, Vance ML, Laws Jr ER. Outcomes after repeat transsphenoidal surgery for recurrent Cushing's disease. Neurosurgery. 2008;63(2):266–70. discussion 70-1.
47. Dickerman RD, Oldfield EH. Basis of persistent and recurrent Cushing disease: an analysis of findings at repeated pituitary surgery. J Neurosurg. 2002;97(6):1343–9.
48. Dimopoulou C, Schopohl J, Rachinger W, Buchfelder M, Honegger J, Reincke M, et al. Long-term remission and recurrence rates after first and second transsphenoidal surgery for Cushing's disease: care reality in the Munich Metropolitan Region. Eur J Endocrinol. 2014;170(2):283–92.
49. Nakane T, Kuwayama A, Watanabe M, Takahashi T, Kato T, Ichihara K, et al. Long term results of transsphenoidal adenomectomy in patients with Cushing's disease. Neurosurgery. 1987;21(2):218–22.
50. Castinetti F, Nagai M, Dufour H, Kuhn JM, Morange I, Jaquet P, et al. Gamma knife radiosurgery is a successful adjunctive treatment in Cushing's disease. Eur J Endocrinol. 2007;156(1):91–8.
51. Reincke M, Ritzel K, Osswald A, Berr C, Stalla G, Hallfeldt K, et al. A critical reappraisal of bilateral adrenalectomy for ACTH-dependent Cushing's syndrome. Eur J Endocrinol. 2015;173(4):M23–32.

Medical Therapies in Cushing's Syndrome

Nicholas A. Tritos and Beverly M.K. Biller

Abstract Medical therapy has an important, albeit secondary, role in patients with Cushing's syndrome. While medications are not currently used as definitive therapy of this condition, they can be very effective in controlling hypercortisolism in patients who fail surgery, those who are not surgical candidates, or those whose tumor location is unknown. Medical therapies can be particularly helpful to control hypercortisolism in patients with Cushing's disease who underwent radiation therapy and are awaiting its salutary effects.

Currently available treatment options include several steroidogenesis inhibitors (ketoconazole, metyrapone, mitotane, etomidate), which block one or several steps in cortisol synthesis in the adrenal glands, centrally acting agents (cabergoline, pasireotide), which decrease ACTH secretion, and glucocorticoid receptor antagonists, which are represented by a single agent (mifepristone). With the exception of pasireotide and mifepristone, available agents are used "off-label" to manage hypercortisolism. Several other medications are at various stages of development and may offer additional options for the management of this serious condition.

As more potential molecular targets become known and our understanding of the pathogenesis of Cushing's syndrome improves, it is anticipated that novel, rationally designed medical therapies may emerge. Clinical trials are needed to further investigate the relative risks and benefits of currently available and novel medical therapies and examine the potential role of combination therapy in the management of Cushing's syndrome.

Keywords Cabergoline • Etomidate • Ketoconazole • Levoketoconazole • Metyrapone • Mifepristone • Mitotane • Osilodrostat • Pasireotide • Pituitary adenoma

N.A. Tritos, M.D., D.Sc. (✉) • B.M.K. Biller, M.D.
Neuroendocrine Unit/Neuroendocrine Clinical Center, Massachusetts General Hospital, Zero Emerson Place, Suite 112, Boston, MA 02114, USA

Harvard Medical School, Boston, MA, USA
e-mail: ntritos@mgh.harvard.edu

© Springer International Publishing Switzerland 2017
E.B. Geer (ed.), *The Hypothalamic-Pituitary-Adrenal Axis in Health and Disease*, DOI 10.1007/978-3-319-45950-9_9

Introduction

Definitive treatment of Cushing's syndrome involves the resection of the underlying lesion driving hypercortisolism [1–3]. However, medical therapy has an important adjunctive role in the management of patients in whom surgery is not effective in controlling cortisol excess or in patients who cannot undergo surgery because of uncertainty about the location of the underlying tumor, the presence of metastatic disease, or very poor general health associated with high surgical risk [3, 4].

In patients with Cushing's disease, pituitary surgery is first-line therapy [3–5]. Medical therapy can be recommended in patients who remain hypercortisolemic after pituitary surgery and are not considered to be good candidates for repeat pituitary surgery or those with persistent cortisol excess after reoperation. In patients who have undergone radiation therapy, medications controlling hypercortisolism are often used as a "bridge" until the radiation therapy takes effect. Anecdotally, preoperative medical therapy has also been implemented in some patients awaiting surgery in order to improve their overall condition and decrease surgical risk [3, 4].

Currently available medications include steroidogenesis inhibitors (which decrease cortisol synthesis), centrally acting agents (which can be effective in patients with Cushing's disease and occasionally ectopic corticotropin secretion), and glucocorticoid receptor antagonists (which are represented by a single available agent, mifepristone) [3, 6]. With the exception of pasireotide and mifepristone, available agents are used "off-label" in patients with Cushing's syndrome (Box 1). The aim of the present chapter is to review the use of current and emerging medical therapies in Cushing's syndrome. A discussion of treatments for comorbidities associated with hypercortisolism is beyond the scope of this chapter [3, 7].

Box 1
Currently Available Medical Therapies for Cushing's Syndrome

Steroidogenesis inhibitors
Ketoconazole
Metyrapone
Mitotane
Etomidate
Centrally acting agents
Cabergoline
Pasireotide
Glucocorticoid receptor antagonist
Mifepristone

Note: With the exception of pasireotide and mifepristone, these agents are used "off-label" in Cushing's syndrome

Steroidogenesis Inhibitors

These agents inhibit one or several enzymatic steps leading to cortisol biosynthesis in the adrenal glands (Table 1). They can be used to control hypercortisolism regardless of the underlying etiology [3, 6]. Measuring 24 h urine free cortisol (UFC) is helpful in dose titration. Two different therapeutic strategies can be employed: either achieving UFC normalization by titrating the dose of medical therapy or completely suppressing endogenous cortisol synthesis with backup glucocorticoid replacement ("block and replace" regimen). The latter regimen can be particularly helpful in patients with cyclic (intermittent or periodic) hypercortisolism but requires meticulous follow-up in order to avoid glucocorticoid excess resulting from residual (incompletely suppressed) endogenous cortisol synthesis.

All steroidogenesis inhibitors may lead to hypoadrenalism as a result of excess enzymatic blockade of cortisol biosynthesis. Therefore, patients on these agents need to be monitored for clinical and biochemical evidence of hypoadrenalism. Pituitary corticotroph tumors, which maintain some degree of feedback regulation by glucocorticoids, may increase their corticotropin (ACTH) output in response to treatment with steroidogenesis inhibitors, potentially overriding enzymatic blockade in some cases [3, 6].

Ketoconazole

Ketoconazole is an imidazole derivative that inhibits several steps in adrenal steroidogenesis, including 11,20-lyase (desmolase), 17-alpha hydroxylase, and 11-beta hydroxylase [8]. Limited data suggested that ketoconazole might also have direct inhibitory effects on ACTH secretion from pituitary corticotrophs, but this is not

Table 1 Currently available steroidogenesis inhibitors

Name	Dose range	Remarks
Ketoconazole	200–600 mg po bid–tid	Rapid onset of action
		Requires regular monitoring of liver chemistries
Metyrapone	250–1000 mg po qid	Rapid onset of action
		Preferred in pregnancy
Mitotane	0.5–3.0 g po tid	Very gradual onset of action
		Adrenolytic in higher doses
		Preferred in adrenocortical carcinoma
Etomidate	0.03 mg/kg iv as a bolus, followed by infusion (0.1–0.3 mg/kg/h)	Useful in patients with severe hypercortisolism
		Use limited by intravenous route and potential for sedation

Abbreviations: bid: twice daily; iv: intravenously; po: by mouth; qid: four times daily; tid: three times daily widely accepted [8]. Ketoconazole was originally licensed as an antifungal agent and has been prescribed "off-label" to control hypercortisolism. Used as monotherapy in patients with Cushing's disease, ketoconazole has been reported to control hyper-cortisolism in 70 % of treated patients based on pooled analyses of 8 small, retrospec-tive studies that included a total of 82 patients [8]. However, a more recent multicenter study found that ketoconazole use led to UFC normalization in approximately 50 % of patients with Cushing's disease [9]. Ketoconazole has also been effective in control-ling hypercortisolism in approximately 50 % of patients with the ectopic ACTH syn-drome [8]. As this medication has a rapid onset of action, it can be particularly helpful among patients with severe manifestations of cortisol excess.

Common, but generally mild, adverse effects associated with ketoconazole use may include gastrointestinal symptoms (nausea, dyspepsia), rash, and headache [3]. Hypogonadism may also develop in men as a result of inhibition of testosterone synthesis. Severe adverse effects are uncommon, including idiosyncratic hepatotox-icity (occurring in approximately 1 in 15,000 treated patients) [10]. Regular moni-toring of liver chemistries is recommended in treated patients, who need to be warned of possible symptoms associated with liver toxicity. Asymptomatic transa-minitis is more common (occurring in approximately 12 % of patients) and gener-ally improves or resolves with a decrease in medication dose [8].

Ketoconazole absorption is significantly higher in the presence of an acidic environment in the stomach. Accordingly, use of medications that raise gastric pH, including proton pump inhibitors or H_2 receptor antagonists, is best avoided in patients receiving ketoconazole therapy. Of note, ketoconazole is metabolized in the liver by the CYP450 3A4 enzyme, raising the potential for drug–drug interactions with other medications (such as several "statins") that are substrates of the same enzyme or those that either inhibit or induce this enzymatic activity [11, 12].

Metyrapone

Metyrapone inhibits 11-beta hydroxylase, which catalyzes the last step in cortisol synthesis [3]. As a corollary, several steroid precursors accumulate in patients receiving this medication, including 11-deoxycortisol, 11-deoxycorticosterone, as well as several androgenic precursors. Since 11-deoxycortisol often cross-reacts with cortisol in immunoassays, serum cortisol levels may be overestimated in patients on this therapy (depending on the assay used).

Metyrapone was reported to control hypercortisolism in up to 75 % of 53 patients with Cushing's disease treated for up to 16 weeks based on serum cortisol data (using cortisol day curves) [13]. A more recent study found that metyrapone use led to UFC normalization in approximately 50 % of patients with Cushing's disease [14]. Escape from its salutary effects may occur in a minority of patients with Cushing's disease. Metyrapone has also been effective in controlling hypercortisolism in substantial proportions (40–75 %) of patients with the ectopic ACTH syndrome, as well as those with benign or malignant adrenal pathologies [14]. Metyrapone has a rapid onset of action, which can be quite helpful when prompt control of severe hypercortisolism is needed. Metyrapone is considered the preferred medical agent to control hypercortisolism during pregnancy, but is not licensed for use specifically for this indication [15].

Common adverse effects (25 %) associated with metyrapone use include nausea, vomiting, and dizziness [14]. In addition, the accumulation of precursors with mineralocorticoid activity (including 11-deoxycortisol and 11-deoxycorticosterone) may lead to hypertension, edema, and hypokalemia. Similarly, androgenic precursors that accumulate as a result of metyrapone therapy may lead to hirsutism and acne in women [3].

Mitotane

Mitotane inhibits several steps in adrenal steroidogenesis, including the cholesterol side-chain cleavage enzyme, 3-beta hydroxysteroid dehydrogenase, and 11-beta hydroxylase. In addition, it is adrenolytic when used long term in higher doses (>4 g/daily) [3]. This latter effect has led to its use in adrenocortical carcinoma, either as adjuvant postoperative therapy or as treatment in patients with advanced disease [16–18]. In patients with adrenocortical carcinoma, monitoring of systemic levels is advisable with a goal to maintain plasma mitotane levels ≥14 mg/L, which correlate with higher likelihood of achieving tumor control [19]. Used as monotherapy, mitotane is effective in controlling hypercortisolemia in 72–83 % of patients with Cushing's disease, but has been used in only a few centers worldwide for this indication [16, 20]. Of note, its onset of action is slow, requiring several weeks to months to reach maximal effect in individual patients. As a consequence, mitotane monotherapy is not appropriate when rapid control of severe hypercortisolism is needed. Adrenal insufficiency may occur over time, necessitating the administration of glucocorticoid replacement in treated patients. Escape from its effects on cortisol synthesis is unlikely with long-term use.

Mitotane use may lead to several adverse effects, including gastrointestinal (nausea, vomiting, dyspepsia, diarrhea) and neurologic (dizziness, ataxia, dysarthria, confusion) symptoms, which may limit its use [3]. Other side effects include rash, gynecomastia, abnormal liver chemistries, and dyslipidemia. Rare adverse events include hemorrhagic cystitis, ophthalmic, and hematologic abnormalities. Mitotane is highly lipophilic and can persist in the adipose tissue for months or years after it is stopped. In view of its long half-life and teratogenicity, pregnancy should be avoided for up to 5 years after mitotane discontinuation [3].

Mitotane increases systemic corticosteroid-binding globulin (CBG) levels and accelerates cortisol clearance. Consequently, glucocorticoid replacement doses need to be higher in patients treated with mitotane therapy.

Etomidate

Etomidate is primarily used in anesthesia induction. However, it also inhibits 11-beta hydroxylase, leading to rapid suppression of cortisol synthesis within hours, even in subhypnotic doses [21]. Etomidate can be particularly helpful when rapid control of severe hypercortisolism is needed, especially in patients unable to take oral medications, but requires careful monitoring to avoid excessive sedation [22]. Etomidate is the only intravenous preparation that can be used to control hypercortisolism [22]. However, its use is limited to hospitalized patients with severe hypercortisolism.

Novel Agents Under Investigation

Osilodrostat (LCI699) is a novel 11-beta hydroxylase and aldosterone synthase inhibitor that is currently under study in patients with Cushing's disease. In a phase II, proof-of-concept trial, osilodrostat administration led to UFC normalization in 92 % of 12 patients with Cushing's disease who were treated for 70 days [23]. Whether escape from its effects may occur remains to be established. Of note, approximately 79 % of patients treated with LCI699 achieved normal UFC in a 6 month extension of the phase II study that included 19 patients [12, 24]. The efficacy and safety of osilodrostat are being investigated in a phase III study.

Osilodrostat appears to be well tolerated in most patients. However, fatigue, headache, gastrointestinal symptoms, and dizziness may occur. Hypertension, edema, and hypokalemia may develop as a consequence of accumulation of precursors with mineralocorticoid activity, and hirsutism or acne may occur as a result of accumulation of androgenic precursors.

Levoketoconazole is a ketoconazole enantiomer that is also under investigation in Cushing's disease. Based on preliminary data, it may have increased potency and duration of action and potentially a lower risk of hepatotoxicity [12].

Abiraterone is an inhibitor of 17-alpha hydroxylase and 17,20-lyase activity and has been used to suppress androgen synthesis in patients with castration-resistant advanced prostate cancer [25]. Based on its mechanism of action, it would be predicted to be potentially efficacious in patients with Cushing's syndrome. However, clinical studies are required to examine this possibility.

Subgroups of adrenal masses in patients with bilateral macronodular adrenal hyperplasia or adrenal adenomas may express a wide variety of receptors, including those engaging glucose-dependent insulinotropic peptide (GIP), luteinizing hormone (LH)/human chorionic gonadotropin (hCG), vasopressin (V1, V2, V3), serotonin (5HT4 and 5HT7), angiotensin (AT1), glucagon, or beta adrenergic receptors [26]. Based on these considerations, medications that inhibit some of these receptors or pathways, including octreotide or pasireotide (inhibiting GIP secretion), leuprolide (inhibiting LH secretion), and propranolol (inhibiting beta adrenergic receptors), have shown at least transient effectiveness in controlling hypercortisolism in small numbers of patients with adrenal masses expressing the respective receptors [26–29].

Table 2 Currently available centrally acting agents

Name	Dose range	Remarks
Cabergoline	0.5–7.0 mg po weekly	Escape (loss of effectiveness) may occur over time
		Potential risk of valvulopathy in high doses
Pasireotide	0.3–0.9 mg sc bid	Hyperglycemia or diabetes mellitus may develop
		Glucose, hepatic function, and electrocardiographic monitoring advised

Abbreviations: bid: twice daily; po: by mouth; sc: subcutaneously

Centrally Acting Agents

These agents are directed at suppressing ACTH synthesis and/or release and may be efficacious in controlling hypercortisolism in patients with ACTH-dependent Cushing's syndrome, primarily those with Cushing's disease (Table 2) [3, 4, 30]. In addition, they might lead to a decrease in pituitary tumor size in patients with Cushing's disease or Nelson's syndrome. Currently, these medications are used primarily for their antisecretory effects, since data on tumor control are limited.

Cabergoline

Cabergoline is a dopamine receptor (type 2 specific) agonist, which is licensed as therapy for hyperprolactinemia, but has also been used "off-label" to treat patients with Cushing's disease [3, 31]. Its potential effectiveness is predicated by the presence of dopamine receptors in the majority of corticotropinomas [32]. Cabergoline administration may control hypercortisolism in 30–40 % of patients with ACTH-secreting pituitary adenomas [33, 34]. However, escape from its effects may occur over time. It should also be noted that cabergoline doses that are required to control hypercortisolism are generally larger (1–7 mg/week) than those that are effective in the majority of patients with hyperprolactinemia (0.5–2.0 mg/week). In contrast to cabergoline, bromocriptine, an older dopamine receptor agonist, is largely ineffective in patients with Cushing's disease.

Cabergoline administration is generally tolerated well. However, nausea, vomiting, and dizziness may occur and are more common among patients receiving high doses. Other less common adverse effects include headache, nasal congestion, constipation, digital vasospasm, anxiety, depression, exacerbation of psychosis, or a variety of manifestations of impulsivity [31, 35]. When administered in high doses in patients with Parkinson's disease, cabergoline use was associated with cardiac valvulopathy, which is presumed to occur as a consequence of serotonin receptor (5HT2B) activation [36, 37]. While cabergoline use in doses typically required to treat hyperprolactinemia (0.5–2.0 mg/week) appears to be safe with regard to cardiac valvulopathy, it is less clear whether its long-term use in higher doses (up to 7.0 mg/week) needed to control hypercortisolism may increase the risk of valvular damage [38]. Periodic echocardiography seems prudent in patients receiving such higher cabergoline doses. However, there are currently no data examining the cost-effectiveness of echocardiography in detecting valvulopathy in this population.

Pasireotide

Pasireotide is a somatostatin receptor agonist with expanded specificity, which activates type 1, 2, 3, and 5 somatostatin receptor isoforms [11, 30]. It is thought that stimulation of the type 5 receptor isoform accounts for its efficacy in patients with Cushing's disease [39]. In contrast, octreotide, which activates type 2 and (weakly) type 5 somatostatin receptors, has very limited efficacy in patients with Cushing's disease. Of note, type 5 and type 2 somatostatin receptor isoforms are expressed by approximately 84% and 74% of corticotropinomas, respectively [32].

The efficacy of pasireotide administration was established in a phase 3, multi-center clinical trial of 162 adults with Cushing's disease, who were randomly allocated to either of two pasireotide starting doses (600 mcg twice daily and 900 mcg twice daily) and were treated for 12 months. Control of hypercortisolism, based on UFC normalization, was reported in 15% and 26% of patients who received the lower and higher pasireotide starting dose without need for dose uptitration, respectively [40]. In addition, pasireotide therapy led to weight loss, decrease in blood pressure, and improved quality of life as well as a decrease in tumor size among patients with measurable tumor mass (by 9.1% and 43.8% in patients receiving the lower and higher pasireotide starting dose, respectively). Pasireotide has been approved by the FDA and EMA for use in patients with Cushing's disease who have failed pituitary surgery or are not surgical candidates. Pasireotide LAR, a long-acting form of pasireotide, is under evaluation in a phase III clinical trial as a possible therapy in patients with Cushing's disease [41].

Similar to octreotide, pasireotide administration is associated with possible gastro-intestinal adverse events (nausea, abdominal pain, diarrhea, gallstones or sludge, mild transaminitis). Asymptomatic sinus bradycardia, QT prolongation, and hair loss may also occur. In addition, pasireotide therapy appears to be associated with the development of hyperglycemia or diabetes mellitus. Indeed, hyperglycemia developed in 73% of patients in the phase 3 trial [40]. The hyperglycemic effects of pasireotide occur as a consequence of inhibition of insulin secretion, which is partly attributable to suppression of incretin secretion from the gastrointestinal tract [42]. Self-monitoring of blood glucose is advisable in patients treated with pasireotide. Hyperglycemia may be treated with metformin therapy with possible stepwise addition of incretin mimetics, dipeptidyl peptidase inhibitors, and/or insulin. In addition to monitoring for hyperglycemia, pasireotide-treated patients are advised to undergo periodic evaluation of serum electrolytes, liver function tests, electrocardiograms, and gallbladder ultrasound examinations.

Novel Agents and Targets Under Investigation

The retinoic acid receptor appears to have a role in the regulation of proopiomelanocortin and ACTH synthesis [43, 44]. Accordingly, cognate retinoic acid receptor agonists may be of potential benefit in patients with Cushing's disease. Preliminary data suggest some evidence of in vitro and in vivo effectiveness of retinoic acid in Cushing's disease, but its clinical use has not been adequately investigated [45].

The epidermal growth factor receptor is often expressed in corticotropinomas [46]. Recent in vitro and preclinical data suggest a role for epidermal growth factor receptor inhibition with gefitinib in controlling tumor size and hypercortisolism [46]. While epidermal growth factor receptor inhibition (with gefitinib) is of potential interest as a treatment strategy in patients with Cushing's disease, its efficacy and safety in this population remain to be explored in clinical studies.

Corticotropinomas may also express growth hormone secretagogue receptors or vasopressin receptors, suggesting that respective receptor antagonists might have a role in the management of Cushing's disease [47–49]. However, clinical data are needed to examine whether medications that inhibit these receptors might be efficacious in Cushing's disease.

Glucocorticoid Receptor Antagonist

Mifepristone

Mifepristone is a glucocorticoid and progesterone receptor antagonist, which has been approved by the FDA as therapy in patients with Cushing's syndrome of diverse etiologies and hyperglycemia, who have failed surgery or are not surgical candidates [50, 51]. Mifepristone administration effectively inhibits glucocorticoid action, leading to a decrease in glycemia, body weight, and improved overall health status based on the findings of an open label forced titration study of 50 patients with Cushing's syndrome (including 43 patients with Cushing's disease), hyperglycemia, or hypertension who were treated with mifepristone for 6 months [51]. Specifically, 60 % of hyperglycemic patients improved with regard to glucose tolerance, and hemoglobin A1c (HbA1c) values declined from 7.4 % (baseline) to 6.3 %. Diastolic blood pressure improved in 38 % of hypertensive patients. Body weight decreased by 5.7 % in the study population. In addition, 87 % of patients showed overall clinical improvement based on the findings of a blinded board [51].

Mifepristone doses, ranging between 300 mg/daily and 1200 mg/daily, must be titrated based on clinical evaluation alone in patients with corticotropinomas. Patients with Cushing's disease on mifepristone therapy generally show an increase in ACTH and cortisol levels in response to mifepristone therapy, which is reversible upon drug discontinuation [51].

Glucocorticoid receptor inhibition may lead to symptoms of hypoadrenalism. Treated patients need to be monitored clinically for suggestive symptoms (headache, nausea, vomiting, dizziness, orthostasis, and arthralgias). If hypoadrenalism is clinically suspected, patients can be treated with dexamethasone, and mifepristone can be temporarily suspended and reintroduced after patients become asymptomatic [52]. Laboratory testing is not helpful in establishing hypoadrenalism in these patients; in fact, cortisol levels are typically elevated in patients with Cushing's disease on mifepristone therapy, but cortisol action is blocked at the glucocorticoid receptor.

Mifepristone does not inhibit the mineralocorticoid receptor, which can be activated by cortisol, thus leading to a potential increase in blood pressure and the frequent development of hypokalemia (34 % of patients) [51]. Regular monitoring of blood pressure and serum potassium levels is advisable in patients on mifepristone therapy. Severe hypokalemia may occur, requiring large doses of potassium replacement and/or spironolactone therapy. In addition, progesterone receptor inhibition will terminate pregnancy and may lead to irregular vaginal bleeding (14 %) as a result of endometrial thickening (28 %), which appears to be pathologically distinct from endometrial hyperplasia [53]. Other possible adverse events associated with mifepristone administration include dyslipidemia and elevated thyrotropin levels. Pituitary tumor progression was noted in 3 patients with macroadenomas and 1 patient with microadenoma [54]. Tumor regression was found in 2 patients out of 43 patients with Cushing's disease, who were treated with mifepristone for 6 months (27 of whom continued into a long-term extension phase and were treated for a median duration of 11.3 months) [54]. More long-term data are needed to examine any possible effects of mifepristone therapy on pituitary adenomas in Cushing's disease.

Mifepristone is metabolized in the liver and inhibits several cytochrome P450 enzymes (including CYP450 3A4), thus leading to possible drug–drug interactions with other medications that influence and/or are metabolized through the same enzymatic activity [3, 52].

Combination Therapy

The use of medical therapies in combination has been reported in several case series but has not been examined in a clinical trial. Patients with severe ACTH-dependent Cushing's syndrome that is not amenable to surgery (including Cushing's disease and ectopic ACTH secretion) may benefit from the combined administration of ketoconazole, metyrapone, and mitotane in order to control hypercortisolism rapidly and avert the need for bilateral adrenalectomy [55]. In another study, the combination of ketoconazole and metyrapone was found to be effective in controlling cortisol excess in, respectively, 73 % and 86 % of patients with severe hypercortisolism at baseline, including 14 with the ectopic ACTH syndrome and 8 patients with adrenocortical carcinoma [56]. In a third case series, pasireotide monotherapy was administered to 17 patients with Cushing's disease who had failed pituitary surgery. Subsequently, cabergoline was added in patients who did not adequately respond to pasireotide and, in a third step, ketoconazole was added to the combination of pasireotide and cabergoline when the two drug combination was not sufficient in controlling hypercortisolism. In this small series, UFC normalization occurred in 88 % of patients treated with 1–3 medications, thus demonstrating the potential role of combination therapy [57]. However, properly designed clinical trials will be needed in order to fully elucidate the risks and benefits of this approach.

Summary and Future Directions

Medical therapy has an important, albeit adjuvant, role in the management of patients with Cushing's syndrome. Several steroidogenesis inhibitors, centrally acting agents, and a glucocorticoid receptor antagonist are currently available or being investigated as potential therapies. It may be noted that the choice between therapies is largely empiric as a consequence of lack of head-to-head clinical trials and depends on several factors, including severity of hypercortisolism and the clinical need to achieve rapid biochemical control, tumor size and location, patient comorbidities, medication tolerance, potential for drug interactions, patient compliance and preference, medication availability, and cost. It is anticipated that better understanding of the molecular underpinnings of Cushing's syndrome will eventually lead to more efficacious, rationally designed therapies for this potentially devastating condition.

Disclosures BMKB has received research grants (to MGH) from Cortendo and Novartis, and has consulted for Cortendo, HRA Pharma, Ipsen and Novartis. NAT has received research grants (to MGH) from Ipsen, Novartis, Novo Nordisk and Pfizer.

References

1. Cushing H. The basophil adenomas of the pituitary body and their clinical manifestations (pituitary basophilism). Bull Johns Hopkins Hosp. 1932;50:137–95.
2. Cushing H. The basophil adenomas of the pituitary body. Ann R Coll Surg Engl. 1969;44(4):180–1.
3. Nieman LK, Biller BM, Findling JW, Murad MH, Newell-Price J, Savage MO, Tabarin A. Treatment of Cushing's syndrome: an endocrine society clinical practice guideline. J Clin Endocrinol Metab. 2015;100(8):2807–31. doi:10.1210/jc.2015-1818.
4. Tritos NA, Biller BM. Medical management of Cushing's disease. J Neurooncol. 2014;117(3):407–14. doi:10.1007/s11060-013-1269-1.
5. Swearingen B, Biller BM, Barker 2nd FG, Katznelson L, Grinspoon S, Klibanski A, Zervas NT. Long-term mortality after transsphenoidal surgery for Cushing disease. Ann Intern Med. 1999;130(10):821–4.
6. Nieman LK, Ilias I. Evaluation and treatment of Cushing's syndrome. Am J Med. 2005; 118(12):1340–6.

7. Arnaldi G, Angeli A, Atkinson AB, Bertagna X, Cavagnini F, Chrousos GP, Fava GA, Findling JW, Gaillard RC, Grossman AB, Kola B, Lacroix A, Mancini T, Mantero F, Newell-Price J, Nieman LK, Sonino N, Vance ML, Giustina A, Boscaro M. Diagnosis and complications of Cushing's syndrome: a consensus statement. J Clin Endocrinol Metab. 2003;88(12): 5593–602.
8. Engelhardt D, Weber MM. Therapy of Cushing's syndrome with steroid biosynthesis inhibitors. J Steroid Biochem Mol Biol. 1994;49(4-6):261–7.
9. Castinetti F, Guignat L, Giraud P, Muller M, Kamenicky P, Drui D, Caron P, Luca F, Donadille B, Vantyghem MC, Bihan H, Delemer B, Raverot G, Motte E, Philippon M, Morange I, Conte-Devolx B, Quinquis L, Martinie M, Vezzosi D, Le Bras M, Baudry C, Christin-Maitre S, Goichot B, Chanson P, Young J, Chabre O, Tabarin A, Bertherat J, Brue T. Ketoconazole in Cushing's disease: is it worth a try? J Clin Endocrinol Metab. 2014;99(5):1623–30. doi:10.1210/jc.2013-3628.
10. McCance DR, Ritchie CM, Sheridan B, Atkinson AB. Acute hypoadrenalism and hepatotoxicity after treatment with ketoconazole. Lancet. 1987;1(8532):573.
11. Nieman LK. Update in the medical therapy of Cushing's disease. Curr Opin Endocrinol Diabetes Obes. 2013;20(4):330–4.
12. Fleseriu M, Petersenn S. Medical therapy for Cushing's disease: adrenal steroidogenesis inhibitors and glucocorticoid receptor blockers. Pituitary. 2015;18(2):245–52. doi:10.1007/s11102-014-0627-0.
13. Verhelst JA, Trainer PJ, Howlett TA, Perry L, Rees LH, Grossman AB, Wass JA, Besser GM. Short and long-term responses to metyrapone in the medical management of 91 patients with Cushing's syndrome. Clin Endocrinol (Oxf). 1991;35(2):169–78.
14. Daniel E, Aylwin S, Mustafa O, Ball S, Munir A, Boelaert K, Chortis V, Cuthbertson DJ, Daousi C, Rajeev SP, Davis J, Cheer K, Drake W, Gunganah K, Grossman A, Gurnell M, Powlson AS, Karavitaki N, Huguet I, Kearney T, Mohit K, Meeran K, Hill N, Rees A, Lansdown AJ, Trainer PJ, Minder AH, Newell-Price J. Effectiveness of metyrapone in treating Cushing's syndrome: a retrospective multicenter study in 195 patients. J Clin Endocrinol Metab. 2015;100(11):4146–54. doi:10.1210/jc.2015-2616.
15. Gormley MJ, Hadden DR, Kennedy TL, Montgomery DA, Murnaghan GA, Sheridan B. Cushing's syndrome in pregnancy—treatment with metyrapone. Clin Endocrinol (Oxf). 1982;16(3):283–93.
16. Luton JP, Cerdas S, Billaud L, Thomas G, Guilhaume B, Bertagna X, Laudat MH, Louvel A, Chapuis Y, Blondeau P, et al. Clinical features of adrenocortical carcinoma, prognostic factors, and the effect of mitotane therapy. N Engl J Med. 1990;322(17):1195–201.
17. Mauclere-Denost S, Leboulleux S, Borget I, Paci A, Young J, Al Ghuzlan A, Deandreis D, Drouard L, Tabarin A, Chanson P, Schlumberger M, Baudin E. High-dose mitotane strategy in adrenocortical carcinoma (ACC): prospective analysis of plasma mitotane measurement during the first three months of follow-up. Eur J Endocrinol. 2012;166(2):261–8. doi:10.1530/EJE-11-0557.
18. Terzolo M, Angeli A, Fassnacht M, Daffara F, Tauchmanova L, Conton PA, Rossetto R, Buci L, Sperone P, Grossrubatscher E, Reimondo G, Bollito E, Papotti M, Saeger W, Hahner S, Koschker AC, Arvat E, Ambrosi B, Loli P, Lombardi G, Mannelli M, Bruzzi P, Mantero F, Allolio B, Dogliotti L, Berruti A. Adjuvant mitotane treatment for adrenocortical carcinoma. N Engl J Med. 2007;356(23):2372–80.
19. Hermsen IG, Fassnacht M, Terzolo M, Houterman S, den Hartigh J, Leboulleux S, Daffara F, Berruti A, Chadarevian R, Schlumberger M, Allolio B, Haak HR, Baudin E. Plasma concentrations of o, p'DDD, o, p'DDA, and o, p'DDE as predictors of tumor response to mitotane in adrenocortical carcinoma: results of a retrospective ENS@T multicenter study. J Clin Endocrinol Metab. 2011;96(6):1844–51.
20. Baudry C, Coste J, Bou Khalil R, Silvera S, Guignat L, Guibourdenche J, Abbas H, Legmann P, Bertagna X, Bertherat J. Efficiency and tolerance of mitotane in Cushing's disease in 76 patients from a single center. Eur J Endocrinol. 2012;167(4):473–81.

21. Schulte HM, Benker G, Reinwein D, Sippell WG, Allolio B. Infusion of low dose etomidate: correction of hypercortisolemia in patients with Cushing's syndrome and dose-response relationship in normal subjects. J Clin Endocrinol Metab. 1990;70(5):1426–30.

22. Preda VA, Sen J, Karavitaki N, Grossman AB. Etomidate in the management of hypercortisolaemia in Cushing's syndrome: a review. Eur J Endocrinol. 2012;167(2):137–43. doi:10.1530/EJE-12-0274.

23. Bertagna X, Pivonello R, Fleseriu M, Zhang Y, Robinson P, Taylor A, Watson CE, Maldonado M, Hamrahian AH, Boscaro M, Biller BM. LCI699, a potent 11beta-hydroxylase inhibitor, normalizes urinary cortisol in patients with Cushing's disease: results from a multicenter, proof-of-concept study. J Clin Endocrinol Metab. 2014;99(4):1375–83. doi:10.1210/jc.2013-2117.

24. Fleseriu M, Pivonello R, Young J, Hamrahian AH, Molitch ME, Shimizu C, Tanaka T, Shimatsu A, White T, Hilliard A, Tian C, Sauter N, Biller BM, Bertagna X. Osilodrostat, a potent oral 11beta-hydroxylase inhibitor: 22-week, prospective, Phase II study in Cushing's disease. Pituitary. 2015;19(2):138–48. doi:10.1007/s11102-015-0692-z.

25. Gartrell BA, Saad F. Abiraterone in the management of castration-resistant prostate cancer prior to chemotherapy. Ther Adv Urol. 2015;7(4):194–202. doi:10.1177/1756287215592288.

26. Ghorayeb NE, Bourdeau I, Lacroix A. Multiple aberrant hormone receptors in Cushing's syndrome. Eur J Endocrinol. 2015;173(4):M45–60. doi:10.1530/EJE-15-0200.

27. Lacroix A, Hamet P, Boutin JM. Leuprolide acetate therapy in luteinizing hormone-dependent Cushing's syndrome. N Engl J Med. 1999;341(21):1577–81. doi:10.1056/NEJM199911183412104.

28. Lacroix A, Tremblay J, Rousseau G, Bouvier M, Hamet P. Propranolol therapy for ectopic beta-adrenergic receptors in adrenal Cushing's syndrome. N Engl J Med. 1997;337(20):1429–34. doi:10.1056/NEJM199711133372004.

29. Preumont V, Mermejo LM, Damoiseaux P, Lacroix A, Maiter D. Transient efficacy of octreotide and pasireotide (SOM230) treatment in GIP-dependent Cushing's syndrome. Horm Metab Res. 2011;43(4):287–91. doi:10.1055/s-0030-1270523.

30. Fleseriu M, Petersenn S. New avenues in the medical treatment of Cushing's disease: corticotroph tumor targeted therapy. J Neurooncol. 2013.

31. Klibanski A. Clinical practice. Prolactinomas. N Engl J Med. 2010;362(13):1219–26.

32. de Bruin C, Feelders RA, Lamberts SW, Hofland LJ. Somatostatin and dopamine receptors as targets for medical treatment of Cushing's syndrome. Rev Endocr Metab Disord. 2009;10(2):91–102.

33. Godbout A, Manavela M, Danilowicz K, Beauregard H, Bruno OD, Lacroix A. Cabergoline monotherapy in the long-term treatment of Cushing's disease. Eur J Endocrinol. 2010;163(5):709–16.

34. Pivonello R, De Martino MC, Cappabianca P, De Leo M, Faggiano A, Lombardi G, Hofland LJ, Lamberts SW, Colao A. The medical treatment of Cushing's disease: effectiveness of chronic treatment with the dopamine agonist cabergoline in patients unsuccessfully treated by surgery. J Clin Endocrinol Metab. 2009;94(1):223–30.

35. Barake M, Evins AE, Stoeckel L, Pachas GN, Nachtigall LB, Miller KK, Biller BM, Tritos NA, Klibanski A. Investigation of impulsivity in patients on dopamine agonist therapy for hyperprolactinemia: a pilot study. Pituitary. 2014;17(2):150–6. doi:10.1007/s11102-013-0480-6.

36. Schade R, Andersohn F, Suissa S, Haverkamp W, Garbe E. Dopamine agonists and the risk of cardiac-valve regurgitation. N Engl J Med. 2007;356(1):29–38.

37. Zanettini R, Antonini A, Gatto G, Gentile R, Tesei S, Pezzoli G. Valvular heart disease and the use of dopamine agonists for Parkinson's disease. N Engl J Med. 2007;356(1):39–46.

38. Valassi E, Klibanski A, Biller BM. Clinical review: potential cardiac valve effects of dopamine agonists in hyperprolactinemia. J Clin Endocrinol Metab. 2010;95(3):1025–33.

39. Ben-Shlomo A, Schmid H, Wawrowsky K, Pichurin O, Hubina E, Chesnokova V, Liu NA, Culler M, Melmed S. Differential ligand-mediated pituitary somatostatin receptor subtype

signaling: implications for corticotroph tumor therapy. J Clin Endocrinol Metab. 2009;94(11): 4342–50.

40. Colao A, Petersenn S, Newell-Price J, Findling JW, Gu F, Maldonado M, Schoenherr U, Mills D, Salgado LR, Biller BM. A 12-month phase 3 study of pasireotide in Cushing's disease. N Engl J Med. 2012;366(10):914–24.

41. Ligueros-Saylan M, Zhang Y, Newell-Price J, Petersenn S, Lymperopoulos S. Evaluation of the efficacy and safety of pasireotide LAR in patients with mild-to-moderate Cushing's disease: a randomized, double-blind, multicenter, phase III study design. Endocr Abstr. 2012;29: P1542.1.

42. Henry RR, Ciaraldi TP, Armstrong D, Burke P, Ligueros-Saylan M, Mudaliar S. Hyperglycemia associated with pasireotide: results from a mechanistic study in healthy volunteers. J Clin Endocrinol Metab. 2013;98(8):3446–53.

43. Castillo V, Giacomini D, Paez-Pereda M, Stalla J, Labeur M, Theodoropoulou M, Holsboer F, Grossman AB, Stalla GK, Arzt E. Retinoic acid as a novel medical therapy for Cushing's disease in dogs. Endocrinology. 2006;147(9):4438–44. doi:10.1210/en.2006-0414.

44. Labeur M, Theodoropoulou M, Sievers C, Paez-Pereda M, Castillo V, Arzt E, Stalla GK. New aspects in the diagnosis and treatment of Cushing disease. Front Horm Res. 2006;35:169–78. doi:10.1159/000094325.

45. Pecori Giraldi F, Ambrogio AG, Andrioli M, Sanguin F, Karamouzis I, Corsello SM, Scaroni C, Arvat E, Pontecorvi A, Cavagnini F. Potential role for retinoic acid in patients with Cushing's disease. J Clin Endocrinol Metab. 2012;97(10):3577–83. doi:10.1210/jc.2012-2328.

46. Fukuoka H, Cooper O, Ben-Shlomo A, Mamelak A, Ren SG, Bruyette D, Melmed S. EGFR as a therapeutic target for human, canine, and mouse ACTH-secreting pituitary adenomas. J Clin Invest. 2011;121(12):4712–21. doi:10.1172/JCI60417.

47. Ghigo E, Arvat E, Ramunni J, Colao A, Gianotti L, Deghenghi R, Lombardi G, Camanni F. Adrenocorticotropin- and cortisol-releasing effect of hexarelin, a synthetic growth hormone-releasing peptide, in normal subjects and patients with Cushing's syndrome. J Clin Endocrinol Metab. 1997;82(8):2439–44. doi:10.1210/jcem.82.8.4132.

48. Dahia PL, Ahmed-Shuaib A, Jacobs RA, Chew SL, Honegger J, Fahlbusch R, Besser GM, Grossman AB. Vasopressin receptor expression and mutation analysis in corticotropin-secreting tumors. J Clin Endocrinol Metab. 1996;81(5):1768–71. doi:10.1210/jcem.81.5. 8626831.

49. Korbonits M, Jacobs RA, Aylwin SJ, Burrin JM, Dahia PL, Monson JP, Honegger J, Fahlbush R, Trainer PJ, Chew SL, Besser GM, Grossman AB. Expression of the growth hormone secretagogue receptor in pituitary adenomas and other neuroendocrine tumors. J Clin Endocrinol Metab. 1998;83(10):3624–30. doi:10.1210/jcem.83.10.5210.

50. Castinetti F, Fassnacht M, Johanssen S, Terzolo M, Bouchard P, Chanson P, Do Cao C, Morange I, Pico A, Ouzounian S, Young J, Hahner S, Brue T, Allolio B, Conte-Devolx B. Merits and pitfalls of mifepristone in Cushing's syndrome. Eur J Endocrinol. 2009;160(6): 1003–10.

51. Fleseriu M, Biller BM, Findling JW, Molitch ME, Schteingart DE, Gross C. Mifepristone, a glucocorticoid receptor antagonist, produces clinical and metabolic benefits in patients with Cushing's syndrome. J Clin Endocrinol Metab. 2012;97(6):2039–49.

52. Fleseriu M, Molitch ME, Gross C, Schteingart DE, Vaughan TB, Biller BM. A new therapeutic approach in the medical treatment of Cushing's syndrome: glucocorticoid receptor blockade with Mifepristone. Endocr Pract. 2013;19(2):313–26.

53. Mutter GL, Bergeron C, Deligdisch L, Ferenczy A, Glant M, Merino M, Williams AR, Blithe DL. The spectrum of endometrial pathology induced by progesterone receptor modulators. Mod Pathol. 2008;21(5):591–8. doi:10.1038/modpathol.2008.19.

54. Fleseriu M, Findling JW, Koch CA, Schlaffer SM, Buchfelder M, Gross C. Changes in plasma ACTH levels and corticotroph tumor size in patients with Cushing's disease during long-term treatment with the glucocorticoid receptor antagonist mifepristone. J Clin Endocrinol Metab. 2014;99(10):3718–27. doi:10.1210/jc.2014-1843.

55. Kamenicky P, Droumaguet C, Salenave S, Blanchard A, Jublanc C, Gautier JF, Brailly-Tabard S, Leboulleux S, Schlumberger M, Baudin E, Chanson P, Young J. Mitotane, metyrapone, and ketoconazole combination therapy as an alternative to rescue adrenalectomy for severe ACTH-dependent Cushing's syndrome. J Clin Endocrinol Metab. 2011;96(9):2796–804.

56. Corcuff JB, Young J, Masquefa-Giraud P, Chanson P, Baudin E, Tabarin A. Rapid control of severe neoplastic hypercortisolism with metyrapone and ketoconazole. Eur J Endocrinol. 2015;172(4):473–81. doi:10.1530/EJE-14-0913.

57. Feelders RA, de Bruin C, Pereira AM, Romijn JA, Netea-Maier RT, Hermus AR, Zelissen PM, van Heerebeek R, de Jong FH, van der Lely AJ, de Herder WW, Hofland LJ, Lamberts SW. Pasireotide alone or with cabergoline and ketoconazole in Cushing's disease. N Engl J Med. 2010;362(19):1846–8.

Mild Adrenal Cortisol Excess

Adina F. Turcu and Richard J. Auchus

Abstract Adrenal subclinical hypercortisolism or mild adrenal cortisol excess has been defined by alterations of the hypothalamic–pituitary–adrenal axis in patients with adrenal adenomas and without overt Cushing syndrome. Mild hypercortisolism is the most common hormonal dysfunction in patients with incidentally diagnosed adrenal masses. Recent reports have linked mild adrenal cortisol excess with several cardiovascular, bone, and metabolic complications, as well as with increased mortality. The pathophysiological mechanisms of mild adrenal cortisol excess are poorly understood, and no consensus exists regarding the appropriate diagnostic criteria of mild adrenal cortisol excess or its management. Existing data have derived predominantly from retrospective or nonrandomized studies. This chapter overviews the most recent progress in the understanding of mild adrenal cortisol excess and highlights remaining gaps to be filled by thoughtfully designed future research.

Keywords Subclinical hypercortisolism • Subclinical Cushing syndrome • Adrenal adenoma • Adrenal incidentaloma • Adrenal • Cortisol • Hypothalamic–pituitary–adrenal axis • Mortality • Cardiovascular risk • Osteoporosis

Introduction

Mild adrenal cortisol excess (MACE) usually arises in the context of incidentally discovered adrenal masses (also called adrenal incidentalomas, AI). With the rising availability and performance of imaging studies applied to routine clinical care, AI

A.F. Turcu
Division of Metabolism, Endocrinology, & Diabetes, Department of Internal Medicine, University of Michigan, Ann Arbor, MI, 48109, USA

R.J. Auchus (✉)
Division of Metabolism, Endocrinology, & Diabetes, Department of Internal Medicine, University of Michigan, Ann Arbor, MI, 48109, USA

Department of Pharmacology, University of Michigan, Room 5560A, 1150 West Medical Center Drive, Ann Arbor, MI 48109, USA
e-mail: rauchus@med.umich.edu

© Springer International Publishing Switzerland 2017
E.B. Geer (ed.), *The Hypothalamic-Pituitary-Adrenal Axis in Health and Disease*, DOI 10.1007/978-3-319-45950-9_10

are found in 4–7 % of cross-sectional studies, and their prevalence increases with age [1, 2]. MACE is uniformly the most frequently reported hormonal abnormality in AI, but the incidence varies, depending on the diagnostic criteria used [3–5]. Overall, MACE has been estimated to affect 0.2–2 % of adults [6, 7]. Along with the mounting frequency of MACE diagnosis, a series of clinical dilemmas have emerged, most of which are interdependent. Debates start with the terminology and definition of this elusive entity, which aim to accurately reflect its clinical implications, and from where, in turn, appropriate management derives. This overview intends to underline the most up-to-date understanding of MACE and to point out aspects that need further clarification by properly designed research.

Definition and Terminology

Mild adrenal or subclinical hypercortisolism is generally defined as autonomous glucocorticoid secretion from an adrenal mass and absence of clinically overt signs and/or symptoms of Cushing syndrome. The terminology used to describe this entity has evolved over time. An initial term used was "preclinical Cushing syndrome." This term was quickly abandoned once longitudinal observational studies demonstrated that progression towards overt hypercortisolism is rather rare [5, 8, 9]. "Subclinical Cushing syndrome" or hypercortisolism has been the most widely used by both clinicians and investigators. While stigmata of Cushing syndrome—such as purple striae, plethora, easy bruising, and proximal muscle weakness—are absent or very subtle, nonspecific comorbidities associated with cortisol excess—including glucose intolerance, hypertension, bone loss, central obesity, and even increased mortality—have been linked with MACE (Box 1) [10, 11], thus further questioning if "subclinical" is an accurate nomenclature. To further complicate matters, diabetes, hypertension, and obesity are common in western populations, and their prevalence increases with age, as does that of AI. Nonetheless, as it will be later detailed in this chapter, recent studies have built strong arguments for a direct impact of even subtle cortisol excess on bone health, cardiometabolic risk factors and related events [9, 12–15]. Adrenal mild hypercortisolism [16] or MACE are terms that avoid the connotation of low clinical relevance and will be used in this chapter, although interchangeably with older terminology.

Box 1: Clinical Implications of Mild Adrenal Cortisol Excess
Dyslipidemia
    Increased fasting glucose and insulin
    Increased visceral adiposity
    Increased waist/hip ratio
    Osteoporosis/fragility fractures
    Increased cardiovascular events
    Increased mortality

Assessment of Dysregulated Cortisol Synthesis

Hormonal Testing

What constitutes adequate hormonal evidence for alterations in the hypothalamic–pituitary–adrenal (HPA) axis has been a subject of debate amongst endocrinologists, and no gold standard for MACE exists. In order to establish adrenal autonomy in cortisol production, clinicians and investigators might use one or multiple tests, in different combinations and with variable cutoffs. Another factor to take into account is that multiple drugs and conditions, including some that can result from hypercortisolism (such as type 2 diabetes mellitus and obesity), can lead to activation of the HPA axis and yield false positive results. All published clinical practice guidelines recommend 1 mg overnight dexamethasone suppression test (DST) for screening of MACE [2, 17–20], but the cutoff defining MACE remains variable. Serum cortisol concentration >1.8 µg/dL (50 nmol/L) after dexamethasone confers a higher sensitivity, while cutoffs >5 µg/dL (138 nmol/L) increase the specificity of the test [21] but encroach upon criteria for overt Cushing syndrome. Some investigators have used an intermediate cutoff of >3 µg/dL (83 nmol/L), or stratified hypercortisolism, by adding an intermediate group (1.8–5 µg/dL). So far, most clinical laboratories have used immunoassays to measure cortisol; the existing cutoffs might experience further transformations in the years to come, particularly with the emergence of liquid chromatography-tandem mass spectrometry (LC-MS/MS), which improves the performance of steroid assays.

Additional proposed tools for diagnosis of MACE include ACTH <10 pg/mL, low dehydroepiandrosterone sulfate (DHEAS) for age, elevated late-night salivary or serum cortisol, and elevated 24-h urinary free cortisol (Box 2), but none of these tests can be used in isolation. As evidenced in a recent systematic review, adrenal insufficiency after surgical resection was more common in patients in whom more indicators of HPA dysregulation were documented [22]. The time to achieve eucortisolemia was shorter in MACE than in overt Cushing syndrome patients (6.5 vs. 11.2 months). Taken together, these data suggest that a continuum of HPA axis disturbances exists.

Box 2: Tests Suggestive of Mild Autonomous Adrenal Cortisol Excess
AM cortisol after 1 mg dexamethasone

- >1.8 µg/dL — Sensitivity 71–100 %, Specificity 24–91 %
- >3 µg/dL — Sensitivity 52–86 %, Specificity 75–96 %
- >5 µg/dL — Sensitivity 22–91 %, Specificity 83–100 %

Suppressed ACTH (<10 pg/mL)
Suppressed DHEAS for age
Increased late-night salivary or serum cortisol
Increased 24-h urinary free cortisol

Partial ACTH suppression is found in many patients with MACE; however, not only is ACTH suppression inconsistent in this group of patients, but ACTH is sometimes normal even in overt adrenal hypercortisolism [23], thus limiting its utility. Peak ACTH values below 30 pg/mL (6.6 pmol/L) after CRH stimulation have been proposed as an additional tool to reveal subtle pituitary suppression by autonomous cortisol production, but CRH testing is rarely helpful [6, 23–26].

Early alterations in the cortisol circadian rhythm might be present in patients with MACE [22, 27, 28]. Late-night serum cortisol has been proposed as the best compromise between sensitivity (64 %) and specificity (81 %) for predicting adrenal insufficiency after adrenalectomy in patients with MACE [22]. However, late-night serum cortisol testing is not usually feasible for ambulatory patients and is physiologically elevated when these patients are hospitalized for unrelated medical problems. While more feasible, midnight salivary cortisol has poor sensitivity for detecting patients with MACE, even when measured by LC-MS/MS [29–33]. Similarly, 24-h urinary free cortisol is only rarely elevated in these patients. Urine cortisol excretion above 70 μg/24 h (193.1 nmol/L) by immunoassays has been used to diagnose MACE, but only in conjunction with other parameters of adrenal autonomy [34–40]. Data from a recent study of patients with AI using multiplex mass spectrometry suggest that urinary cortisol metabolites might become abnormal before cortisol does [41], explaining the poor sensitivity of urine testing. LC-MS/MS assays for urine cortisol are not likely to improve sensitivity [42].

Both dehydroepiandrosterone (DHEA) and DHEAS are regarded as ACTH-dependent hormones [43]. Therefore, some investigators have proposed DHEAS suppression as a useful indicator of HPA dysregulation. Immunohistochemical studies in patients with cortisol-producing adenomas showed suppression of sulfotransferase 2A1 (SULT2A1) expression in the adjacent zona reticularis tissue [44]. Although suppressed DHEAS as an indicator of subclinical hypercortisolism was first proposed a decade ago, supporting data have remained inconsistent [44–46]. In a Japanese study of AI, only 27 % of patients with serum cortisol ≥1.8 μg/dL after dexamethasone had low serum DHEAS, as assessed by immunoassay [46]. Conversely, another group, which defined MACE more stringently with at least two of three criteria: serum cortisol after dexamethasone >3 μg/dL, urinary free cortisol >70 μg/24 h, and ACTH <10 pg/mL, found that DHEAS was significantly lower in patients with subclinical hypercortisolism (27.95 μg/dL, $n=38$) compared to nonfunctioning AI (65.90 μg/dL, $n=141$) [47]. In a recent cross-sectional study, Di Dalmazi and colleagues used LC-MS/MS to measure a panel of steroids in 28 patients with MACE, 66 patients with nonsecretory adrenal adenomas, and 188 age- and sex-matched controls [48]. Patients with MACE had lower DHEA and androstenedione than those with non-secreting adenomas and controls, both at baseline and after cosyntropin stimulation. The advent of gas chromatography (GC)- and LC-MS/MS will help characterize the hormonal signature of both MACE and nonsecretory adenomas in greater detail. More importantly, the interplay between secreted compounds and their activation of gluco- and mineralocorticoid receptors to yield the resultant clinical outcomes has not been carefully studied.

Imaging

Autonomous cortisol synthesis from the adrenal typically correlates with the size of the nodules. In a multicenter longitudinal Italian study of 206 patients with AI followed for a median of six years, an adenoma size >2.4 cm predicted conversion to subclinical hypercortisolism with a sensitivity of 73.3 % and a specificity of 60.5 % [15]. Another imaging finding suggestive of autonomous adrenal cortisol excess is atrophy of the contralateral adrenal gland (Fig. 1).

Some investigators explored the idea of assessing the adrenal function by tracking the incorporation of radiolabeled cholesterol derivatives within the gland. Using this principle, scintigraphic uptake exclusively to an adrenal adenoma indicates autonomous cortisol production, while symmetrical incorporation of the tracer supports an ACTH-responsive cortisol synthesis. As an example, Valli and colleagues used [^{131}I]-6β-iodomethyl norcholesterol scintigraphy (IMS) in 31 patients with benign cortical adenomas and found that the sensitivity and specificity of the test in detecting MACE was 58 % and 83 %, respectively, if referenced to a dexamethasone-suppressed cortisol of 5 µg/dL (138 nmol/L) and 100 % and 67 %, respectively, for a dexamethasone-suppressed cortisol of 2.2 µg/dL (60 nmol/L) [36]. Barzon and colleagues obtained similar results with [^{75}Se]-selenio-6α-methyl-19-norcholesterol [49]. These studies, however, are limited by burdensome protocols, scarce availability of the tracers, and high cost. Furthermore, because in contrast to primary aldosteronism, adrenal cortisol excess correlates closely with the size of the adenomas, there is little value of the scintigraphic studies over routine hormonal tests and cross-sectional imaging [50].

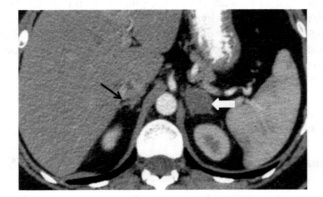

Fig. 1 Left adrenal incidentaloma with subtle autonomous cortisol secretion in a 37-year-old woman (white arrow); the right adrenal gland is partially atrophied (black arrow). Over 1 year, the cortisol after 1 mg dexamethasone rose from 1.4 to 2.1 µg/dL, the AM ACTH fell from 14 to 2 pg/mL, and the DHEAS fell from 126 to 41 µg/dL. Improvement in weight and blood pressure was noted after laparoscopic adrenalectomy and several weeks of partial cortisol deficiency

Pathogenesis of Dysregulated Adrenal Cortisol Synthesis

The molecular pathogenesis of adrenal Cushing syndrome is covered in detail in another chapter. To summarize, Assie and colleagues identified inactivating mutations of armadillo repeat containing 5 (ARMC5) in 18 of 33 patients with macronodular adrenal hyperplasia and hypercortisolism, 3 of which presented with MACE [51]. ARMC5 mutations have been found in patients with both familial and sporadic macronodular adrenal hyperplasia with a range of hypercortisolism [52–54]. Beuschlein and colleagues found somatic mutations in PRKACA, which encodes the main catalytic subunit of protein kinase A (PKA), in cortisol-producing adenomas associated with overt Cushing syndrome but not in 40 patients with MACE [55]. Germline PRKACA duplications were identified in 14% of patients with Cushing syndrome due to bilateral adrenal hyperplasia, and three other groups reported PRKACA mutations in 35–69% of cortisol-producing adrenal adenomas [56–58], primarily mutation L206R, which leads to constitutive PKA activation [57, 58]. Of all the patients with PRKACA mutations across these four studies, only four patients had MACE [57, 58]. Thus, ARMC5 and PRKACA mutations are found in some patients with MACE but more often in those with overt hypercortisolism.

Beyond the genetic and epigenetic aspects contributing to excessive ACTH-independent adrenal cortisol synthesis, several additional factors modulate the effects of excessive glucocorticoids in the target tissues. These include the cortisol binding globulins, tissue-specific glucocorticoid activating and inactivating enzymes, and glucocorticoid receptor (NR3C1) polymorphisms. Recent studies have identified polymorphisms in the 11β-hydroxysteroid dehydrogenase type 1 (11βHSD1) and glucocorticoid receptor genes that are protective against a Cushingoid phenotype, including cognitive impairment [59] and diabetes [60]. 11βHSD1 knockout mice with circulating glucocorticoid excess were protected from the glucose intolerance, hyperinsulinemia, hepatic steatosis, hypertension, myopathy, and dermal atrophy of Cushing syndrome [61]. The intricate interplay between various factors that constitute the basis of a specific phenotype remains elusive and deserves to be dissected further.

Clinical Consequences of MACE

Cardiometabolic Profile and Related Outcomes in MACE

Research conducted over a decade ago found that surrogates of cardiovascular risk, including blood pressure, fasting glucose, homeostatic assessment model-insulin resistance (HOMA-IR) index, lipoproteins and triglycerides, fibrinogen, waist-to-hip ratio, and mean carotid artery intima-media thickness, were significantly worse in patients with MACE than in age-, sex-, and body mass index (BMI)-matched controls [62]. More recent studies have linked mild hypercortisolism with cardiometabolic morbidity and mortality. In a first large cross-sectional study, Di Dalmazi and colleagues stratified patients with AI and hypercortisolism in an intermediate group, with a cortisol after dexamethasone between 1.8 and 5 µg/dL, or >5 µg/dL, respectively. They found that the prevalence of type 2 diabetes mellitus and coronary heart disease increased in parallel with progressively higher degrees of hypercortisolism, as compared with patients with nonfunctioning adrenal adenomas [13]. The same group longitudinally followed a cohort of 198 patients with AI (mean follow-up, 7.5±3.2 years), and they found that the incidence of cardiovascular events and related mortality was higher in patients with subclinical hypercortisolism (cortisol >1.8 µg/dL) [9]. Worsening hypercortisolism during follow-up was independently associated with cardiovascular events and mortality.

Another Italian multicenter study retrospectively analyzed the outcomes of 206 patients with AI followed for a median of 6 years. Of these, 11.6% patients were classified to have subclinical hypercortisolism, based on a cortisol after dexamethasone >5 µg/dL, or at least two other indicators of altered HPA axis (low ACTH, increased urinary free cortisol, and cortisol >3 µg/dL after dexamethasone) [15]. Subclinical hypercortisolism was associated with a higher incidence of cardiovascular events and worsening of at least two metabolic parameters (weight, glycemic, lipid, and blood pressure control), independent of age. Debono and colleagues retrospectively studied a similar size cohort of patients with AI followed for 4.2±2.3 years in the UK [12]. During the time interval studied, 18/206 patients died, and of these, 17 patients had a cortisol >1.8 µg/dL after dexamethasone. Mortality was higher in patients with a cortisol after dexamethasone >5 µg/dL vs. 1.8–5 µg/dL, and half of the deaths were attributed to cardiovascular causes. The mean time to death was 3.2 years, and the age of death was lower than the life expectancy at birth for the general population in the same area. Taken together, these studies strongly suggest that chronic hypercortisolism is a direct contributor to cardiovascular events and related mortality even when subtle, and that the impact directly increases with the degree of hypercortisolism.

Metabolic Bone Disease

The deleterious effects on bone metabolism of overt glucocorticoid excess, both endogenous and exogenous, have been widely documented [63]. Evidence that mild hypercortisolism leads to osteoporosis and fragility fractures emerged predominantly from Italian cohorts [13, 14, 34, 64–67]. In a cross-sectional study of 219 patients evaluated for osteoporosis without any known secondary causes, subclinical hypercortisolism was present in 5% of patients and in 10% of the subset who also had vertebral fractures [64]. Similarly, a 2-year longitudinal study of 103 consecutive patients with AI documented a higher incidence of vertebral fractures in patients with MACE [14]. Patients with MACE experienced worsening of their spinal deformity index, independent of age, gender, BMI, bone mineral density, baseline spinal deformity index (SDI), and menopause duration. In another cohort including 287 patients with AI, both bone mineral density and bone quality, as measured by the SDI, were significantly worse in patients with MACE [34]. The trabecular bone score, another index of bone quality, was found to be worse amongst patients with AI who had MACE, and this parameter was proposed to be a useful predictor of fractures [68].

Pathophysiology of MACE

Pathogenic Mechanisms of MACE on Cardiovascular and Glucose Metabolism

While solid evidence exists to explain the deleterious effects of overt cortisol excess, the pathogenic mechanisms derived from mild chronic hypercortisolism remain speculative. The most commonly entertained hypothesis is that even subtle cortisol excess over time has cumulative effects, leading to clinical consequences similar to overt Cushing syndrome, but at a smaller scale. This hypothesis is supported by evidence of an incremental effect of cortisol on cardiovascular events and mortality [9, 12, 13]. In addition, an increased vascular mortality rate has been observed in patients with primary adrenal insufficiency on various glucocorticoid replacement regimens and has been attributed to chronic overtreatment [69]. Metabolic components associated with an increased cardiovascular risk, such as high blood pressure, fasting glucose and insulin, cholesterol, fibrinogen and waist to hip ration, are common to mild and overt hypercortisolism [62]. Cortisol-induced visceral adiposity might explain both the increased insulin resistance and cardiovascular risk in these patients. A retrospective study of 125 patients with AI conducted by Debono and team found that patients with a cortisol >1.8 µg/dL following dexamethasone had significantly more visceral fat than those with nonsecretory adenomas [70]. Furthermore, there was no difference in visceral fat between patients with subclinical and overt hypercortisolism, although only nine women in the latter group were included.

Beyond patients with MACE, data have emerged to support that even apparently nonfunctioning adrenal adenomas are associated with increased cardiovascular risk. In 2009, Yener and colleagues proposed that the increased carotid intima-media thickness was a consequence of insulin resistance associated with subtle cortisol autonomy [71]. The same group later suggested that impaired arterial flow-mediated dilatation and elevated IL-18 might underlie the endothelial alterations in patients with adrenal adenomas and early cortisol autonomy [72]. To eliminate the confounding effects of comorbidities associated with increased cardiovascular risk frequently present in hypercortisolism, Androulakis and colleagues studied a group of 60 normotensive and normoglycemic patients with apparently nonfunctioning AI and 32 healthy controls [73]. Besides absence of clinically overt Cushing syndrome, patients were enrolled if they had normal early morning basal serum ACTH and cortisol levels, preservation of ACTH and cortisol circadian rhythm, and normal 24-h urinary free cortisol excretion. Subsequently, a group of 26 patients was classified as cortisol-secreting, based on low-dose 2-day DST greater than 1.09 µg/dL, cutoff derived from the mean +2 SD values of the control group. The authors found that carotid intima-media thickness measurements were higher and that flow-mediated dilatation was lower in the cortisol-secreting group compared with both nonfunctioning and control groups. In addition, they found that intima-media thickness correlated with cortisol, urinary free cortisol, and cortisol after dexamethasone. The authors concluded that this disproportionate cortisol secretion might potentially lead to microvasculature damage [73].

Another hypothesis to explain some of the cardiovascular profile and outcomes in patients with MACE is that before the cortisol excess becomes apparent, other alterations in steroidogenesis and HPA axis might occur. In support of this hypothesis stand two lines of evidence: (1) cardiovascular risk factors appear to be increased in patients with so-called "nonfunctioning" adrenal adenomas [71, 73–77], terminology that only excludes cortisol, aldosterone, and catecholamines excess; (2) cortisol secretion typically becomes apparent in large adenomas [15], suggesting that intrinsic enzymatic alterations in the steroid biosynthesis within the tumor might lead to an atypical steroid profile prior to the development of clinical manifestations. Using LC-MS/MS, Di Dalmazi and colleagues have recently measured ten steroids, both at baseline and after cosyntropin stimulation, in patients with adrenal adenomas (66 nonfunctional and 28 subclinical hypercortisolism) and in 188 age- and sex-matched controls [48]. Basal and cosyntropin-stimulated DHEA, androstenedione and, in women, basal testosterone concentrations were lower in patients with MACE than in those with non-secreting adenomas and controls. Increased cortisol and reduced DHEA levels were independently associated with increased waist circumference. Cortisol, but not androstenedione, was independently associated with increased number of cardiovascular risk factors in patients with MACE. Patients with MACE also demonstrated increased production of 21-deoxycortisol and the mineralocorticoid 11-deoxycorticosterone after cosyntropin stimulation. In addition, the ratio between 17α-hydroxyprogesterone and androstenedione was higher in the MACE than in nonfunctioning adenomas group, suggesting alterations in P450c17 and P450c21 activities. A second hypothesis postulated by the authors was that the cortisol excess secreted from an adenoma suppresses ACTH, and this in turn leads to decreased adrenal androgen synthesis from the remaining adrenal tissue. However, although a positive correlation with ACTH was noted for both DHEA and androstenedione, DHEA was also reduced in patients with nonfunctioning adenomas, despite normal ACTH levels. Further studies to assess the common and unusual steroids synthesized both in vivo and in vitro are needed as an initial step; subsequently, it would be important to establish the function of steroids other than cortisol and their links with clinical outcomes.

Pathogenic Mechanisms of MACE on Bone Metabolism

Cortisol alters bone metabolism by decreasing bone formation and increasing bone resorption [78, 79]. The magnitude at which cortisol excess starts to affect bone metabolism and the relationship between time and degree of cortisol excess remain unclear. Tauchmanova and colleagues assessed the bone density and vertebral fractures in 71 consecutive women with either overt ($n = 36$) or subclinical ($n = 35$) hypercortisolism and corresponding controls [65]. Interestingly, bone mineral density and prevalence of any vertebral fractures did not differ between women with overt and subclinical hypercortisolism, defined by a cortisol after dexamethasone >3 µg/dL. Di Dalmazi et al. found that osteoporosis was independently associated with subclinical hypercortisolism, as defined by a cortisol after dexamethasone >5 µg/dL [13], while an intermediate group, with a cortisol after dexamethasone of 1.8–5 µg/dL, was no different than the non-AI group.

The interrelation between sex steroids and cortisol on bone metabolism has been explored in both men and women [65, 67]. In women with MACE, eugonadism was partially protective, but this effect was lost in patients with overt Cushing syndrome [65]. MACE was associated with low bone mineral density and high prevalence of vertebral fractures in eugonadal men [67]; however, a direct comparison with hypogonadal men has not been done. A more intriguing aspect is the role of DHEAS in bone health. Beyond cortisol itself, Tauchmanova et al. found that the cortisol/DHEAS ratio was a predictor of fractures in all patients [65], but to what degree this association is reflective of the cortisol excess alone remains unclear. Other authors have suggested a benefit of DHEAS on bone density. Studies investigating the association between DHEAS and bone mineral density in postmenopausal women have found conflicting results [80–84]. In placebo-controlled studies of DHEA administration in elderly men and women, DHEA was found to have a positive effect on bone mineral density only in women in one study [85] and in both sexes in another [86].

Management of MACE

So far, the evidence to guide appropriate treatment of subclinical hypercortisolism has been modest. The few studies that have looked at management of MACE have used different diagnostic criteria and have enrolled small numbers of patients. In addition, most studies are retrospective and prone to selection bias. As an example, surgery could have been more frequently offered to patients with subtle comorbidities typically associated to hypercortisolism. One prospective study randomized 45 patients with MACE to either laparoscopic surgery ($n = 23$) or observation ($n = 24$) and followed them for a mean of 7.7 years [87]. In the surgical group, diabetes mellitus, hypertension, hyperlipidemia, and obesity normalized or improved in 62.5 %, 67 %, 37.5 %, and 50 %, respectively. In contrast, some worsening of diabetes, hypertension, and hyperlipidemia was noted in conservatively managed patients. Similar outcomes were reported by smaller retrospective studies [88–93]. Surgery has also been proposed for patients with MACE and bilateral adrenal nodules, by selectively removing the gland with the largest nodules. In a retrospective study of 33 patients with bilateral AI and MACE followed for up to 4.5 years, Perogamvros and colleagues found that markers of HPA axis dysregulation were significantly improved in the 14 patients who underwent unilateral adrenalectomy [94].

In addition, comorbidities associated with hypercortisolism, such as hypertension, impaired glucose tolerance or diabetes mellitus, dyslipidemia, and osteoporosis, improved in the surgical group, while no changes were noted in the observational group [94]. A recent systematic review of outcomes of adrenalectomy for MACE concluded that, compared with conservative management, surgery cured or improved blood pressure, glucometabolic control, and obesity in 72, 46, and 39 % of patients, respectively [95]. The main limitations to this analysis were the heterogeneity of diagnosis and outcomes followed and the retrospective nature of all but one of the studies included. Furthermore, the interventions in the nonsurgical groups were often poorly defined, and no studies have evaluated the outcomes of MACE in patients with intensive comorbidity-specific medical therapy.

Medical treatment for MACE has not been much assessed. One open-label pilot study observed a reduction in insulin AUC in 4/6 patients with MACE treated with the glucocorticoid receptor antagonist mifepristone for 4 weeks [96]. Other strategies to decrease cortisol synthesis have been tried in adrenal tumors with aberrant receptor expression, such as glucose-dependent insulinotropic peptide, catecholamine, serotonin, vasopressin, angiotensin II, leptin, and luteinizing hormone/chorionic gonadotropin receptors [97–102]. Examples of such successful therapies include somatostatin analogs [99, 100], propranolol [101], and leuprolide acetate [102]. Inhibitors of cortisol synthetic enzymes, such as ketoconazole, metyrapone, or LCI699 (osilodrostat), have not yet been formally studied in MACE. While medical treatment for overt Cushing syndrome is reserved for inoperable cases, emerging medical therapies with a favorable safety profile might offer a safe and effective alternative to surgery in MACE, especially if low doses could successfully inhibit the hormone synthesis in these already inefficient adenomas.

Conclusion

It is clear that cortisol excess spans a spectrum of severities and, not surprisingly, establishing rigid lines to define clinically important disease becomes an unrealistic task. Solid evidence has emerged that even mild hypercortisolism has important clinical consequences, including deleterious effects on cardiovascular risk, glucose, lipid and bone metabolism, and even survival. However, numerous aspects remain to be clarified in order to best guide clinical practice for MACE. Beyond the mere association of MACE and unfavorable outcomes, the responsible mechanisms remain speculative. Do other steroid precursors produced by apparently nonfunctioning and/or cortisol-producing adrenal adenomas have direct clinical impact, either by activating nuclear hormone receptors or by different mechanisms? Is it reasonable to conclude that surgery should be offered to all patients with MACE? If not, how should we best follow these patients and when should we recommend treatment? A future research agenda aiming to answer some of these questions should include prospective studies of large cohorts, to characterize detailed steroid profiles and autonomy in these patients and to assess clinical outcomes in three distinctive arms: surgical treatment, steroid synthesis or action blockade, and intensive comorbidity-specific interventions. Until then, defining clinically important hypercortisolism and appropriate management remain rather arbitrary, and decisions must be individualized empirically.

References

1. Bovio S, Cataldi A, Reimondo G, Sperone P, Novello S, Berruti A, et al. Prevalence of adrenal incidentaloma in a contemporary computerized tomography series. J Endocrinol Invest. 2006;29(4):298–302.
2. Terzolo M, Stigliano A, Chiodini I, Loli P, Furlani L, Arnaldi G, et al. AME position statement on adrenal incidentaloma. Eur J Endocrinol. 2011;164(6):851–70.
3. Mantero F, Terzolo M, Arnaldi G, Osella G, Masini AM, Ali A, et al. A survey on adrenal incidentaloma in Italy. Study Group on Adrenal Tumors of the Italian Society of Endocrinology. J Clin Endocrinol Metab. 2000;85(2):637–44.
4. Kim J, Bae KH, Choi YK, Jeong JY, Park KG, Kim JG, et al. Clinical characteristics for 348 patients with adrenal incidentaloma. Endocrinol Metab (Seoul). 2013;28(1):20–5.
5. Barzon L, Sonino N, Fallo F, Palu G, Boscaro M. Prevalence and natural history of adrenal incidentalomas. Eur J Endocrinol. 2003;149(4):273–85.
6. Chiodini I. Clinical review: Diagnosis and treatment of subclinical hypercortisolism. J Clin Endocrinol Metab. 2011;96(5):1223–36.
7. Reincke M. Subclinical Cushing's syndrome. Endocrinol Metab Clin North Am. 2000;29(1):43–56.
8. Barzon L, Scaroni C, Sonino N, Fallo F, Paoletta A, Boscaro M. Risk factors and long-term follow-up of adrenal incidentalomas. J Clin Endocrinol Metab. 1999;84(2):520–6.
9. Di Dalmazi G, Vicennati V, Garelli S, Casadio E, Rinaldi E, Giampalma E, et al. Cardiovascular events and mortality in patients with adrenal incidentalomas that are either non-secreting or associated with intermediate phenotype or subclinical Cushing's syndrome: a 15-year retrospective study. Lancet Diabetes Endocrinol. 2014;2(5):396–405.
10. Di Dalmazi G, Pasquali R. Adrenal adenomas, subclinical hypercortisolism, and cardiovascular outcomes. Curr Opin Endocrinol Diabetes Obes. 2015;22(3):163–8.
11. Di Dalmazi G, Pasquali R, Beuschlein F, Reincke M. Subclinical hypercortisolism: a state, a syndrome, or a disease? Eur J Endocrinol. 2015;173(4):M61–71.
12. Debono M, Bradburn M, Bull M, Harrison B, Ross RJ, Newell-Price J. Cortisol as a marker for increased mortality in patients with incidental adrenocortical adenomas. J Clin Endocrinol Metab. 2014;99(12):4462–70.
13. Di Dalmazi G, Vicennati V, Rinaldi E, Morselli-Labate AM, Giampalma E, Mosconi C, et al. Progressively increased patterns of subclinical cortisol hypersecretion in adrenal incidentalomas differently predict major metabolic and cardiovascular outcomes: a large cross-sectional study. Eur J Endocrinol. 2012;166(4):669–77.
14. Morelli V, Eller-Vainicher C, Salcuni AS, Coletti F, Iorio L, Muscogiuri G, et al. Risk of new vertebral fractures in patients with adrenal incidentaloma with and without subclinical hypercortisolism: a multicenter longitudinal study. J Bone Miner Res. 2011;26(8):1816–21.
15. Morelli V, Reimondo G, Giordano R, Della Casa S, Policola C, Palmieri S, et al. Long-term follow-up in adrenal incidentalomas: an Italian multicenter study. J Clin Endocrinol Metab. 2014;99(3):827–34.
16. Goddard GM, Ravikumar A, Levine AC. Adrenal mild hypercortisolism. Endocrinol Metab Clin North Am. 2015;44(2):371–9.
17. Nieman LK, Biller BM, Findling JW, Newell-Price J, Savage MO, Stewart PM, et al. The diagnosis of Cushing's syndrome: an Endocrine Society Clinical Practice Guideline. J Clin Endocrinol Metab. 2008;93(5):1526–40.

18. Tabarin A, Bardet S, Bertherat J, Dupas B, Chabre O, Hamoir E, et al. Exploration and management of adrenal incidentalomas. French Society of Endocrinology Consensus. Ann Endocrinol. 2008;69(6):487–500.

19. Zeiger MA, Thompson GB, Duh QY, Hamrahian AH, Angelos P, Elaraj D, et al. The American Association of Clinical Endocrinologists and American Association of Endocrine Surgeons medical guidelines for the management of adrenal incidentalomas. Endocr Pract. 2009;15 Suppl 1:1–20.

20. NIH state-of-the-science statement on management of the clinically inapparent adrenal mass ("incidentaloma"). NIH Consens State Sci Statements. 2002;19(2):1–25.

21. Ceccato F, Barbot M, Zilio M, Frigo AC, Albiger N, Camozzi V, et al. Screening tests for Cushing's syndrome: urinary free cortisol role measured by LC-MS/MS. J Clin Endocrinol Metab. 2015;100(10):3856–61.

22. Di Dalmazi G, Berr CM, Fassnacht M, Beuschlein F, Reincke M. Adrenal function after adrenalectomy for subclinical hypercortisolism and Cushing's syndrome: a systematic review of the literature. J Clin Endocrinol Metab. 2014;99(8):2637–45.

23. Hong AR, Kim JH, Hong ES, Kim IK, Park KS, Ahn CH, et al. Limited diagnostic utility of plasma adrenocorticotropic hormone for differentiation between adrenal Cushing syndrome and Cushing disease. Endocrinol Metab (Seoul). 2015;30(3):297–304.

24. Reincke M, Nieke J, Krestin GP, Saeger W, Allolio B, Winkelmann W. Preclinical Cushing's syndrome in adrenal "incidentalomas": comparison with adrenal Cushing's syndrome. J Clin Endocrinol Metab. 1992;75(3):826–32.

25. Osella G, Terzolo M, Borretta G, Magro G, Ali A, Piovesan A, et al. Endocrine evaluation of incidentally discovered adrenal masses (incidentalomas). J Clin Endocrinol Metab. 1994;79(6):1532–9.

26. Ambrosi B, Peverelli S, Passini E, Re T, Ferrario R, Colombo P, et al. Abnormalities of endocrine function in patients with clinically "silent" adrenal masses. Eur J Endocrinol. 1995;132(4):422–8.

27. Terzolo M, Bovio S, Pia A, Conton PA, Reimondo G, Dall'Asta C, et al. Midnight serum cortisol as a marker of increased cardiovascular risk in patients with a clinically inapparent adrenal adenoma. Eur J Endocrinol. 2005;153(2):307–15.

28. Tanabe A, Naruse M, Nishikawa T, Yoshimoto T, Shimizu T, Seki T, et al. Autonomy of cortisol secretion in clinically silent adrenal incidentaloma. Horm Metab Res. 2001; 33(7):444–50.

29. Sereg M, Toke J, Patocs A, Varga I, Igaz P, Szucs N, et al. Diagnostic performance of salivary cortisol and serum osteocalcin measurements in patients with overt and subclinical Cushing's syndrome. Steroids. 2011;76(1-2):38–42.

30. Kidambi S, Raff H, Findling JW. Limitations of nocturnal salivary cortisol and urine free cortisol in the diagnosis of mild Cushing's syndrome. Eur J Endocrinol. 2007;157(6): 725–31.

31. Nunes ML, Vattaut S, Corcuff JB, Rault A, Loiseau H, Gatta B, et al. Late-night salivary cortisol for diagnosis of overt and subclinical Cushing's syndrome in hospitalized and ambulatory patients. J Clin Endocrinol Metab. 2009;94(2):456–62.

32. Masserini B, Morelli V, Bergamaschi S, Ermetici F, Eller-Vainicher C, Barbieri AM, et al. The limited role of midnight salivary cortisol levels in the diagnosis of subclinical hypercortisolism in patients with adrenal incidentaloma. Eur J Endocrinol. 2009;160(1):87–92.

33. Palmieri S, Morelli V, Polledri E, Fustinoni S, Mercadante R, Olgiati L, et al. The role of salivary cortisol measured by liquid chromatography-tandem mass spectrometry in the diagnosis of subclinical hypercortisolism. Eur J Endocrinol. 2013;168(3):289–96.

34. Chiodini I, Morelli V, Masserini B, Salcuni AS, Eller-Vainicher C, Viti R, et al. Bone mineral density, prevalence of vertebral fractures, and bone quality in patients with adrenal incidentalomas with and without subclinical hypercortisolism: an Italian multicenter study. J Clin Endocrinol Metab. 2009;94(9):3207–14.

35. Torlontano M, Chiodini I, Pileri M, Guglielmi G, Cammisa M, Modoni S, et al. Altered bone mass and turnover in female patients with adrenal incidentaloma: the effect of subclinical hypercortisolism. J Clin Endocrinol Metab. 1999;84(7):2381–5.

36. Valli N, Catargi B, Ronci N, Vergnot V, Leccia F, Ferriere JM, et al. Biochemical screening for subclinical cortisol-secreting adenomas amongst adrenal incidentalomas. Eur J Endocrinol. 2001;144(4):401–8.
37. Chiodini I, Torlontano M, Carnevale V, Guglielmi G, Cammisa M, Trischitta V, et al. Bone loss rate in adrenal incidentalomas: a longitudinal study. J Clin Endocrinol Metab. 2001;86(11):5337–41.
38. Chiodini I, Tauchmanova L, Torlontano M, Battista C, Guglielmi G, Cammisa M, et al. Bone involvement in eugonadal male patients with adrenal incidentaloma and subclinical hypercortisolism. J Clin Endocrinol Metab. 2002;87(12):5491–4.
39. Eller-Vainicher C, Morelli V, Salcuni AS, Battista C, Torlontano M, Coletti F, et al. Accuracy of several parameters of hypothalamic-pituitary-adrenal axis activity in predicting before surgery the metabolic effects of the removal of an adrenal incidentaloma. Eur J Endocrinol. 2010;163(6):925–35.
40. Reimondo G, Allasino B, Coletta M, Pia A, Peraga G, Zaggia B, et al. Evaluation of midnight salivary cortisol as a predictor factor for common carotid arteries intima media thickness in patients with clinically inapparent adrenal adenomas. Int J Endocrinol. 2015;2015:674734.
41. Brossaud J, Ducint D, Corcuff JB. Urinary glucocorticoid metabolites: biomarkers to classify adrenal incidentalomas? Clin Endocrinol (Oxf). 2015 Jan 8. doi: 10.1111/cen.12717.
42. Raff H, Auchus RJ, Findling JW, Nieman LK. Urine free cortisol in the diagnosis of Cushing's syndrome: is it worth doing, and if so, how? J Clin Endocrinol Metab. 2015;100(2):395–7.
43. Parker LN. Control of adrenal androgen secretion. Endocrinol Metab Clin North Am. 1991;20(2):401–21.
44. Morio H, Terano T, Yamamoto K, Tomizuka T, Oeda T, Saito Y, et al. Serum levels of dehydroepiandrosterone sulfate in patients with asymptomatic cortisol producing adrenal adenoma: comparison with adrenal Cushing's syndrome and non-functional adrenal tumor. Endocr J. 1996;43(4):387–96.
45. Bencsik Z, Szabolcs I, Kovacs Z, Ferencz A, Voros A, Kaszas I, et al. Low dehydroepiandrosterone sulfate (DHEA-S) level is not a good predictor of hormonal activity in nonselected patients with incidentally detected adrenal tumors. J Clin Endocrinol Metab. 1996;81(5):1726–9.
46. Akehi Y, Kawate H, Murase K, Nagaishi R, Nomiyama T, Nomura M, et al. Proposed diagnostic criteria for subclinical Cushing's syndrome associated with adrenal incidentaloma. Endocr J. 2013;60(7):903–12.
47. Yener S, Yilmaz H, Demir T, Secil M, Comlekci A. DHEAS for the prediction of subclinical Cushing's syndrome: perplexing or advantageous? Endocrine. 2015;48(2):669–76.
48. Di Dalmazi G, Fanelli F, Mezzullo M, Casadio E, Rinaldi E, Garelli S, et al. Steroid profiling by LC-MS/MS in nonsecreting and subclinical cortisol-secreting adrenocortical adenomas. J Clin Endocrinol Metab. 2015;100(9):3529–38.
49. Barzon L, Fallo F, Sonino N, Boscaro M. Overnight dexamethasone suppression of cortisol is associated with radiocholesterol uptake patterns in adrenal incidentalomas. Eur J Endocrinol. 2001;145(2):223–4.
50. Fagour C, Bardet S, Rohmer V, Arimone Y, Lecomte P, Valli N, et al. Usefulness of adrenal scintigraphy in the follow-up of adrenocortical incidentalomas: a prospective multicenter study. Eur J Endocrinol. 2009;160(2):257–64.
51. Assie G, Libe R, Espiard S, Rizk-Rabin M, Guimier A, Luscap W, et al. ARMC5 mutations in macronodular adrenal hyperplasia with Cushing's syndrome. N Engl J Med. 2013;369(22):2105–14.
52. Elbelt U, Trovato A, Kloth M, Gentz E, Finke R, Spranger J, et al. Molecular and clinical evidence for an ARMC5 tumor syndrome: concurrent inactivating germline and somatic mutations are associated with both primary macronodular adrenal hyperplasia and meningioma. J Clin Endocrinol Metab. 2015;100(1):E119–28.
53. Suzuki S, Tatsuno I, Oohara E, Nakayama A, Komai E, Shiga A, et al. Germline deletion of Armc5 in familial primary macronodular adrenal hyperplasia. Endocr Pract. 2015; 21(10):1152–60.

54. Espiard S, Drougat L, Libe R, Assie G, Perlemoine K, Guignat L, et al. ARMC5 mutations in a large cohort of primary macronodular adrenal hyperplasia: clinical and functional consequences. J Clin Endocrinol Metab. 2015;100(6):E926–35.
55. Beuschlein F, Fassnacht M, Assie G, Calebiro D, Stratakis CA, Osswald A, et al. Constitutive activation of PKA catalytic subunit in adrenal Cushing's syndrome. N Engl J Med. 2014;370(11):1019–28.
56. Cao Y, He M, Gao Z, Peng Y, Li Y, Li L, et al. Activating hotspot L205R mutation in PRKACA and adrenal Cushing's syndrome. Science. 2014;344(6186):913–7.
57. Goh G, Scholl UI, Healy JM, Choi M, Prasad ML, Nelson-Williams C, et al. Recurrent activating mutation in PRKACA in cortisol-producing adrenal tumors. Nat Genet. 2014;46(6):613–7.
58. Sato Y, Maekawa S, Ishii R, Sanada M, Morikawa T, Shiraishi Y, et al. Recurrent somatic mutations underlie corticotropin-independent Cushing's syndrome. Science. 2014;344(6186):917–20.
59. Ragnarsson O, Glad CA, Berglund P, Bergthorsdottir R, Eder DN, Johannsson G. Common genetic variants in the glucocorticoid receptor and the 11β-hydroxysteroid dehydrogenase type 1 genes influence long-term cognitive impairments in patients with Cushing's syndrome in remission. J Clin Endocrinol Metab. 2014;99(9):E1803–7.
60. Trementino L, Appolloni G, Concettoni C, Cardinaletti M, Boscaro M, Arnaldi G. Association of glucocorticoid receptor polymorphism A3669G with decreased risk of developing diabetes in patients with Cushing's syndrome. Eur J Endocrinol. 2012;166(1):35–42.
61. Morgan SA, McCabe EL, Gathercole LL, Hassan-Smith ZK, Larner DP, Bujalska IJ, et al. 11β-HSD1 is the major regulator of the tissue-specific effects of circulating glucocorticoid excess. Proc Natl Acad Sci U S A. 2014;111(24):E2482–91.
62. Tauchmanova L, Rossi R, Biondi B, Pulcrano M, Nuzzo V, Palmieri EA, et al. Patients with subclinical Cushing's syndrome due to adrenal adenoma have increased cardiovascular risk. J Clin Endocrinol Metab. 2002;87(11):4872–8.
63. Toth M, Grossman A. Glucocorticoid-induced osteoporosis: lessons from Cushing's syndrome. Clin Endocrinol (Oxf). 2013;79(1):1–11.
64. Chiodini I, Mascia ML, Muscarella S, Battista C, Minisola S, Arosio M, et al. Subclinical hypercortisolism among outpatients referred for osteoporosis. Ann Intern Med. 2007;147(8):541–8.
65. Tauchmanova L, Pivonello R, De Martino MC, Rusciano A, De Leo M, Ruosi C, et al. Effects of sex steroids on bone in women with subclinical or overt endogenous hypercortisolism. Eur J Endocrinol. 2007;157(3):359–66.
66. Chiodini I, Guglielmi G, Battista C, Carnevale V, Torlontano M, Cammisa M, et al. Spinal volumetric bone mineral density and vertebral fractures in female patients with adrenal incidentalomas: the effects of subclinical hypercortisolism and gonadal status. J Clin Endocrinol Metab. 2004;89(5):2237–41.
67. Chiodini I, Viti R, Coletti F, Guglielmi G, Battista C, Ermetici F, et al. Eugonadal male patients with adrenal incidentalomas and subclinical hypercortisolism have increased rate of vertebral fractures. Clin Endocrinol (Oxf). 2009;70(2):208–13.
68. Eller-Vainicher C, Morelli V, Ulivieri FM, Palmieri S, Zhukouskaya VV, Cairoli E, et al. Bone quality, as measured by trabecular bone score in patients with adrenal incidentalomas with and without subclinical hypercortisolism. J Bone Miner Res. 2012;27(10):2223–30.
69. Johannsson G, Ragnarsson O. Cardiovascular and metabolic impact of glucocorticoid replacement therapy. Front Horm Res. 2014;43:33–44.
70. Debono M, Prema A, Hughes TJ, Bull M, Ross RJ, Newell-Price J. Visceral fat accumulation and postdexamethasone serum cortisol levels in patients with adrenal incidentaloma. J Clin Endocrinol Metab. 2013;98(6):2383–91.
71. Yener S, Genc S, Akinci B, Secil M, Demir T, Comlekci A, et al. Carotid intima media thickness is increased and associated with morning cortisol in subjects with non-functioning adrenal incidentaloma. Endocrine. 2009;35(3):365–70.

72. Yener S, Baris M, Secil M, Akinci B, Comlekci A, Yesil S. Is there an association between non-functioning adrenal adenoma and endothelial dysfunction? J Endocrinol Invest. 2011;34(4):265–70.
73. Androulakis II, Kaltsas GA, Kollias GE, Markou AC, Gouli AK, Thomas DA, et al. Patients with apparently nonfunctioning adrenal incidentalomas may be at increased cardiovascular risk due to excessive cortisol secretion. J Clin Endocrinol Metab. 2014;99(8):2754–62.
74. Erbil Y, Ozbey N, Barbaros U, Unalp HR, Salmaslioglu A, Ozarmagan S. Cardiovascular risk in patients with nonfunctional adrenal incidentaloma: myth or reality? World J Surg. 2009;33(10):2099–105.
75. Tuna MM, Imga NN, Dogan BA, Yilmaz FM, Topcuoglu C, Akbaba G, et al. Non-functioning adrenal incidentalomas are associated with higher hypertension prevalence and higher risk of atherosclerosis. J Endocrinol Invest. 2014;37(8):765–8.
76. Sereg M, Szappanos A, Toke J, Karlinger K, Feldman K, Kaszper E, et al. Atherosclerotic risk factors and complications in patients with non-functioning adrenal adenomas treated with or without adrenalectomy: a long-term follow-up study. Eur J Endocrinol. 2009;160(4):647–55.
77. Ermetici F, Dall'Asta C, Malavazos AE, Coman C, Morricone L, Montericcio V, et al. Echocardiographic alterations in patients with non-functioning adrenal incidentaloma. J Endocrinol Invest. 2008;31(6):573–7.
78. O'Brien CA, Jia D, Plotkin LI, Bellido T, Powers CC, Stewart SA, et al. Glucocorticoids act directly on osteoblasts and osteocytes to induce their apoptosis and reduce bone formation and strength. Endocrinology. 2004;145(4):1835–41.
79. Canalis E. Clinical review 83: Mechanisms of glucocorticoid action in bone: implications to glucocorticoid-induced osteoporosis. J Clin Endocrinol Metab. 1996;81(10):3441–7.
80. Ghebre MA, Hart DJ, Hakim AJ, Kato BS, Thompson V, Arden NK, et al. Association between DHEAS and bone loss in postmenopausal women: a 15-year longitudinal population-based study. Calcif Tissue Int. 2011;89(4):295–302.
81. Tok EC, Ertunc D, Oz U, Camdeviren H, Ozdemir G, Dilek S. The effect of circulating androgens on bone mineral density in postmenopausal women. Maturitas. 2004;48(3):235–42.
82. Szathmari M, Szucs J, Feher T, Hollo I. Dehydroepiandrosterone sulphate and bone mineral density. Osteoporos Int. 1994;4(2):84–8.
83. Zofkova I, Bahbouh R, Hill M. The pathophysiological implications of circulating androgens on bone mineral density in a normal female population. Steroids. 2000;65(12):857–61.
84. Guthrie JR, Lehert P, Dennerstein L, Burger HG, Ebeling PR, Wark JD. The relative effect of endogenous estradiol and androgens on menopausal bone loss: a longitudinal study. Osteoporos Int. 2004;15(11):881–6.
85. Baulieu EE, Thomas G, Legrain S, Lahlou N, Roger M, Debuire B, et al. Dehydroepiandrosterone (DHEA), DHEA sulfate, and aging: contribution of the DHEAge Study to a sociobiomedical issue. Proc Natl Acad Sci U S A. 2000;97(8):4279–84.
86. Nair KS, Rizza RA, O'Brien P, Dhatariya K, Short KR, Nehra A, et al. DHEA in elderly women and DHEA or testosterone in elderly men. N Engl J Med. 2006;355(16):1647–59.
87. Toniato A, Merante-Boschin I, Opocher G, Pelizzo MR, Schiavi F, Ballotta E. Surgical versus conservative management for subclinical Cushing syndrome in adrenal incidentalomas: a prospective randomized study. Ann Surg. 2009;249(3):388–91.
88. Iacobone M, Citton M, Viel G, Boetto R, Bonadio I, Mondi I, et al. Adrenalectomy may improve cardiovascular and metabolic impairment and ameliorate quality of life in patients with adrenal incidentalomas and subclinical Cushing's syndrome. Surgery. 2012;152(6):991–7.
89. Akaza I, Yoshimoto T, Iwashima F, Nakayama C, Doi M, Izumiyama H, et al. Clinical outcome of subclinical Cushing's syndrome after surgical and conservative treatment. Hypertens Res. 2011;34(10):1111–5.
90. Guerrieri M, Campagnacci R, Patrizi A, Romiti C, Arnaldi G, Boscaro M. Primary adrenal hypercortisolism: minimally invasive surgical treatment or medical therapy? A retrospective study with long-term follow-up evaluation. Surg Endosc. 2010;24(10):2542–6.

91. Tsuiki M, Tanabe A, Takagi S, Naruse M, Takano K. Cardiovascular risks and their long-term clinical outcome in patients with subclinical Cushing's syndrome. Endocr J. 2008; 55(4):737–45.
92. Chiodini I, Morelli V, Salcuni AS, Eller-Vainicher C, Torlontano M, Coletti F, et al. Beneficial metabolic effects of prompt surgical treatment in patients with an adrenal incidentaloma causing biochemical hypercortisolism. J Clin Endocrinol Metab. 2010;95(6):2736–45.
93. Perysinakis I, Marakaki C, Avlonitis S, Katseli A, Vassilatou E, Papanastasiou L, et al. Laparoscopic adrenalectomy in patients with subclinical Cushing syndrome. Surg Endosc. 2013;27(6):2145–8.
94. Perogamvros I, Vassiliadi D, Karapanou O, Botoula E, Tzanela M, Tsagarakis S. Biochemical and clinical benefits of unilateral adrenalectomy in patients with subclinical hypercortisolism and bilateral adrenal incidentalomas. Eur J Endocrinol. 2015;173(6):719–25.
95. Iacobone M, Citton M, Scarpa M, Viel G, Boscaro M, Nitti D. Systematic review of surgical treatment of subclinical Cushing's syndrome. Br J Surg. 2015;102(4):318–30.
96. Debono M, Chadarevian R, Eastell R, Ross RJ, Newell-Price J. Mifepristone reduces insulin resistance in patient volunteers with adrenal incidentalomas that secrete low levels of cortisol: a pilot study. PLoS One. 2013;8(4), e60984.
97. Lacroix A, Ndiaye N, Tremblay J, Hamet P. Ectopic and abnormal hormone receptors in adrenal Cushing's syndrome. Endocr Rev. 2001;22(1):75–110.
98. Reznik Y, Lefebvre H, Rohmer V, Charbonnel B, Tabarin A, Rodien P, et al. Aberrant adrenal sensitivity to multiple ligands in unilateral incidentaloma with subclinical autonomous cortisol hypersecretion: a prospective clinical study. Clin Endocrinol (Oxf). 2004;61(3):311–9.
99. Libe R, Coste J, Guignat L, Tissier F, Lefebvre H, Barrande G, et al. Aberrant cortisol regulations in bilateral macronodular adrenal hyperplasia: a frequent finding in a prospective study of 32 patients with overt or subclinical Cushing's syndrome. Eur J Endocrinol. 2010;163(1):129–38.
100. de Herder WW, Hofland LJ, Usdin TB, de Jong FH, Uitterlinden P, van Koetsveld P, et al. Food-dependent Cushing's syndrome resulting from abundant expression of gastric inhibitory polypeptide receptors in adrenal adenoma cells. J Clin Endocrinol Metab. 1996;81(9):3168–72.
101. Lacroix A, Tremblay J, Rousseau G, Bouvier M, Hamet P. Propranolol therapy for ectopic β-adrenergic receptors in adrenal Cushing's syndrome. N Engl J Med. 1997;337(20):1429–34.
102. Lacroix A, Hamet P, Boutin JM. Leuprolide acetate therapy in luteinizing hormone--dependent Cushing's syndrome. N Engl J Med. 1999;341(21):1577–81.

Long-Term Effects of Prior Cushing's Syndrome

Anna Aulinas, Elena Valassi, Eugenia Resmini, Alicia Santos, Iris Crespo, María-José Barahona, and Susan M. Webb

Abstract Cushing's syndrome, and its most frequent cause pituitary-dependent Cushing's disease, is a rare disease due to excessive glucocorticoid (GC) secretion. Chronic exposure to GC excess determines a large number of deleterious effects leading to increased morbidity (i.e., cardiovascular complications, psychiatric symptoms, osteoporosis, cognitive impairment, hormonal dysfunctions after surgery) and mortality. Although most of these effects improve after normalization of cortisol, not all are completely reversible after remission of hypercortisolism and negatively impact on health-related quality of life. Therefore, there is a need for both greater diagnostic suspicion and improved diagnostic tools to hasten the delay to diagnosis and effective therapy aimed at improving long-term prognosis. The lack of systematic data analysis and prospective longitudinal studies is due to low prevalence and orphan disease status of CS. Multicenter registries collecting longitudinal data on these patients would contribute to further knowledge on the natural history and long-term outcome data in these patients.

Keywords Hypercortisolism • Cushing syndrome • Morbidity • Mortality • Cardiovascular risk • Quality of life • Remission

A. Aulinas • E. Valassi • E. Resmini • A. Santos • I. Crespo • S.M. Webb (✉)
Endocrinology/Medicine Department, Hospital Sant Pau, IIB-Sant Pau,
Pare Claret 167, 08025 Barcelona, Spain

Centro de Investigación Biomédica en Red de Enfermedades Raras (CIBERER, Unidad 747),
ISCIII and Universitat Autònoma de Barcelona (UAB), Barcelona, Spain
e-mail: swebb@santpau.cat

M.-J. Barahona
Centro de Investigación Biomédica en Red de Enfermedades Raras (CIBERER, Unidad 747),
ISCIII and Universitat Autònoma de Barcelona (UAB), Barcelona, Spain

Department of Endocrinology, Hospital Universitari Mútua de Terrassa, Barcelona, Spain

© Springer International Publishing Switzerland 2017
E.B. Geer (ed.), *The Hypothalamic-Pituitary-Adrenal Axis in Health and Disease*, DOI 10.1007/978-3-319-45950-9_11

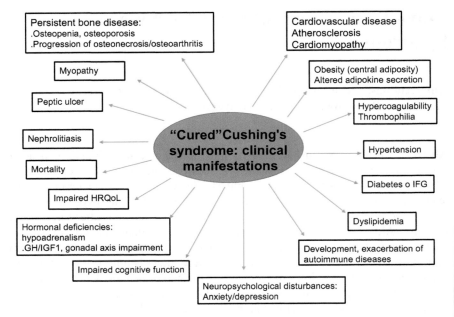

Fig. 1 Main clinical manifestations after remission of Cushing's syndrome

Introduction

Cushing's syndrome is a rare and severe disease due to excessive cortisol secretion. Chronic exposure to high glucocorticoid (GC) levels has been associated with increased morbidity and mortality. Metabolic and cardiovascular complications, osteoporosis, psychiatric symptoms, and cognitive impairments are the most common. Additionally, increased nephrolithiasis and hormonal dysfunctions after surgery (i.e., growth hormone deficiency or adrenal insufficiency) together lead to health-related quality of life (HRQoL) impairment and increased mortality (Fig. 1). Although most of these comorbidities improve after initial therapy, not all are completely reversible in spite of being biochemical "cure" of hypercortisolism. Therefore, long-term follow-up is mandatory to foresee and control complications due to prior, chronic exposure to high cortisol levels. Data regarding final outcome after complete resolution are lacking and need further study on survival and natural history of the affected subjects.

This chapter addresses current information on the main long-term/persistent effects of prior Cushing's disease/glucocorticoid exposure.

oh, hello.

I ♡ U

-E

from the desk of
David Behar
213 593 6704

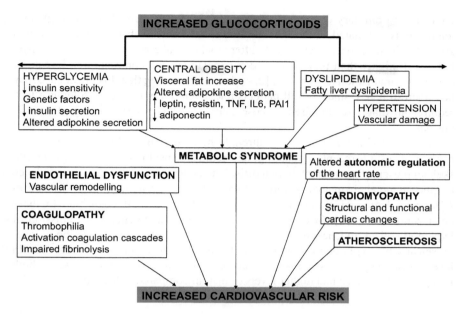

Fig. 2 Cardiovascular risk in Cushing's syndrome

Cardiovascular and Metabolic Comorbidities

Hypercortisolism enhances cardiovascular risk factors such as glucose intolerance, central obesity, hypertension, and dyslipidemia. All are linked to an increased incidence of atherosclerosis and coronary disease, and impact on morbidity, cardiovascular disease being the leading cause of death in patients with Cushing's syndrome (CS). However, this cardiovascular risk profile is not completely explained by conventional cardiovascular risk factors; other still inadequately defined disease-specific factors, partially related to the hypercoagulable and inflammatory state with an unfavorable adipokine profile, have also been observed [1]. Although most of the risk factors improve, cardiovascular risk is clearly increased in CS patients even years after remission (Fig. 2) [2].

Glucose Metabolism

Glucose metabolism abnormalities are common in CS; in fact diabetes is one of the most common metabolic complications of CS. The prevalence of these abnormalities varies depending on the series and the etiology of CS (higher in ectopic CS compared to pituitary or adrenal adenomas) [3]. Prevalence of impaired glucose tolerance is estimated around 21–64 % and of overt diabetes mellitus around 20–47 %; the latter decreases by 40 % after biochemical control of hypercortisolism, but is still higher than body mass index (BMI)-matched controls after 5 years of cortisol normalization (33 vs. 7 %) [2, 4]. It is worth noting that this prevalence may be underestimated, since not all patients underwent an oral glucose tolerance test, required to diagnose impaired glucose tolerance (IGT) when fasting glucose is normal.

GCs affect glucose homeostasis through the induction of gluconeogenesis, disruption of insulin receptor signaling, and reducing insulin sensitivity in liver and skeletal muscle [5]. Although hypercortisolism is involved in this higher prevalence of glucose metabolism abnormalities, it seems that age, genetic predisposition, lifestyle, and degree of exposure to hypercortisolism may all contribute to these deleterious effects [6]. Insulin resistance persists after biochemical remission of hypercortisolism, independently of body weight, suggesting that reduction in insulin sensitivity is not due to obesity but to hypercortisolism per se. Although insulin levels in patients in remission were observed to be lower than in active disease, both groups of CS patients had higher levels of insulin compared to healthy controls [7].

Obesity, Central Adiposity, and Chronic Inflammatory State

Chronic hypercortisolism determines a redistribution of body fat leading to increased abdominal fat and reduced peripheral subcutaneous adipose depots, with the related metabolic consequences.

Several studies have observed a higher body mass index and waist/hip ratio in CS patients compared to an age- and sex-matched controls. Persistently increased abdominal circumference was seen in CS patients (irrespective of the cause) 1 year after hormonal remission [8]. In a recent published study evaluating cardiovascular risk factors after remission of hypercortisolism, the authors observed that all the risk factors returned to a level comparable to the control subjects, except for obesity and triglyceridemia (related directly to central obesity) [1]. When comparing body composition before surgery and in remission (mean of 20 months after surgery) using whole body magnetic resonance imaging, although an important part of the fat depots had decreased and reverted fat to a distribution more consistent with favorable cardiovascular risk, most patients with Cushing's disease (CD) in remission continued to have overweight, obese, and had persistence of cardiovascular risk [9]. A case–control study showed that patients with CS after a mean of 11 years in remission continued to have greater total fat and central obesity as compared to age- and sex-matched controls [7]. In the same line, in a group of 50 women with CS in remission (median time 13 years), abdominal fat mass was increased compared to matched controls. The authors also observed that increased abdominal obesity was associated to ongoing GC replacement therapy, as well to polymorphism rs1045642 in a ABCB1 gene, related to GC sensitivity [10].

Although correction of hypercortisolism is generally associated with a reduction of visceral and subcutaneous fat mass, body cell mass loss does not recover after remission, indicating true protein loss in these patients [11].

Moreover, it seems that the effects of exogenous hypercortisolism on body composition is different from those seen in endogenous CS, where the increase in total body fat and trunk fat is higher [12]. Recently, glucocorticoid-induced obesity has been evaluated among different diagnostic groups of CS. Interestingly, patients with primary pigmented nodular adrenocortical disease who presented a PRKAR1A gene mutation (increased cAMP-dependent protein kinase levels) were less obese than other patients with CS [13, 14].

Altered Adipokine Secretion

This increased central obesity and visceral adiposity characteristic of CS induces impaired adipokine production. The persistence of central adiposity and an unfavorable adipokine profile may link metabolic alterations and cardiovascular morbidity in CS after biochemical remission. Some adipokines may contribute to the pathogenesis of vascular, metabolic and inflammatory complications such as endothelial damage, high blood pressure, impaired bone remodeling, atherosclerosis, and low grade inflammation [15].

Increased levels of leptin, resistin, and proinflammatory cytokines such as tumor necrosis factor alpha (TNF-alfa) and interleukin-6 observed both in active CS and even years after biochemical remission are associated with greater cardiovascular risk [7, 15, 16]. These and other adipokines and humoral factors may stimulate circulating cortisol levels (activating 11ß-hydroxysteroid dehydrogenase type 1 11ß-HSD1), contributing to the typical characteristics of metabolic syndrome and visceral obesity in CS [17].

Leptin, an anorexigenic hormone, in general is elevated in active CS. It decreases after correction of hypercortisolism, depending on the timed evaluation and changes in body fat [15]. Leptin elevation persists 10 days after surgery for CD despite a drop in cortisol levels, suggesting that factors other than cortisol, such as persistently abnormal fat distribution, play a role in leptin hypersecretion [18]. In long-term remission of hypercortisolism, leptin gradually decreases in parallel to a decrease in BMI, fat mass, and insulin levels [9]. Also, a decrease in leptin concentrations, 9 months after curative surgery in CD patients, was observed, similar to findings in obese patients following bariatric surgery [16, 19].

On the other hand, adiponectin (an adipokine with antiatherogenic and anti-inflammatory properties) is decreased in patients with active and cured CS after 11 years of biochemical control compared to controls; however, the differences were no longer significant when patients were stratified based on their estrogen status [7]. Nonobese CS patients had lower adiponectin concentrations compared to non-obese controls, but this difference was not present when comparing obese CS patients and obese controls. This suggests that obesity is crucial when considering adiponectin levels in CS patients [20]. Another peptide with anti-inflammatory, as well as anti-fibrotic effects (although not an adipokine), is ghrelin; its levelsy have been found to be higher 24 months after successful surgical correction of hypercortisolism compared with values before surgery, together with an improvement in glucose and lipid homeostasis and a progressive weight loss [21, 22].

If concomitant growth hormone deficiency exists after pituitary surgery, cardio-vascular risk and metabolic and body composition abnormalities worsen even more, all of which may improve after GH replacement therapy [23, 24].

To summarize, imbalance of adipokine production is associated with increased cardiovascular risk and central fat accumulation in CS. Persistent impairment of adipokine secretion may also contribute to the increased long-term cardiovascular risk in patients cured of CS.

Dyslipidemia in CS

According to different series, lipid abnormalities have been observed in 37–71 % of patients with CS, mainly hypercholesterolemia in 16–60 % and hypertriglyceridemia in 1–36 % of patients [25]. Improvements of dyslipidemia after cure/remission occur, but an adverse lipid profile (higher total/HDL cholesterol ratio) can persist in around 30 % of patients, probably due to GC-induced modifications of adipose tissue [2]. However, in a subgroup of subclinical CS patients due to adrenal adenoma, no significant improvements in lipid profile was observed after adrenalectomy [26].

Although the pathogenetic mechanisms of dyslipidemia are multifactorial; insulin resistance and growth hormone deficiency combined with impaired gonadal function can contribute to lipid abnormalities [27]. Given the increased cardiovascular mortality in CS, treatment of dyslipidemia is strongly recommended.

Hypertension and Vascular Damage

Hypertension is one of the most prevalent cardiovascular risk factors in CS, reported in 55–85 % of CS patients, and is associated with the duration of hypercortisolism [4]. Moderately high blood pressure persists despite effective treatment of CS in around 24–56 % of cured CS, mainly when patients are older, had a longer exposure to high levels of GCs, and longer duration of hypertension in the active phase of hypercortisolism. Removal of the source of hypercortisolism led to improvement of hypertension in a significant proportion of patients but not all [28, 29]. Although with a lower prevalence, hypertension, impaired glucose tolerance, and dyslipid-emia were still present in a group of cured CS patients; furthermore, a more marked decrease was observed in adrenal adenomas compared to pituitary adenomas [8]. CS patients in remission with persistently high blood pressure have more structural and functional cardiac changes as compared to control hypertensive subjects [30].

Hypertension has also been associated with brain white matter lesions in CS patients in remission [31]. Therefore, it is strongly recommended and often required to prescribe antihypertensive treatment while hypercortisolism exists, as well as in cases in which hypertension persists despite control of hypercortisolism. ACE inhibitors and angiotensin receptor blockers, with their cardioprotective effects, have been recently proposed as a first line treatment [32].

Pathogenesis of hypertension appears to be multifactorial: inhibition of the vasodilating system, activation of the renin–angiotensin–aldosterone system, inhibition of peripheral catecholamine catabolism, increased cardiac output, total peripheral resistance, and renovascular resistance. All these factors together with concomitant insulin resistance and/or sleep apnea are the main contributors to hypertension in CS [32, 33]. Moreover, increased cortisol levels may override the capacity of 11ß-HSD2 (which inactivates cortisol), facilitating cortisol binding to the mineralocorticoid receptor, resulting in an increased effect of aldosterone, that has growth-promoting and profibrotic activities, leading to remodeling and fibrosis of both small vessels and the myocardium [34].

Increased oxidative stress and inflammatory markers (soluble receptor of tumor necrosis factor type 1 (sTNFR1), interleukin-6, interleukin-8, glutathione peroxidase, thromboxaneB2, 15-F2t-isoprostane) and decreased antioxidants levels (vitamin E) have been observed in CS compared to controls. These prooxidative processes induced by GCs in combination with metabolic comorbidities lead to a worsening oxidant–antioxidant balance and an increased cardiovascular morbimortality [35]. sTNFR1 has been found to correlate with the Agatston score and to be a predictor of coronary calcifications in a cohort of active and cured CS patients [36]. Also, sTNFR1 has been found to be the strongest predictor of carotid intima media thickness in females with CS [37]. Moreover, endothelin, homocysteine, VEGF, and cell adhesion molecules are increased in active CS patients, while taurine, a suggested protective factor, is decreased. Most of these molecules improved after successful normalization of cortisol levels [38, 39].

An increased carotid intima media thickness and a lower distensibility coefficient were observed in CS after 1 year of remission compared to a BMI-matched control group [2, 40]. The same group observed that atherosclerotic plaques were present in 26.7 % of CD patients compared to <4 % of controls 5 years after remission [2]. Cardiovascular disease was more prevalent in CS patients even after long-term remission (mean time of 11 years); a greater prevalence of coronary calcifications (31 % vs. 21 %) and noncalcified atheroma plaques (20 % vs. 7.8 %), quantified by cardiac multidetector computed tomography (MDCT) angiogram scan, were observed in cured CS compared to age- and gender-matched controls, even after excluding patients with hypopituitarism or dyslipidemia [41]. Also by MDCT, increased coronary calcifications and noncalcified coronary plaque volumes were present in patients with active or previous hypercortisolism, in a small series of mostly ectopic CS [42]. In the same line, atherosclerotic plaques were more prevalent in CS compared to populations matched for similar cardiovascular risk factors, even long-term after remission and they correlated with insulin resistance and central adiposity [43].

Cardiac Morphology: Cardiomyopathy

Several groups have reported functional and structural cardiac lesions such as left ventricular hypertrophy, diastolic dysfunction, and decreased systolic performance in patients with active CS. With remission of hypercortisolemia, cardiac alterations significantly improve, but may not normalize. Myocardial fibrosis has been observed in active CS compared to healthy controls and controls with high blood pressure. Fibrosis appears to be one of the greatest determinants for the degree of regression of cardiomyopathy seen in CS. Nevertheless, successful treatment of CS normalized the extent of myocardial fibrosis, suggesting that hypercortisolism may have a direct effect on myocardial fibrosis independent of left ventricular hypertrophy and high blood pressure [44]. Eighteen months after successful treatment of CS, improvement in left ventricular systolic and diastolic function in parallel to a reduction in myocardial fibrosis was found [45]. In the same line, echocardiographic abnormalities in left ventricular mass parameters were seen in around 70 % of active CS. These abnormalities substantially improved during a mean follow-up of 4 years after the remission of hypercortisolism, although they continued to be more marked as compared to controls [46]. Using cardiac MRI, subclinical systolic biventricular dysfunction together with increased left ventricular mass was found in CS patients compared to controls [47]. After effective treatment of hypercortisolism, an improvement of the systolic performance of both ventricles and reduced left ventricular mass were observed together with a regression of the concentric left ventricular remodeling pattern. This reduction in left ventricular mass was independently associated with changes in glucose metabolism and BMI. Moreover, on the basis of the absence of late gadolinium myocardial enhancement, dense replacement myocardial fibrosis was ruled out in uncomplicated CS [47].

On the other hand, prolonged QTcd (QTc dispersion) in association with ECG evidence of left ventricular hypertrophy seems to be specific features of CD patients and to correlate with hypercortisolemia independently of other cardiovascular risk factors, suggesting a cardiotoxic effect of cortisol excess per se [48]. Also, reduced heart rate variability, an abnormality in cardiovascular autonomic regulation, has been observed in patients with CS; hypercortisolism and disease duration were found to be the main causative factors [49].

Thus, both excess cortisol and high blood pressure contribute to alter cardiac mass and increase the prevalence of damage in target organs. The importance of controlling high blood pressure and other cardiovascular risk factors before surgery to improve long-term prognosis should be emphasized.

In summary, although there is a reduction of fat mass and central obesity after normalization of cortisol, adverse metabolic profile, overweight, and increased cardiovascular risk still persist after remission.

Coagulopathy, Thrombophilia

Cortisol excess induces a procoagulative phenotype (activation of coagulation cascades and impaired fibrinolysis), so that patients with CS have a greater predisposition to thromboembolic events, especially in the perioperative period. This hypercoagulable state in CS is explained by higher levels of procoagulant factors, mainly factors VIII, IX, and von Willebrand factor, as well as an impaired fibrinolytic capacity, due to increase synthesis of the plasminogen activator inhibitor type 1 (the main inhibitor of the fibrinolytic system) [15]. Consequently, there is a shortening of activated partial thromboplastin time and increased thrombin generation [50, 51]. Moreover, both a rise in platelets and endothelial dysfunction observed in patients with CS predispose to increased cardiovascular risk and play a role in the pathogenesis of the prothrombotic state in patients with CS [52, 53]. The incidence of venous thromboembolism (VTE) in CS is higher than in the general population (2.5–3.1 vs. 1.0–2.0 per 1000 persons/year, respectively) [51, 54]. Patients who undergo transsphenoidal surgery for CD have greater risk of thromboembolism than those for a nonfunctional pituitary adenoma, suggesting a role of cortisol (or ACTH) inducing changes in hemostatic factors [54]. Hemostatic and fibrinolytic parameters did not normalize 80 days after biochemical remission with medical therapy [55]. In the same line, in a systematic review the authors observed that even after remission of hypercortisolism, v Willebrand Factor, VII, and IX factors remained high [51]. An improvement in hemostatic parameters after one year of successful surgery has been described, but complete normalization of hemostasis does not occur [56].

In a recent study, an increase risk for VTE (Hazard Ratio, HR 2.6) in patients with CS was found to be already present 3 years before diagnosis, being highest the first year after diagnosis (HR 20.6) and still remained elevated from 1 to 30 years after diagnosis, although most of the cases occurred during persistent hypercortisolism [28].

Although it is still a matter of debate whether systematic antithrombotic prophylaxis in CS should be used, it seems that thromboprophylaxis could be recommended in patients with CS undergoing surgery. However, there is no consensus on the dose or duration of use of prophylactic anticoagulant therapy. Prospective placebo-controlled trials to evaluate the effects of thromboprophylaxis in patients with CD are still lacking.

Additional Hormonal Dysfunction

Remission rates after pituitary surgery can be achieved for 65–100 % of patients. These percentages are lower in patients with a non-visible adenoma, microadenoma with unfavorable localization or macroadenomas and recurrence rates can reach 5–36 % [4]. Secondary hypothyroidism and hypogonadism are common in patients with CS, due to the functional suppression of thyrotropin and gonadotropin secretion by GC excess. After normalization of cortisol secretion, these endocrine abnormalities usually recover, as well as normal menstrual cycles and sexual activity. However, due to structural damage of the residual pituitary gland after surgical removal of the tumor or prolonged inhibition of the hypothalamic–pituitary–adrenal axis, permanent hormone deficiency may occur (hypopituitarism or adrenal insufficiency) [57].

After surgery the most common pituitary insufficiency observed is GH deficiency (which is not always evaluated), followed by thyrotropin and gonadotropin deficiencies [58]. Some patients require life-long replacement with exogenous GC.

GH/IGF1 Axis Impairment: GH Deficiency

GCs are important regulators for GH secretion and action. Prolonged GC excess is a well-known negative regulator of GH secretion. Short stature and delayed linear growth are typical features of pediatric CS, and slowed growth is common in children undergoing long-term high-dose GC therapy. Spontaneous catch-up growth is unlikely even after successful treatment in pediatric CS [59].

There is also evidence supporting the negative impact of hypercortisolism on GH secretion in adult patients. In a group of 34 patients with CD evaluated after long-term remission (median 3.3 years), 65 % presented abnormal GH secretion [60]. The GH/IGF-1 axis recovered at 6 months after successful treatment in half of these patients and was more commonly observed in those patients in whom the HPA axis recovered as well [58].

Interestingly, patients with subclinical hypercortisolism due to adrenal adenomas had a reduced GH secretion reserve compared to patients with nonfunctioning adrenal adenomas after adjusting for age and BMI. In these patients, GH secretion improved after normalization of hypercortisolism [61].

A 3-year follow-up study of GH-treated CD and nonfunctioning pituitary adenomas (NFPA) patients found that in spite of similar prevalence of metabolic syndrome at baseline, metabolic syndrome and cardio- and cerebrovascular disease were significantly higher in treated CD than NFPA patients, suggesting that GHD CD subjects were more predisposed to adverse metabolic features and increased cardiovascular risk [23]. Comparing the effect of GH treatment on lean body mass in cured CD and NFPA patients, NFPA patients showed greater improvement of lean body mass than cured CD after GH treatment, indicating that CD patients could be resistant to the anabolic effect of GH on protein, even years after remission [62].

Assessment of GH secretion is therefore recommended for patients cured from CD, even if not submitted to radiotherapy. Studies on the clinical impact of GH deficiency and the use of GH replacement therapy seem warranted in patients cured from CD.

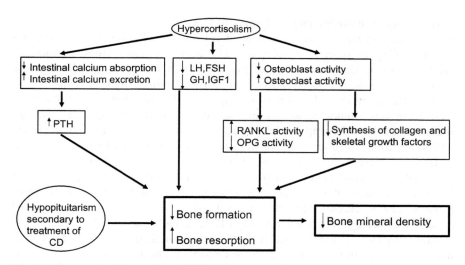

Fig. 3 Pathogenesis of bone disease in CS. (CD: Cushing's Disease, LH: luteinizing hormone, FSH: follicle-stimulating hormone, PTH: parathyroid hormone, GH: growth hormone, IGF: Insulin-like Growth Factor I, RANKL: receptor activator of NF-Kappa B-Rank-ligand, OPG: Osteoprotegerin)

Bone: Osteoporosis

The prevalence of bone disease, mainly osteoporosis, is high and often underestimated in patients with CS, since not all patients undergo DXA scans, and asymptomatic vertebral and rib fractures can remain undiagnosed. Approximately 30–50 % of CS patients present with fractures, particularly vertebral fractures [3]. Additionally, osteoarthritis and osteonecrosis have been reported mainly in patients with iatrogenic CS, but rarely in patients with endogenous hypercortisolism.

GCs have direct and indirect effects on bone, including decrease osteoblastic and increased osteoclastic activity, reduced intestinal calcium absorption, and increased urinary calcium excretion which induces in both cases a modest increase in parathyroid hormone levels [63]. Deleterious effects on bone, especially on cortical bone microstructure, have been observed using a high-resolution peripheral quantitative computed tomography in patients with active CS [64]. Furthermore, secondary hypogonadism and/or decreases in GH or IGF1 levels induced by excessive amounts of cortisol contribute to the loss of bone mineral density (BMD). The pathophysiology of bone disease in CS is detailed in (Fig. 3) [65].

Studies evaluating bone status after biochemical control of hypercortisolism, however, are often conflicting. While some observed a reversal of GC-induced osteoporosis, others showed an incomplete recovery of BMD and quality of bone after remission. Reversal of GC-induced osteoporosis after long-term remission of CS (mean 72 months) has been described in parallel with increased osteocalcin levels [66]. In the same line, after remission of hypercortisolism, bone mass changes were reversible, probably due to the fact that prior exposure time to endogenous hypercortisolism was shorter than in other studies [67]. The mechanisms causing BMD recovery are speculative. They could be attributed to an increase in osteocalcin levels and to the preservation of trabecular architecture despite the thinning induced by GCs, so osteoblasts may continue synthesizing new bone [68].

A partial recovery of BMD and bone quality after treatment for CS has been reported in most studies (in adolescents and adult patients), although the series are small and median follow-up is relatively short (less than 2 years) [69]. In the series with longer follow-up after remission of hypercortisolism (mean 11 years), decreased BMD values were seen in estrogen-sufficient women as compared to age-, sex- and BMI-matched controls, but not in women with estrogen deficiency, suggesting that the protective effect of estrogens on bone mass is lost with hypercortisolism. Prior exposure time to excess endogenous cortisol and the duration of postoperative GC replacement therapy were predictors of low BMD in these patients [70].

In a group of 50 cured CS, with a median remission time of 13 years, BMD was not significantly different at any site between patients and age- and gender-matched controls. The authors observed that the NR3C1 Bcl1 polymorphism of the GC receptor was associated with reduced total and femoral neck BMD, and patients with ongoing GC replacement presented worse skeletal health (reduced total and lumbar spine BMD) [10].

In summary, BMD recovery appears to be only partial in most patients with "cured" CS.

Myopathy

Around 60–80 % of CS patients present with proximal muscle atrophy and weakness, more frequently in males [3]. GC-induced changes in muscle are evident after a few days of GC exposure or administration, with a more prominent effect on proximal muscles [71]. In aging subjects without CS, muscle mass loss were not associated to circulating or urinary cortisol, but muscle strength correlated with quadriceps expression of 11ß-HSD1, supporting the importance of tissue-specific cortisol metabolism and conversion, rather than overall circulating levels in determining negative effects of GCs [72]. Muscle damage can persist both short- and long-term after cure; it has been related to protein synthesis inhibition and increased rate of protein degradation of myofibrillar and extracellular matrix proteins. Indeed, reduced arm muscle area showed no relevant improvement 6 months after successful treatment, despite a reduction of body fat mass [11]. Reduced lean mass due to muscle loss of limb muscle was observed in CS compared to obese controls with same total body fat mass [73]. In a long-term follow-up, patients with CS had reduced limb skeletal muscle mass, but similar lean body mass compared to age- and gender-matched controls [10]. MRI body composition assessment of CD patients 20 months after remission showed that total and limb skeletal muscle is actually reduced compared to active disease, probably due to the GC replacement therapy after cure [9].

Moreover, postmenopausal women in remission presented with similar muscle mass as active disease patients, suggesting a role of estrogen deprivation in muscle mass as well [9]. Creatinin kinase, plasma myoglobin, and muscle fiber conduction velocity were reduced in the active phase of the disease compared to healthy age-, sex-, and BMI-matched controls and correlated with disease duration [74]. It has been suggested that aerobic and resistance exercises could probably be effective in attenuating GC-induced muscle atrophy [75].

Nephrolithiasis

Nephrolithiasis has been reported in 50% of active CD and 30% of cured CD patients compared to 6.5% in age- and gender-matched controls [4, 76]. The pathogenesis of nephrolithiasis in CS is not fully elucidated. There is probably a synergistic effect of different metabolic changes (hypercalciuria, hypocitraturia, hyperuricosuria, and hyperoxalaturia) together with hemodynamic changes caused by hypercortisolism. In fact, obesity, hypertension, and diabetes, common features of CS, have been seen more frequently in patients with kidney stones. It seems that normalization of cortisol levels can restore the amino acid profile in urine. In a large series investigating the role of different lithogenic factors in CS, high blood pressure and excessive excretion of uric acid were found to be independent risk factors for the recurrence of nephrolithiasis [76, 77].

Cognitive Function and Behavior

Chronic hypercortisolism has been related to changes in memory, behavior, verbal learning, neuronal activity, and other processes of the central nervous system. Psychiatric disturbances have been reported in 54–81% in different series, major depression and irritability being the most common psychiatric disorders. Emotional lability, mania, paranoia, acute psychosis, anxiety, and panic attacks may also occur in CS [78, 79]. Few reports assess psychopathology after effective surgery; although most of these symptoms and changes improve one year after remission, many persist and do not appear to be fully reversible in the long-term follow-up. An increased prevalence of psychopathology and maladaptive personality traits compared to patients with nonfunctioning pituitary adenomas (NFPAs) and matched controls have been found, indicating that cortisol excess has irreversible effects on the central nervous system, rather than any effect of the pituitary tumor itself [80]. Recently, a retrospective study in a group of patients with CD who underwent bilateral adrenalectomy, with a median follow-up of 11 years, observed improvements in almost all Cushing-specific comorbidities, except for psychiatric morbidities (which included self-reported anxiety, depression, panic attacks, and psychosis) [81].

On the other hand, the hippocampus, amygdala, and cerebral cortex are important structures involved in cognitive and emotional functions. These structures are rich in GC receptors and, therefore, particularly vulnerable to hypercortisolism. Moreover, 11ß-HSD2 (which inactivates cortisol to cortisone) is not expressed in the hippocampus or limbic structures, which allows the sustained activation of mineralocorticoid receptors by GCs. Since common genetic polymorphism variants in the GC receptor and the 11ß-HSD1 have been recently associated with long-term cognitive impairments in CS in remission (for a median time of 13 years), these results indicate that GC sensitivity and prereceptor regulation of GC action may play a role in the etiology of cognitive dysfunction in these patients [82].

After successful biochemical treatment of CD, psychiatric symptoms may decrease, but patients still show cognitive impairment, decreased quality of life, and a higher prevalence of affective disorders and apathy compared to healthy controls [83–86]. Long-lasting impairments have been reported in several domains of cognitive (attention, visuospatial orienting, alerting, working, verbal and visual memory, verbal fluency, reading speed) and executive functions [83–85]. Higher prevalence of "maladaptive" personality traits in CD, even after long-term cure, has been described [80]. Impaired decision-making together with decreased cortical thickness in selective frontal areas irrespective of the activity of disease has also been observed in CS patients compared to healthy controls, suggesting that chronic hypercortisolemia promotes brain changes which are not reversible after endocrine remission [86]. In the same line, mental fatigue, characterized by mental exhaustion and long recovery time following mentally strenuous tasks, is more common in patients with CS in remission compared to healthy education-, age-, gender-matched controls, according to a very recent study [87].

Decreased hippocampal volume (HV) assessed by 3-T cerebral MRI was seen in CS patients with severe memory impairments compared to controls [83]. Both brain atrophy and reduction in total and cortical grey matter volumes have also been observed in CS compared to controls, but subcortical gray matter reduction has only been seen in those patients with severe memory impairments in parallel to the findings of reduced HV. The negative effects of GC excess on memory and HV seem to be not totally reversible after biochemical cured, since no differences, either in HV or in memory performance between active and cured CS, were found [83]. Brain volumes and cognitive functions have been associated with cardiovascular risk in CS patients in remission [31]. Furthermore, using a proton magnetic resonance spectroscopy, lower N-acetyl-aspartate in the hippocampus (suggesting neuronal dysfunction/loss) and higher levels of glutamate (suggesting glial proliferation as a repair mechanism after neuronal dysfunction) have been observed in cured CS patients compared to matched controls [88]. The authors suggest that these persistently abnormal metabolites could be early markers of GC neurotoxicity, preceding HV reduction [88]. In major depressive disorder patients, similar patterns of reduced HV and reversibility of hippocampal atrophy after treatment have been observed [89]. Moreover, widespread reductions of white matter integrity (reflecting a structural abnormality of white brain matter, like demyelination or loss of axonal integrity) in CD patients with long-term remission (mean 11.9 years) compared with matched controls have also been observed, together with abnormalities in the integrity of the uncinate fasciculus being related to the severity of depressive symptoms [90]. Similarly, structural abnormalities of the brain white matter have been identified with diffusion tensor imaging (DTI) in the brains of CS patients, again suggesting a widespread loss of axonal integrity and demyelination compared to controls [91]. Once present, these alterations seem to be independent of concomitant hypercortisolism, persisting after remission, since a more of these white matter lesions have also been found in CS in remission compared to healthy controls [31, 91]. Moreover, reduced anterior cingulate cortex grey matter volumes and greater volume of the left posterior lobe of the cerebellum in patients with long-term remission of CD (mean 11.2 years) have been observed compared to matched controls [92]. However, another study observed a smaller bilateral cerebellar cortex volume in active CS compared to matched controls [93]. Recently, aberrant resting-state functional connectivity of the brain with the limbic network (responsible for emotional processing and regulation, as well as encoding of memories) and executive control network has been observed in CD patients with long-term remission, suggesting that hypercortisolism may lead to persistent changes in brain functional connectivity (involving episodic memories, semantic knowledge, prospective memory, attention demands, working memory, and cognitive control) [94]. In the same line, altered neural processing of emotional faces after long-term remission of hypercortisolism has been recently reported in CD compared to matched healthy controls [95].

To summarize all these findings, a recent systematic review was performed, including 19 studies using MRI in a total of 339 unique patients with CS (active and in remission). Smaller hippocampal volumes, enlarged ventricles, and cerebral atrophy, as well as alterations in neurochemical concentrations and functional activity, were observed. The reversibility of structural and neurochemical alterations were incomplete after long-term remission. These findings are related to clinical characteristics (cortisol levels, duration of hypercortisolism, age at diagnosis, current age, and triglyceride levels) and behavioral outcome (cognitive and emotional functioning, mood, and quality of life) [96].

In general, active CS demonstrates brain abnormalities, which only partly recover after biochemical cure, because these still occur even after long-term remission. All these functional alterations observed may, together with abnormalities in brain structure, be related to the persisting psychological morbidity in patients with CD after long-term remission.

Autoimmune Diseases and Infections

GCs have an inhibitory action on the immune system, inducing a state of immunosuppression. Lymphoid tissue involution and lymphopenia lead to an increased susceptibility to infections and an improvement of autoimmune diseases during the active phase of CS. The opposite situation has been reported after remission of hypercortisolism, where new onset or exacerbation of previously existing

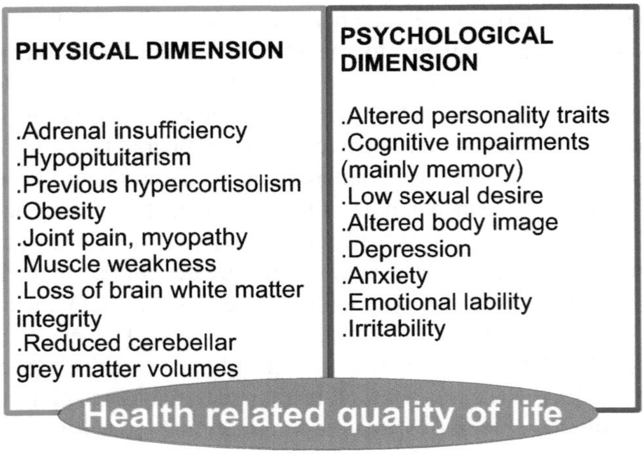

PHYSICAL DIMENSION	PSYCHOLOGICAL DIMENSION
.Adrenal insufficiency .Hypopituitarism .Previous hypercortisolism .Obesity .Joint pain, myopathy .Muscle weakness .Loss of brain white matter integrity .Reduced cerebellar grey matter volumes	.Altered personality traits .Cognitive impairments (mainly memory) .Low sexual desire .Altered body image .Depression .Anxiety .Emotional lability .Irritability

Health related quality of life

Fig. 4 Features believed to negatively impact on quality of life in Cushing's syndrome

autoimmune diseases are common. The exacerbation of autoimmune diseases appears to be related to an improvement in immune activity, suppressed by endogenous hypercortisolism during the active phase of the disease. Celiac disease, rheumatoid arthritis, the Sjögren-like sicca syndrome of dry eyes, development of sarcoidosis, or lupus erythematous has been reported in different forms of CS after the correction of hypercortisolism [97]. Nevertheless, the most common reported autoimmune disease is autoimmune thyroiditis. Thyroid autoimmunity was found in 35 % of patients "cured" of CD as compared to 10 % of controls. Thyroid autoimmunity appears to occur more frequently in patients with multinodular goiter or positive antithyroid antibodies during the active phase of the disease, suggesting that preexistent thyroid abnormalities and genetic predisposition to autoimmunity are factors for the development of autoimmune thyroid disorders after cortisol normalization [98]. The titers of autoantibodies tend to increase after surgery [99].

In conclusion, it should be borne in mind that an immune disease, which is silent during the active phase of CS may "reappear" after remission of CS; these patients with positive autoantibodies should be followed closely after remission of hypercortisolism in order to identify the eventual onset of subclinical or overt post-CS reappearance of thyroiditis and hypothyroidism.

Health-Related Quality of Life

HRQoL is significantly impaired in patients with CS of any etiology, specially in active hypercortisolism but also after endocrine cure. Considering all the systemic and neuropsychiatric complications associated with hypercortisolism this impairment in HRQoL is not unexpected in CS patients. Features believed to negatively impact on quality of life in CS are summarized in Fig. 4. In a recent cross-sectional study of CD patients with a mean time of 7.4 years since surgery, 92 % met biochemical remission criteria; however, only 80 % felt that they had been cured, reflecting the discordance between biochemical and self-assessed disease status and its impact on HRQoL in CD patients [100]. Impairment in HRQoL has been demonstrated by both generic and disease-specific questionnaires CushingQoL and Tuebingen CD-25, the most appropriate tools to assess the impact of the disease and its treatment on HRQoL [101]. CushingQoL was validated in a study with a large series of 125 CS patients (active, cured, with adrenal insufficiency secondary to treatment); the authors observed that active hypercortisolism (elevated urinary free cortisol, UFC) and female sex were the main predictors of low HRQoL. Good psychometric properties and sensitivity to change in conditions of real clinical practice were confirmed with the CushingQoL questionnaire, demonstrating that it is a valid, reliable, and responsive tool to assess HRQoL in CS [102]. Results from ERCUSYN (European Registry on Cushing's syndrome) showed that depression was an independent predictor of a lower CushingQoL score, suggesting that psychiatric disorders play a pivotal role in affecting HRQoL. They also observed that transsphenoidal surgery only improved CushingQoL several months after surgery [3]. In fact, improvements in HRQoL often take several months or even years to appear, and long-term impairments are still present when compared to normal healthy populations [103, 104]. Residual impairment of HRQoL may persist after long-term disease remission, in terms of fatigue, physical aspects, anxiety, depression, and perception of well-being according to different studies [105–107]. These impairments in HRQoL are even worse if pituitary deficiencies coexist [108]. Somatic factors (including hypopituitarism), psychological factors (illness perceptions), and health care environment were identified as factors influencing improvement in HRQoL after remission of CD, compared to other pituitary adenomas [109]. Since drawings can be used to assess perceptions of patients on their disease, the utility of a drawing test and its relation to HRQoL in CS patients in long-term remission was explored. The authors observed that drawings did not share common properties with parameters of QoL or illness perceptions, but did represent the clinical severity of disease, suggesting that drawings could reflect a new dimension of the psychological impact on these patients [110].

The Tuebingen CD-25 questionnaire was developed and validated in 63 CD patients and 1784 healthy controls; female patients scored worse HRQoL than men in the domains of depressive symptoms and social environment. They also observed that preoperative UFC levels correlated significantly with cognition [111, 112]. On average 42 months after remission of active hypercortisolism, both genders presented similar psychopathological profiles; however, in males prolonged time to diagnosis and in females the presence of comorbidities/stressors were the strongest predictive factors for worse psychopathological status [113]. Recently, the authors have provided evidence for the construct and criterion validity of the Tuebingen CD-25 in a group of 176 patients with CD [114].

In children, CS is also associated with residual impairment of HRQoL even after remission of hypercortisolism. Optimization of growth and pubertal development, normalization of body composition, and promotion of psychological health and cognitive maturation, are the specific challenges that affect children and adolescents that can severely impact on HRQoL [115].

To conclude, there appears to be some evidence that elevated, uncontrolled levels of UFC are associated with poorer HRQoL and improvements in UFC leads to better HRQoL, but not always normalization, and also that depression favors poorer HRQoL in these patients.

Peptic Ulcer

Since steroids inhibit the synthesis of prostaglandins, impair gastric bicarbonate secretion, and disturb angiogenesis and epithelial protection, they have been considered to increase the incidence of peptic ulcer disease [116]. However, the ulcerogenic and other upper gastrointestinal system effects of endogenous hypercortisolism are yet to be confirmed. Studies have reported conflicting results concerning the risk of peptic ulcer in patients receiving exogenous GCs. Until 2015, there was no controlled study assessing the frequency of peptic ulcer disease in the presence of high endogenous cortisol levels. Recently, a study evaluating the relationship of endogenous CS with helicobacter pylori infection and peptic ulcer disease was published [117]. All 20 CS patients included were in the active phase of the disease; no differences in the frequency of stomach and duodenal ulcers and Helicobacter pylori infection were observed compared to the control group (who received exogenous GCs). Endoscopic appearance of pangastritis was more common in CS, but it was not histopathologically confirmed. Candida esophagitis was more frequent in cases of CS compared to healthy controls. The authors suggested that prophylactic use of proton pump inhibitors was not compulsory for hypercortisolism of any type [117]. We are unaware of any studies to evaluate the incidence of peptic ulcers in CS in remission.

Mortality

Several studies show that mortality is increased in CS (due to nonmalignant causes), especially in patients with persistent hypercortisolism (Standard Mortality Ratio, SMR 3.7–4.2) compared to those in remission (SMR 1.8–3.17) [3, 118]. Also, SMR is higher in those undergoing transsphenoidal surgery for CD than for nonfunctioning pituitary macroadenomas [28, 119]. Cardiovascular and cerebrovascular events are the most common cause of death in CD [120, 121]. Duration of GC exposure, older age at diagnosis, and preoperative ACTH concentration were identified as independent predictors of mortality in a long-term follow-up of a large cohort of treated patients with CD [29]. A recent meta-analysis revealed that mortality remains increased in patients with CD even after initial biochemical cure or remission. Hypopituitarism after surgery may also contribute to the increased mortality risk [122].

No increased mortality was observed in CS due to a benign adrenal adenoma, but another meta-analysis restricted increased mortality in CS to pituitary CD with persistent hypercortisolism, after surgical failure [121]. Nevertheless, in a large cohort of patients with adrenal incidentalomas, a postdexamethasone serum cortisol >1.8 μg/dL was associated to increased risk for mortality (HR 12; 95%CI 1.6–92.6) mainly related to cardiovascular disease and infection [123]. Mortality in CS was also evaluated after bilateral adrenalectomy in those patients with active disease when all other treatment options failed. Surgical mortality was <1 %; at a follow-up of median 41 months 17 % of patients died with a remarkable excess of mortality within the first year after surgery (46 %) in spite of a clear improvement of symptoms of hypercortisolism, suggesting that intensive clinical care should focus on patients in this period. The main causes of death were stroke, myocardial infarction, and septicemia [124].

A most relevant population-based cohort study including the entire population of Denmark (1980 to 2010) compared 343 benign CS of adrenal or pituitary origin and 34300 matched population [28]. Both morbidity and mortality were assessed during complete follow-up after diagnosis and treatment. Furthermore, morbidity was investigated in the 3 years before diagnosis. Mortality was twice as high in CS patients (HR 2.3, 95%CI 1.8–2.9) compared with controls. Patients with CS were at increased risk for venous thromboembolism (HR 2.6, 95%CI 1.5–4.7), myocardial infarction (HR 3.7, 95%CI 2.4–5.5), stroke (HR 2.0, 95%CI 1.3–3.2), peptic ulcers (HR 2.0, 95%CI 1.1–3.6), fractures (HR 1.4, 95%CI 1.0–1.9), and infections (HR 4.9, 95%CI 3.7–6.4). Importantly, increased multimorbidity risk was present before diagnosis, similarly in adrenal and pituitary CS, reflecting most probably the deleterious effect of undiagnosed hypercortisolism, prior to diagnosis. Mortality and risk of myocardial infarction remained elevated during long-term follow-up. Thus, despite the apparently benign character of the disease, CS is associated with clearly increased mortality and multisystem morbidity, even before diagnosis and treatment.

Conclusions

GC excess determines a large number of deleterious effects. Although most improve after normalization of cortisol, the evidence detailed above highlights significant persistent comorbidities in CS after remission. There is a need for both improved diagnostic tools to reduce the time to diagnosis and effective therapy aimed at improving long-term prognosis. The lack of systematic data analysis and prospective studies is due to the orphan disease status of CS. Multicenter registries collecting data on these patients would provide essential data to answer the remaining questions.

References

1. Terzolo M, Allasino B, Pia A, Peraga G, Daffara F, Laino F, et al. Surgical remission of Cushing's syndrome reduces cardiovascular risk. Eur J Endocrinol. 2014;171(1):127–36.
2. Colao A, Pivonello R, Spiezia S, Faggiano A, Ferone D, Filippella M, et al. Persistence of increased cardiovascular risk in patients with Cushing's disease after five years of successful cure. J Clin Endocrinol Metab. 1999;84(8):2664–72.
3. Valassi E, Santos A, Yaneva M, Tóth M, Strasburger CJ, Chanson P, et al. The European Registry on Cushing's syndrome: 2-year experience. Baseline demographic and clinical characteristics. Eur J Endocrinol. 2011;165(3):383–92.
4. Feelders RA, Pulgar SJ, Kempel A, Pereira AM. Management of endocrine disease: the burden of Cushing's disease: clinical and health-related quality of life aspects. Eur J Endocrinol. 2012;167(3):311–26.
5. Ferraù F, Korbonits M. Metabolic comorbidities in Cushing's syndrome. Eur J Endocrinol. 2015;173(4):M133–57.
6. Giordano C, Guarnotta V, Pivonello R, Amato MC, Simeoli C, Ciresi A, et al. Is diabetes in Cushing's syndrome only a consequence of hypercortisolism? Eur J Endocrinol. 2014;170(2):311–9.
7. Barahona M-J, Sucunza N, Resmini E, Fernández-Real J-M, Ricart W, Moreno-Navarrete J-M, et al. Persistent body fat mass and inflammatory marker increases after long-term cure of Cushing's syndrome. J Clin Endocrinol Metab. 2009;94(9):3365–71.
8. Giordano R, Picu A, Marinazzo E, D'Angelo V, Berardelli R, Karamouzis I, et al. Metabolic and cardiovascular outcomes in patients with Cushing's syndrome of different aetiologies during active disease and 1 year after remission. Clin Endocrinol (Oxf). 2011;75(3):354–60.
9. Geer EB, Shen W, Strohmayer E, Post KD, Freda PU. Body composition and cardiovascular risk markers after remission of Cushing's disease: a prospective study using whole-body MRI. J Clin Endocrinol Metab. 2012;97(5):1702–11.
10. Ragnarsson O, Glad CAM, Bergthorsdottir R, Almqvist EG, Ekerstad E, Widell H, et al. Body composition and bone mineral density in women with Cushing's syndrome in remission and the association with common genetic variants influencing glucocorticoid sensitivity. Eur J Endocrinol. 2015;172(1):1–10.
11. Pirlich M, Biering H, Gerl H, Ventz M, Schmidt B, Ertl S, et al. Loss of body cell mass in Cushing's syndrome: effect of treatment. J Clin Endocrinol Metab. 2002;87(3):1078–84.
12. Resmini E, Farkas C, Murillo B, Barahona MJ, Santos A, Martínez-Momblán MA, et al. Body composition after endogenous (Cushing's syndrome) and exogenous (rheumatoid arthritis) exposure to glucocorticoids. Horm Metab Res. 2010;42(8):613–8.
13. London E, Rothenbuhler A, Lodish M, Gourgari E, Keil M, Lyssikatos C, et al. Differences in adiposity in Cushing syndrome caused by PRKAR1A mutations: clues for the role of cyclic AMP signaling in obesity and diagnostic implications. J Clin Endocrinol Metab. 2014;99(2):E303–10.
14. London E, Lodish M, Keil M, Lyssikatos C, de la Luz SM, Nesterova M, et al. Not all glucocorticoid-induced obesity is the same: differences in adiposity among various diagnostic groups of Cushing syndrome. Horm Metab Res. 2014;46(12):897–903.
15. Valassi E, Biller BMK, Klibanski A, Misra M. Adipokines and cardiovascular risk in Cushing's syndrome. Neuroendocrinology. 2012;95(3):187–206.
16. Krsek M, Silha JV, Jezková J, Hána V, Marek J, Weiss V, et al. Adipokine levels in Cushing's syndrome; elevated resistin levels in female patients with Cushing's syndrome. Clin Endocrinol (Oxf). 2004;60(3):350–7.
17. Iwasaki Y, Takayasu S, Nishiyama M, Tsugita M, Taguchi T, Asai M, et al. Is the metabolic syndrome an intracellular Cushing state? Effects of multiple humoral factors on the transcriptional activity of the hepatic glucocorticoid-activating enzyme (11beta-hydroxysteroid dehydrogenase type 1) gene. Mol Cell Endocrinol. 2008;285(1-2):10–8.

18. Cizza G, Lotsikas AJ, Licinio J, Gold PW, Chrousos GP. Plasma leptin levels do not change in patients with Cushing's disease shortly after correction of hypercortisolism. J Clin Endocrinol Metab. 1997;82(8):2747–50.

19. van Dielen FMH, van't Veer C, Buurman WA, Greve JWM. Leptin and soluble leptin receptor levels in obese and weight-losing individuals. J Clin Endocrinol Metab. 2002;87(4):1708–16.

20. Fallo F, Scarda A, Sonino N, Paoletta A, Boscaro M, Pagano C, et al. Effect of glucocorticoids on adiponectin: a study in healthy subjects and in Cushing's syndrome. Eur J Endocrinol. 2004;150(3):339–44.

21. Libè R, Morpurgo PS, Cappiello V, Maffini A, Bondioni S, Locatelli M, et al. Ghrelin and adiponectin in patients with Cushing's disease before and after successful transsphenoidal surgery. Clin Endocrinol (Oxf). 2005;62(1):30–6.

22. Otto B, Tschöp M, Heldwein W, Pfeiffer AFH, Diederich S. Endogenous and exogenous glucocorticoids decrease plasma ghrelin in humans. Eur J Endocrinol. 2004;151(1):113–7.

23. Webb SM, Mo D, Lamberts SWJ, Melmed S, Cavagnini F, Pecori Giraldi F, et al. Metabolic, cardiovascular, and cerebrovascular outcomes in growth hormone-deficient subjects with previous Cushing's disease or non-functioning pituitary adenoma. J Clin Endocrinol Metab. 2010;95(2):630–8.

24. Höybye C, Ragnarsson O, Jönsson PJ, Koltowska-Häggström M, Trainer P, Feldt-Rasmussen U, et al. Clinical features of GH deficiency and effects of 3 years of GH replacement in adults with controlled Cushing's disease. Eur J Endocrinol. 2010;162(4):677–84.

25. Mancini T, Kola B, Mantero F, Boscaro M, Arnaldi G. High cardiovascular risk in patients with Cushing's syndrome according to 1999 WHO/ISH guidelines. Clin Endocrinol (Oxf). 2004;61(6):768–77.

26. Giordano R, Marinazzo E, Berardelli R, Picu A, Maccario M, Ghigo E, et al. Long-term morphological, hormonal, and clinical follow-up in a single unit on 118 patients with adrenal incidentalomas. Eur J Endocrinol. 2010;162(4):779–85.

27. Arnaldi G, Scandali VM, Trementino L, Cardinaletti M, Appolloni G, Boscaro M. Pathophysiology of dyslipidemia in Cushing's syndrome. Neuroendocrinology. 2010;92 Suppl 1:86–90.

28. Dekkers OM, Horváth-Puhó E, Jørgensen JOL, Cannegieter SC, Ehrenstein V, Vandenbroucke JP, et al. Multisystem morbidity and mortality in Cushing's syndrome: a cohort study. J Clin Endocrinol Metab. 2013;98(6):2277–84.

29. Lambert JK, Goldberg L, Fayngold S, Kostadinov J, Post KD, Geer EB. Predictors of mortality and long-term outcomes in treated Cushing's disease: a study of 346 patients. J Clin Endocrinol Metab. 2013;98(3):1022–30.

30. De Leo M, Pivonello R, Auriemma RS, Cozzolino A, Vitale P, Simeoli C, et al. Cardiovascular disease in Cushing's syndrome: heart versus vasculature. Neuroendocrinology. 2010;92 Suppl 1:50–4.

31. Santos A, Resmini E, Gómez-Ansón B, Crespo I, Granell E, Valassi E, et al. Cardiovascular risk and white matter lesions after endocrine control of Cushing's syndrome. Eur J Endocrinol. 2015 [Epub ahead of print].

32. Isidori AM, Graziadio C, Paragliola RM, Cozzolino A, Ambrogio AG, Colao A, et al. The hypertension of Cushing's syndrome: controversies in the pathophysiology and focus on cardiovascular complications. J Hypertens. 2015;33(1):44–60.

33. Cicala MV, Mantero F. Hypertension in Cushing's syndrome: from pathogenesis to treatment. Neuroendocrinology. 2010;92 Suppl 1:44–9.

34. Rizzoni D, Porteri E, De Ciuceis C, Rodella LF, Paiardi S, Rizzardi N, et al. Hypertrophic remodeling of subcutaneous small resistance arteries in patients with Cushing's syndrome. J Clin Endocrinol Metab. 2009;94(12):5010–8.

35. Karamouzis I, Berardelli R, D'Angelo V, Fussotto B, Zichi C, Giordano R, et al. Enhanced oxidative stress and platelet activation in patients with Cushing's syndrome. Clin Endocrinol (Oxf). 2015;82(4):517–24.

36. Barahona M-J, Resmini E, Viladés D, Fernández-Real J-M, Ricart W, Moreno-Navarrete J-M, et al. Soluble TNFα-receptor 1 as a predictor of coronary calcifications in patients after long-term cure of Cushing's syndrome. Pituitary. 2015;18(1):135–41.

37. Shivaprasad K, Kumar M, Dutta D, Sinha B, Mondal SA, Maisnam I, et al. Increased soluble TNF receptor-1 and glutathione peroxidase may predict carotid intima media thickness in females with Cushing syndrome. Endocr Pract. 2015;21(3):286–95.

38. Kristo C, Ueland T, Godang K, Aukrust P, Bollerslev J. Biochemical markers for cardiovascular risk following treatment in endogenous Cushing's syndrome. J Endocrinol Invest. 2008;31(5):400–5.

39. Faggiano A, Melis D, Alfieri R, De Martino M, Filippella M, Milone F, et al. Sulfur amino acids in Cushing's disease: insight in homocysteine and taurine levels in patients with active and cured disease. J Clin Endocrinol Metab. 2005;90(12):6616–22.

40. Faggiano A, Pivonello R, Spiezia S, De Martino MC, Filippella M, Di Somma C, et al. Cardiovascular risk factors and common carotid artery caliber and stiffness in patients with Cushing's disease during active disease and 1 year after disease remission. J Clin Endocrinol Metab. 2003;88(6):2527–33.

41. Barahona M-J, Resmini E, Viladés D, Pons-Lladó G, Leta R, Puig T, et al. Coronary artery disease detected by multislice computed tomography in patients after long-term cure of Cushing's syndrome. J Clin Endocrinol Metab. 2013;98(3):1093–9.

42. Neary NM, Booker OJ, Abel BS, Matta JR, Muldoon N, Sinaii N, et al. Hypercortisolism is associated with increased coronary arterial atherosclerosis: analysis of noninvasive coronary angiography using multidetector computerized tomography. J Clin Endocrinol Metab. 2013;98(5):2045–52.

43. Albiger N, Testa RM, Almoto B, Ferrari M, Bilora F, Petrobelli F, et al. Patients with Cushing's syndrome have increased intimal media thickness at different vascular levels: comparison with a population matched for similar cardiovascular risk factors. Horm Metab Res. 2006;38(6):405–10.

44. Yiu KH, Marsan NA, Delgado V, Biermasz NR, Holman ER, Smit JWA, et al. Increased myocardial fibrosis and left ventricular dysfunction in Cushing's syndrome. Eur J Endocrinol. 2012;166(1):27–34.

45. Pereira AM, Delgado V, Romijn JA, Smit JWA, Bax JJ, Feelders RA. Cardiac dysfunction is reversed upon successful treatment of Cushing's syndrome. Eur J Endocrinol. 2010;162(2):331–40.

46. Toja PM, Branzi G, Ciambellotti F, Radaelli P, De Martin M, Lonati LM, et al. Clinical relevance of cardiac structure and function abnormalities in patients with Cushing's syndrome before and after cure. Clin Endocrinol (Oxf). 2012;76(3):332–8.

47. Kamenický P, Redheuil A, Roux C, Salenave S, Kachenoura N, Raissouni Z, et al. Cardiac structure and function in Cushing's syndrome: a cardiac magnetic resonance imaging study. J Clin Endocrinol Metab. 2014;99(11):E2144–53.

48. Alexandraki KI, Kaltsas GA, Vouliotis A-I, Papaioannou TG, Trisk L, Zilos A, et al. Specific electrocardiographic features associated with Cushing's disease. Clin Endocrinol (Oxf). 2011;74(5):558–64.

49. Chandran DS, Ali N, Jaryal AK, Jyotsna VP, Deepak KK. Decreased autonomic modulation of heart rate and altered cardiac sympathovagal balance in patients with Cushing's syndrome: role of endogenous hypercortisolism. Neuroendocrinology. 2013;97(4):309–17.

50. Coelho MCA, Dos Santos CV, Vieira Neto L, Gadelha MR. Adverse effects of glucocorticoids: coagulopathy. Eur J Endocrinol. 2015;173(4):M11–21.

51. Van Zaane B, Nur E, Squizzato A, Dekkers OM, Twickler MTB, Fliers E, et al. Hypercoagulable state in Cushing's syndrome: a systematic review. J Clin Endocrinol Metab. 2009;94(8):2743–50.

52. Prázný M, Jezková J, Horová E, Lazárová V, Hána V, Kvasnicka J, et al. Impaired microvascular reactivity and endothelial function in patients with Cushing's syndrome: influence of arterial hypertension. Physiol Res. 2008;57(1):13–22.

53. Isidori AM, Minnetti M, Sbardella E, Graziadio C, Grossman AB. Mechanisms in endocrinology: the spectrum of haemostatic abnormalities in glucocorticoid excess and defect. Eur J Endocrinol. 2015;173(3):R101–13.
54. Stuijver DJF, van Zaane B, Feelders RA, Debeij J, Cannegieter SC, Hermus AR, et al. Incidence of venous thromboembolism in patients with Cushing's syndrome: a multicenter cohort study. J Clin Endocrinol Metab. 2011;96(11):3525–32.
55. van der Pas R, Leebeek FWG, Hofland LJ, de Herder WW, Feelders RA. Hypercoagulability in Cushing's syndrome: prevalence, pathogenesis and treatment. Clin Endocrinol (Oxf). 2013;78(4):481–8.
56. Manetti L, Bogazzi F, Giovannetti C, Raffaelli V, Genovesi M, Pellegrini G, et al. Changes in coagulation indexes and occurrence of venous thromboembolism in patients with Cushing's syndrome: results from a prospective study before and after surgery. Eur J Endocrinol. 2010;163(5):783–91.
57. Pivonello R, De Martino MC, De Leo M, Tauchmanovà L, Faggiano A, Lombardi G, et al. Cushing's syndrome: aftermath of the cure. Arq Bras Endocrinol Metabol. 2007;51(8): 1381–91.
58. Tzanela M, Karavitaki N, Stylianidou C, Tsagarakis S, Thalassinos NC. Assessment of GH reserve before and after successful treatment of adult patients with Cushing's syndrome. Clin Endocrinol (Oxf). 2004;60(3):309–14.
59. Storr HL, Chan LF, Grossman AB, Savage MO. Paediatric Cushing's syndrome: epidemiology, investigation and therapeutic advances. Trends Endocrinol Metab. 2007;18(4):167–74.
60. Giraldi FP, Andrioli M, Marinis LD, Bianchi A, Giampietro A, Martin MD, et al. Significant GH deficiency after long-term cure by surgery in adult patients with Cushing's disease. Eur J Endocrinol. 2007;156(2):233–9.
61. Palmieri S, Morelli V, Salcuni AS, Eller-Vainicher C, Cairoli E, Zhukouskaya VV, et al. GH secretion reserve in subclinical hypercortisolism. Pituitary. 2014;17(5):470–6.
62. Johannsson G, Sunnerhagen KS, Svensson J. Baseline characteristics and the effects of two years of growth hormone replacement therapy in adults with growth hormone deficiency previously treated for Cushing's disease. Clin Endocrinol (Oxf). 2004;60(5):550–9.
63. Shaker JL, Lukert BP. Osteoporosis associated with excess glucocorticoids. Endocrinol Metab Clin North Am. 2005;34(2):341–56. viii–ix.
64. Dos Santos CV, Vieira Neto L, Madeira M, Alves Coelho MC, de Mendonça LMC, Paranhos-Neto F de P, et al. Bone density and microarchitecture in endogenous hypercortisolism. Clin Endocrinol (Oxf). 2015; 83(4):468-74
65. Aulinas A, Valassi E, Webb SM. Prognosis of patients treated for Cushing syndrome. Endocrinol Nutr. 2014;61(1):52–61.
66. Kristo C, Jemtland R, Ueland T, Godang K, Bollerslev J. Restoration of the coupling process and normalization of bone mass following successful treatment of endogenous Cushing's syndrome: a prospective, long-term study. Eur J Endocrinol. 2006;154(1):109–18.
67. Manning PJ, Evans MC, Reid IR. Normal bone mineral density following cure of Cushing's syndrome. Clin Endocrinol (Oxf). 1992;36(3):229–34.
68. Mancini T, Doga M, Mazziotti G, Giustina A. Cushing's syndrome and bone. Pituitary. 2004;7(4):249–52.
69. Di Somma C, Pivonello R, Loche S, Faggiano A, Klain M, Salvatore M, et al. Effect of 2 years of cortisol normalization on the impaired bone mass and turnover in adolescent and adult patients with Cushing's disease: a prospective study. Clin Endocrinol (Oxf). 2003;58(3):302–8.
70. Barahona M-J, Sucunza N, Resmini E, Fernández-Real J-M, Ricart W, Moreno-Navarrete J-M, et al. Deleterious effects of glucocorticoid replacement on bone in women after long-term remission of Cushing's syndrome. J Bone Miner Res. 2009;24(11):1841–6.
71. Alshekhlee A, Kaminski HJ, Ruff RL. Neuromuscular manifestations of endocrine disorders. Neurol Clin. 2002;20(1):35–58, v–vi.

72. Kilgour AHM, Gallagher IJ, MacLullich AMJ, Andrew R, Gray CD, Hyde P, et al. Increased skeletal muscle 11βHSD1 mRNA is associated with lower muscle strength in ageing. PLoS One. 2013;8(12), e84057.
73. Wajchenberg BL, Bosco A, Marone MM, Levin S, Rocha M, Lerário AC, et al. Estimation of body fat and lean tissue distribution by dual energy X-ray absorptiometry and abdominal body fat evaluation by computed tomography in Cushing's disease. J Clin Endocrinol Metab. 1995;80(9):2791–4.
74. Minetto MA, Lanfranco F, Botter A, Motta G, Mengozzi G, Giordano R, et al. Do muscle fiber conduction slowing and decreased levels of circulating muscle proteins represent sensitive markers of steroid myopathy? A pilot study in Cushing's disease. Eur J Endocrinol. 2011;164(6):985–93.
75. Schakman O, Kalista S, Barbé C, Loumaye A, Thissen JP. Glucocorticoid-induced skeletal muscle atrophy. Int J Biochem Cell Biol. 2013;45(10):2163–72.
76. Faggiano A, Pivonello R, Melis D, Filippella M, Di Somma C, Petretta M, et al. Nephrolithiasis in Cushing's disease: prevalence, etiopathogenesis, and modification after disease cure. J Clin Endocrinol Metab. 2003;88(5):2076–80.
77. Faggiano A, Pivonello R, Melis D, Alfieri R, Filippella M, Spagnuolo G, et al. Evaluation of circulating levels and renal clearance of natural amino acids in patients with Cushing's disease. J Endocrinol Invest. 2002;25(2):142–51.
78. Pereira AM, Tiemensma J, Romijn JA. Neuropsychiatric disorders in Cushing's syndrome. Neuroendocrinology. 2010;92 Suppl 1:65–70.
79. Sonino N, Fava GA. Psychiatric disorders associated with Cushing's syndrome. Epidemiology, pathophysiology and treatment. CNS Drugs. 2001;15(5):361–73.
80. Tiemensma J, Biermasz NR, Middelkoop HAM, van der Mast RC, Romijn JA, Pereira AM. Increased prevalence of psychopathology and maladaptive personality traits after long-term cure of Cushing's disease. J Clin Endocrinol Metab. 2010;95(10):E129–41.
81. Oßwald A, Plomer E, Dimopoulou C, Milian M, Blaser R, Ritzel K, et al. Favorable long-term outcomes of bilateral adrenalectomy in Cushing's disease. Eur J Endocrinol. 2014;171(2):209–15.
82. Ragnarsson O, Glad CAM, Berglund P, Bergthorsdottir R, Eder DN, Johannsson G. Common genetic variants in the glucocorticoid receptor and the 11β-hydroxysteroid dehydrogenase type 1 genes influence long-term cognitive impairments in patients with Cushing's syndrome in remission. J Clin Endocrinol Metab. 2014;99(9):E1803–7.
83. Resmini E, Santos A, Gómez-Anson B, Vives Y, Pires P, Crespo I, et al. Verbal and visual memory performance and hippocampal volumes, measured by 3-Tesla magnetic resonance imaging, in patients with Cushing's syndrome. J Clin Endocrinol Metab. 2012;97(2):663–71.
84. Ragnarsson O, Berglund P, Eder DN, Johannsson G. Long-term cognitive impairments and attentional deficits in patients with Cushing's disease and cortisol-producing adrenal adenoma in remission. J Clin Endocrinol Metab. 2012;97(9):E1640–8.
85. Tiemensma J, Kokshoorn NE, Biermasz NR, Keijser B-JSA, Wassenaar MJE, Middelkoop HAM, et al. Subtle cognitive impairments in patients with long-term cure of Cushing's disease. J Clin Endocrinol Metab. 2010;95(6):2699–714.
86. Crespo I, Granell-Moreno E, Santos A, Valassi E, Vives-Gilabert Y, De Juan-Delago M, et al. Impaired decision making and selective cortical frontal thinning In Cushing's syndrome. Clin Endocrinol (Oxf). 2014;81(6):826–33.
87. Papakokkinou E, Johansson B, Berglund P, Ragnarsson O. Mental fatigue and executive dysfunction in patients with Cushing's syndrome in remission. Behav Neurol. 2015;2015:173653.
88. Resmini E, Santos A, Gómez-Anson B, López-Mourelo O, Pires P, Vives-Gilabert Y, et al. Hippocampal dysfunction in cured Cushing's syndrome patients, detected by 1H-MR-spectroscopy. Clin Endocrinol (Oxf). 2013;79(5):700–7.
89. Tendolkar I, van Beek M, van Oostrom I, Mulder M, Janzing J, Voshaar RO, et al. Electroconvulsive therapy increases hippocampal and amygdala volume in therapy refractory depression: a longitudinal pilot study. Psychiatry Res. 2013;214(3):197–203.

90. van der Werff SJA, Andela CD, Nienke Pannekoek J, Meijer OC, van Buchem MA, Rombouts SARB, et al. Widespread reductions of white matter integrity in patients with long-term remission of Cushing's disease. Neuroimage Clin. 2014;4:659–67.

91. Pires P, Santos A, Vives-Gilabert Y, Webb SM, Sainz-Ruiz A, Resmini E, et al. White matter alterations in the brains of patients with active, remitted, and cured Cushing syndrome: a DTI study. AJNR Am J Neuroradiol. 2015;36(6):1043–8.

92. Andela CD, van der Werff SJA, Pannekoek JN, van den Berg SM, Meijer OC, van Buchem MA, et al. Smaller grey matter volumes in the anterior cingulate cortex and greater cerebellar volumes in patients with long-term remission of Cushing's disease: a case-control study. Eur J Endocrinol. 2013;169(6):811–9.

93. Santos A, Resmini E, Crespo I, Pires P, Vives-Gilabert Y, Granell E, et al. Small cerebellar cortex volume in patients with active Cushing's syndrome. Eur J Endocrinol. 2014;171(4):461–9.

94. van der Werff SJA, Pannekoek JN, Andela CD, Meijer OC, van Buchem MA, Rombouts SARB, et al. Resting-state functional connectivity in patients with long-term remission of Cushing's disease. Neuropsychopharmacology. 2015;40(8):1888–98.

95. Bas-Hoogendam JM, Andela CD, van der Werff SJA, Pannekoek JN, van Steenbergen H, Meijer OC, et al. Altered neural processing of emotional faces in remitted Cushing's disease. Psychoneuroendocrinology. 2015;59:134–46.

96. Andela CD, van Haalen FM, Ragnarsson O, Papakokkinou E, Johannsson G, Santos A, et al. Mechanisms in endocrinology: Cushing's syndrome causes irreversible effects on the human brain: a systematic review of structural and functional magnetic resonance imaging studies. Eur J Endocrinol. 2015;173(1):R1–14.

97. da Mota F, Murray C, Ezzat S. Overt immune dysfunction after Cushing's syndrome remission: a consecutive case series and review of the literature. J Clin Endocrinol Metab. 2011;96(10):E1670–4.

98. Colao A, Pivonello R, Faggiano A, Filippella M, Ferone D, Di Somma C, et al. Increased prevalence of thyroid autoimmunity in patients successfully treated for Cushing's disease. Clin Endocrinol (Oxf). 2000;53(1):13–9.

99. Takasu N, Komiya I, Nagasawa Y, Asawa T, Yamada T. Exacerbation of autoimmune thyroid dysfunction after unilateral adrenalectomy in patients with Cushing's syndrome due to an adrenocortical adenoma. N Engl J Med. 1990;322(24):1708–12.

100. Carluccio A, Sundaram NK, Chablani S, Amrock LG, Lambert JK, Post KD, et al. Predictors of quality of life in 102 patients with treated Cushing's disease. Clin Endocrinol (Oxf). 2015;82(3):404–11.

101. Webb SM, Badia X, Barahona MJ, Colao A, Strasburger CJ, Tabarin A, et al. Evaluation of health-related quality of life in patients with Cushing's syndrome with a new questionnaire. Eur J Endocrinol. 2008;158(5):623–30.

102. Santos A, Resmini E, Martínez-Momblán MA, Crespo I, Valassi E, Roset M, et al. Psychometric performance of the Cushing QoL questionnaire in conditions of real clinical practice. Eur J Endocrinol. 2012;167(3):337–42.

103. van der Pas R, de Bruin C, Pereira AM, Romijn JA, Netea-Maier RT, Hermus AR, et al. Cortisol diurnal rhythm and quality of life after successful medical treatment of Cushing's disease. Pituitary. 2013;16(4):536–44.

104. Santos A, Crespo I, Aulinas A, Resmini E, Valassi E, Webb SM. Quality of life in Cushing's syndrome. Pituitary. 2015;18(2):195–200.

105. Lindsay JR, Nansel T, Baid S, Gumowski J, Nieman LK. Long-term impaired quality of life in Cushing's syndrome despite initial improvement after surgical remission. J Clin Endocrinol Metab. 2006;91(2):447–53.

106. Wagenmakers MA. EM, Netea-Maier RT, Prins JB, Dekkers T, den Heijer M, Hermus ARMM. Impaired quality of life in patients in long-term remission of Cushing's syndrome of both adrenal and pituitary origin: a remaining effect of long-standing hypercortisolism? Eur J Endocrinol. 2012;167(5):687–95.

107. van der Klaauw AA, Kars M, Biermasz NR, Roelfsema F, Dekkers OM, Corssmit EP, et al. Disease-specific impairments in quality of life during long-term follow-up of patients with different pituitary adenomas. Clin Endocrinol (Oxf). 2008;69(5):775–84.
108. van Aken MO, Pereira AM, Biermasz NR, van Thiel SW, Hoftijzer HC, Smit JWA, et al. Quality of life in patients after long-term biochemical cure of Cushing's disease. J Clin Endocrinol Metab. 2005;90(6):3279–86.
109. Andela CD, Scharloo M, Pereira AM, Kaptein AA, Biermasz NR. Quality of life (QoL) impairments in patients with a pituitary adenoma: a systematic review of QoL studies. Pituitary. 2015;18(5):752–76.
110. Tiemensma J, Daskalakis NP, van der Veen EM, Ramondt S, Richardson SK, Broadbent E, et al. Drawings reflect a new dimension of the psychological impact of long-term remission of Cushing's syndrome. J Clin Endocrinol Metab. 2012;97(9):3123–31.
111. Milian M, Teufel P, Honegger J, Gallwitz B, Schnauder G, Psaras T. The development of the Tuebingen Cushing's disease quality of life inventory (Tuebingen CD-25). Part I: construction and psychometric properties. Clin Endocrinol (Oxf). 2012;76(6):851–60.
112. Milian M, Teufel P, Honegger J, Gallwitz B, Schnauder G, Psaras T. The development of the Tuebingen Cushing's disease quality of life inventory (Tuebingen CD-25). Part II: normative data from 1784 healthy people. Clin Endocrinol (Oxf). 2012;76(6):861–7.
113. Milian M, Honegger J, Gerlach C, Hemeling X, Psaras T. Similar psychopathological profiles in female and male Cushing's disease patients after treatment but differences in the pathogenesis of symptoms. Neuroendocrinology. 2014;100(1):9–16.
114. Milian M, Kreitschmann-Andermahr I, Siegel S, Kleist B, Führer-Sakel D, Honegger J, et al. Validation of the Tuebingen CD-25 Inventory as a measure of postoperative health-related quality of life in patients treated for Cushing's disease. Neuroendocrinology. 2015;102(1-2):60–7.
115. Keil MF. Quality of life and other outcomes in children treated for Cushing syndrome. J Clin Endocrinol Metab. 2013;98(7):2667–78.
116. Guslandi M. Steroid ulcers: Any news? World J Gastrointest Pharmacol Ther. 2013;4(3):39–40.
117. Hatipoglu E, Caglar AS, Caglar E, Ugurlu S, Tuncer M, Kadioglu P. Peptic ulcer disease in endogenous hypercortisolism: myth or reality? Endocrine. 2015;50(2):489–95.
118. Etxabe J, Vazquez JA. Morbidity and mortality in Cushing's disease: an epidemiological approach. Clin Endocrinol (Oxf). 1994;40(4):479–84.
119. Dekkers OM, Biermasz NR, Pereira AM, Roelfsema F, van Aken MO, Voormolen JHC, et al. Mortality in patients treated for Cushing's disease is increased, compared with patients treated for nonfunctioning pituitary macroadenoma. J Clin Endocrinol Metab. 2007;92(3):976–81.
120. Clayton RN, Raskauskiene D, Reulen RC, Jones PW. Mortality and morbidity in Cushing's disease over 50 years in Stoke-on-Trent, UK: audit and meta-analysis of literature. J Clin Endocrinol Metab. 2011;96(3):632–42.
121. Graversen D, Vestergaard P, Stochholm K, Gravholt CH, Jørgensen JOL. Mortality in Cushing's syndrome: a systematic review and meta-analysis. Eur J Intern Med. 2012;23(3):278–82.
122. van Haalen FM, Broersen LHA, Jorgensen JO, Pereira AM, Dekkers OM. Management of endocrine disease: mortality remains increased in Cushing's disease despite biochemical remission: a systematic review and meta-analysis. Eur J Endocrinol. 2015;172(4):R143–9.
123. Debono M, Bradburn M, Bull M, Harrison B, Ross RJ, Newell-Price J. Cortisol as a marker for increased mortality in patients with incidental adrenocortical adenomas. J Clin Endocrinol Metab. 2014;99(12):4462–70.
124. Ritzel K, Beuschlein F, Mickisch A, Osswald A, Schneider HJ, Schopohl J, et al. Clinical review: outcome of bilateral adrenalectomy in Cushing's syndrome: a systematic review. J Clin Endocrinol Metab. 2013;98(10):3939–48.

Cushing's Disease, Refining the Definition of Remission and Recurrence

Jeremy N. Ciporen, Justin S. Cetas, Shirley McCartney, and Maria Fleseriu

Abstract More than ten decades have passed since the first case of Cushing's disease (CD) was presented and documented; yet CD remains one of the most challenging diseases to diagnose, and treat, in medicine today. Patients frequently have musculoskeletal weakness, hypertension, diabetes, cardiovascular disease, and infectious and psychiatric complications at diagnosis. These symptoms and co-morbidities present more commonly as a continuum rather than all at once, thus making an initial diagnosis more difficult and often, there is a significant delay in diagnosis. Primary CD treatment is surgical in most cases, generally through a transsphenoidal approach; however, there are many challenges in defining disease remission after surgery. Recurrent disease has been shown to be more frequent than previously thought, occurring in approximately a quarter of patients; thus early diagnosis of disease recurrence is essential.

In this chapter, we review the complex evaluation needed for defining CD remission vs. persistent disease after surgery, challenges in how to diagnose early recurrent CD and furthermore discuss the assessment criteria used for remission when patients are treated with medical therapy.

Keywords Cushing's disease • Cushing syndrome • Remission of Cushing's • Recurrence of Cushing • Hypercortisolemia • Urinary-free cortisol • Salivary cortisol • Overnight dexamethasone suppression test

Introduction

Cushing's disease (CD) is the most common cause of endogenous hypercortisolism, Cushing's syndrome (CS), and causes significant morbidity and mortality in those affected. In 1932, Harvey Cushing described CD as a condition of chronic glucocorticoid excess resulting from elevated adrenocorticotrophic hormone (ACTH)

J.N. Ciporen, M.D. • J.S. Cetas, M.D., Ph.D. • S. McCartney, Ph.D.
M. Fleseriu, M.D., F.A.C.E. (✉)
Department of Neurological Surgery and Northwest Pituitary Center,
Oregon Health & Science University, Mail Code CH8N, 3303 SW
Bond Avenue, Portland, OR 97239, USA
e-mail: ciporen@ohsu.edu; cetasj@ohsu.edu; mccartns@ohsu.edu; fleseriu@ohsu.edu

© Springer International Publishing Switzerland 2017   225
E.B. Geer (ed.), *The Hypothalamic-Pituitary-Adrenal Axis in Health and Disease*, DOI 10.1007/978-3-319-45950-9_12

secretion [1]. Approximately 60–80 % of ACTH dependent cases are caused by CD, overproduction of ACTH by the corticotroph cells of the pituitary gland, which stimulates the adrenal gland to produce cortisol. The prevalence of CD is approximately 40 cases per million [2] and comes with significant morbidity and a 50 % mortality at 5 years in untreated patients [3, 4]. In addition to timely diagnosis and treatment, ensuring disease remission after either surgery or medical treatment and detecting recurrence(s) is essential.

Patients with hypercortisolemia experience various symptoms and signs noted on physical examination, consistent with persistent elevation in glucocorticoids: moon face, supraclavicular or dorsoclavicular fat, truncal obesity, skin striae, thinning and easily bruisable skin, and proximal muscle weakness [5]. Hypertension, glucose intolerance or diabetes, cardiomyopathy, and osteoporosis are very frequently observed. Patients with CS may also suffer from depression, cognitive deficits, sleep deprivation, and emotional lability [6, 7]. Cushing's disease patients, in the majority of cases, have small pituitary tumors, microadenomas (<10 mm). Large corticotroph pituitary macroadenomas (>10 mm) are rare, but present an additional treatment challenge [8].

Clinical Assessment

Diagnostic Biochemical Testing and Imaging

Despite advances in biochemical testing, imaging, medical therapies, neurosurgical, and radiosurgical techniques, CD continues to challenge patients and physicians alike. The first step in clinically assessing a patient who potentially has CS is to examine medication usage to definitively rule out exogenous steroid usage. Subsequently, there are different biochemical tests and criteria that can be used in an effort to define presence of CS, and measure disease remission, persistence or recurrence. However, controversy surrounds testing sensitivity and specificity variability, and the effects of variations in diurnal elevated cortisol secretion levels, when used as a disease measure. The most frequently used biochemical tests are 24 hours (h) urine-free cortisol (UFC) levels, low-dose dexamethasone suppression test (LDDST), overnight dexamethasone suppression test (ODST) [9, 10], serum midnight (late-night) cortisol, adrenocorticotropic hormone (ACTH) levels, and a midnight or late-night salivary cortisol concentration test (LNSC).

Screening for newly diagnosed Cushing's patient usually starts with two or three of the following tests: UFC (usually performed twice) test, a 1 mg ODST, a longer low-dose DST (2 mg) over 48 h, LDDST (if the 1 mg DST is equivocal or non-diagnostic) or a LNSC. Urine-free cortisol measures cortisol that is not bound to cortisol-binding globulin and represents integrated adrenal cortisol secretion over 24 h; in CS, the proportion of free cortisol increases, thus the urinary cortisol also increases. A low-dose dexamethasone suppression test and ODST will detect loss of normal feedback (failure to suppress cortisol to low-dose glucocorticoids) while a high LNSC reflects loss of normal diurnal variation of cortisol.

If any of these tests is abnormal, other potential causes of hypercortisolism should be ruled out such as pregnancy, alcohol dependence, morbid obesity, depression, and poorly controlled diabetes. Confirmatory tests (which are not needed if UFC is 3–4×the upper limit of normal; ULN) [9, 10] include a dexamethasone corticotropin-releasing hormone (CRH) test or a desmopressin test [10].

If ACTH is elevated, localization tests will determine if the ACTH source is pituitary (~80 %) vs. an ectopic ACTH secreting tumor (~20 %). If ACTH is low, then the adrenal gland is the source of cortisol excess [10, 11]. Imaging modalities include brain magnetic resonance (MR) imaging with a pituitary gland protocol, computed tomography (CT) with and without contrast of the chest and abdomen/pelvis to evaluate for an adrenal mass or an ectopic source of ACTH. An octreotide scan can be also used to identify the location of ectopic ACTH-producing cells; however, sensitivity is low overall. Of note, one should be aware that one or all of these imaging modalities can be normal [10]. Petrosal sinus or cavernous sinus sampling is best used to confirm that the pituitary gland is the source of abnormal ACTH production. If the ACTH ratio is 2:1 before and 3:1 after CRH administration, this is considered diagnostic for localization; however, accuracy is lower (approximately 40 %) in determining the lateralization side of the lesion within the sella or supra sellar region [12, 13].

Treatment of Cushing's Disease

Transphenoidal surgery (TSS) is the first-line treatment for CD [14]. There is debate as to which clinical method and values best define CD remission and predicts outcome. Reversal of clinical features and normalization of biochemical changes with long-term control are the goals of treatment. Patients with longstanding CD in remission after surgery will initially experience adrenal insufficiency (AI) due to suppression of normal ACTH and it may require months to years for the hypothalamic–pituitary–adrenal axis (HPA) to fully recover. Supraphysiologic doses of glucocorticoids are required in the immediate postoperative period with subsequent tapering to normal physiologic doses. Criteria for disease remission vary significantly, but include resolution of clinical symptoms related to hypercortisolism [15, 16], need for corticosteroid replacement for greater than 6 months after TSS [17], hypocortisolemia/eucortisolemia [18], and presence of clinical and laboratory signs of low cortisol and AI [19, 20].

Remission rates after surgery for microadenomas range from 65 to 90 % [5] with low surgical morbidity in the hands of an experienced neurosurgeon. However, a disease recurrence rate at 10 years might reach approximately 20–25 % [21]. Remission rates for macroadenomas are much lower than for microadenomas [5].

For patients with recurrent or persistent disease after first surgery, a second surgery is sometimes recommended [14]; however, less than 50 % of these patients achieve disease remission. In cases of persistent or recurrent disease after TSS, stereotactic radiosurgery either single- or multi-staged can be used; however, the therapeutic effects of radiation can take 3–5 years to be realized [11, 22–24]. In severe, refractory cases a bilateral adrenalectomy can be performed, but there are substantial risks related to permanent hypocortisolemia, adrenal insufficiency, and risk of Nelson's Syndrome [11, 22]. Medical therapy is now used more frequently in the treatment algorithm [25–29].

Criteria for evaluation of disease remission and recurrence in CD patients are reviewed below.

Evaluating and Defining CD Remission vs. Persistent Disease After Surgery

Transsphenoidal surgery is the mainstay of treatment in patients with CD [14]. After surgery, it is essential to determine which patients are in remission and which patients have persistent disease and will require further treatment. Overall, micro-surgical case series publications detailing outcomes of TSS are more common given historical use; presently, there are few endoscopic endonasal approach (EEA) publications addressing CD outcomes [30–35]. Using heterogeneous criteria for defining remission after TSS, a remission rate of 72–90 % [35] is reported and after EEA reports range from 60–90 % for microadenomas [35, 36]. Studies are largely published by centers with an experienced pituitary surgeon and it can be envisioned that overall rates might be lower in general practice. Remission rates for macroadenomas are lower than for microadenomas [5, 37]. Interestingly, microsurgical TSS vs. EEA for CD have not yielded a significant difference in biochemical outcomes, yet.

Criteria used to define remission (as well as recurrence) of CD after TSS are heterogeneously described in the literature, and report use of various single or combined biochemical markers, with or without assessment of clinical features. Follow-up duration to determine both remission and recurrence is also variable, however similar rates are usually reported for early remission (≤6 months) and late remission (>6 months) [23].

Follow-up is an essential criterion for detecting recurrence; for example, it has been shown that in some cases disease can recur even two decades later [38, 39]. There is no single test that has proven a true litmus to define remission or recurrence of CD after TSS. However, low serum cortisol immediately after surgery, 24 h UFC, and LNSC appear to have a higher sensitivity and specificity in comparison to other biochemical markers such as serum morning cortisol, ACTH levels, and LDDST.

In most studies that used biochemical markers to determine remission and recurrence rates, 24 h UFC was the most common test performed either alone or in combination with serum cortisol with or without LDDST. However, over the last decade it has been suggested that LNSC may more accurately establish remission and recurrence after TSS for CD than 24 h UFC [40–42].

Timeline of Biochemical Testing

The timeline of evaluation is also important. While serial serum cortisol in the 24–48 h after surgery is now considered "the norm" in determining an initial response to TSS, there is a subset of patients (5.6%) with delayed remission [43]. These data support the notion that critical evaluation of a subset of patients without immediate postoperative remission is essential before making definitive treatment recommendations.

The risk of short- and long-term recurrence has been clearly demonstrated [37, 38, 44–51]. Studies that report longer follow-up exhibit higher recurrence rate after a previously documented remission. Patil et al. (2008) noted that CD recurrence incidence was 25.5% in patients followed for 5 years [52]. Interestingly, the 3-year actuarial recurrence rates of patients with postoperative cortisol of >2 and ≤2 μg/dL were 14.1 and 7.0%, respectively [52].

Patients with macroadenomas reportedly have a higher incidence of recurrence than those with microadenomas [37, 53]. This is thought to be due to adenoma size and involvement of surrounding critical structures limiting the extent of safe resection. Another factor that might contribute to recurrence, regardless of tumor size, is tumor invasiveness, most commonly cavernous sinus involvement [35].

A CD patient's status post TSS needs to be followed for their lifetime and an individualized management approach should be based on whether postoperative serum cortisol values are low, normal, or high. Especially in patients who have been treated with medical therapy before surgery, measuring late-night salivary or serum cortisol to exclude hypercortisolemia is needed, in the absence of AI [14].

Standardizing biochemical endpoints, duration of follow-up, size of the tumor and score, would further assist in defining remission and recurrence, which would guide further treatment and potentially improve outcomes.

Biochemical Testing Used to Define Remission After Surgery

As mentioned previously, there is significant heterogeneity in type of tests used (serum cortisol, UFC, and/or LDDST, LNSC) in addition to clinical evaluation, cut-offs considered normal, and timeline of assessments [2, 5, 14, 23]. Among the definitions for remission of CD are reversal of the following: clinical signs and symptoms of CD [54], hypercortisolemia [55], and clinical features [15, 56]. Laboratory evidence of AI [19, 20], the need for corticosteroid replacement for >6 months after TSS [17], and normalization of morning cortisol levels and UFC levels have also been reported to define remission.

The 2008 Endocrine Society Consensus Statement recommends assessing remission by the measurement of morning serum cortisol during the first postoperative week, either by withholding treatment with glucocorticoids or by using low doses of dexamethasone suppression (<1 mg) [10]. When serum cortisol levels are between 2 and 5 μg/dL, the patient can be considered in remission and observed without additional CD treatment [5, 10].

A recent large review [23] highlights the use of different parameters to assess remission and recurrence and summarizes the postoperative remission and recurrence rates of over 6000 patients. The data are somewhat limited because clinical improvements with or without biomarker changes are reported with various follow-up durations: 22 studies reported remission using biochemical evaluation only and 16 used a combination of biochemical and clinical parameters. For the serum morning cortisol postoperatively, the cut-off of 1.8 μg/dL (50 nmol/L) was most consistently used to define remission, but ranged from 50 to 275.9 nmol/L in some studies.

Analyzing all studies that used all of the following parameters, midnight serum cortisol, UFC, and LDDST, a 75.8 % rate of remission over 76.5 months was reported, while studies that used morning serum cortisol, LDDST with variable cut-offs and UFC reported a 71.7 % rate of remission over 67.2 months. Using LDDST achieved similar rates, 77.37 % remission over 55.2 months vs. UFC 77.4 % remission over 55.2 months of remission (Table 1) [18, 41–43, 46, 47, 49, 51, 52, 56–68].

An overall remission rate in 74 studies published between 1976 and 2014 [2], involving 6091 patients CD after TSS was reportedly 25–100 %, (mean 77.7 % and median 78.2 %). Recurrence rates ranged from 0 to 65.5 % (mean 13.4 % and median 10.6 %). This review included studies that overlapped significantly with those studies analyzed by Petersenn et al. [23]. Similar to previous data, the studies included were heterogeneous with a wide number of patients reported (range 6–668) and large variations in follow-up duration (Table 1). Furthermore, the criteria used to define remission and recurrence were not uniformly reported.

A subanalysis of studies with 30 or more patients and a minimum mean/median follow-up of 6 months reported that the percentage of failed pituitary surgeries for CD ranged between 5.7and 63 % with a mean of 31.4 % and median of 29.4 % [2]. In studies that further stratified results by adenoma size (micro vs. macro) the mean rates of remission were 85 % for microadenomas and 58 % for macroadenomas (Table 1) [18, 41–43, 46, 47, 49, 51, 52, 56–68].

Serial Serum Cortisol and ACTH in the Immediate Postoperative Period

In one study, Swearingen et al. looked at factors associated with remission and determined cure by using both fasting serum cortisol levels less than 138 nmol/L and UFC less than 55 nmol/day [49]. The 5-year cure rate for patients was lower than the 10-year one: 96 % for microadenomas, 96 % for macroadenomas vs. 93 % and 55 %, respectively.

In other studies, a serum cortisol of 50 nmol/L (1.8 μg/dL at ODST after surgery) and normal 24 h UFC was most frequently used to define remission [2]. Overall, patients with serum cortisol levels of <2 μg/dL in the immediate postoperative period achieved long-term remission at 10 years in approximately 90 % of cases [5, 15, 18, 21, 45, 51, 56, 61, 69, 70].

Table 1 Remission: clinical and biochemical assessments

	Author	No. of patients (n)	Follow-up (months)	Remission rate (%)
Overnight dexa	Bochicchio et al. [41]	668	Mean 24; median 13	76.3
	Chee et al. [56]	61	Median 88	78.7 (at 12 months) 67.2 (at 14–211 months)
	Hofmann et al. [65]	426 369 (adenomectomy)	73.2±43.2 (range, 13–207)	75.9
UFC	Fomekong et al. [59]	40	Mean 84±44	80
	Kim et al. [47]	54	Mean 50.7	70.4
	Patil et al. [52]	215	Mean 45	85.6
	Prevedello et al. [64]	167	Mean 39 (range, 6–157)	88.6 (100 % microadenoma)
mSC + UFC	Acebes et al. [57]	44	Mean 49 (range, 19–102)	89
	Invitti et al. [66]	288	Not available (for n = 288) Range 6–180 (for n = 129)	69
	Jagannathan et al. [68]	261	Mean 84 (range, 12–215)	96.5
	Salenave et al. [62]	54	Mean 19.9±22.7 (range, 1–89)	66.5
	Swearingen et al. [49]	161 154 (TSS)	Mean 104.4; median 96 (range, 12–240)	90 (microadenoma) 65 (macroadenoma) 85 (total)
	Valassi et al. [43]	620	Range, 1–300	70.5
	Yap et al. [51]	97	Mean 92 (range, 6–348)	68.5 (immediate)

(continued)

Table 1 (continued)

	Author	No. of patients (n)	Follow-up (months)	Remission rate (%)
Morning SC	Bou Khalil et al. [58]	127	Mean 48.8 (range, 3.7–148.7)	79.5 (early) 62.9 (maintained)
	Carrasco et al. [42]	68	Range, 6–12	74
	Esposito et al. [18]	40	Mean 33 (minimum 14)	79.5
	Flitsch et al. [46]	147	61	93
	Hassan-Smith et al. [60]	80	Median 55.2	83 (early remission) 72 (cure)
	Rees et al. [61]	54	Median 72 (range, 6–252)	77
	Storr et al. [63]	155 (microadenoma) 28 (macroadenoma)	Not available	Overall 61 (microadenoma) 32 (macroadenoma)
	Trainer et al. [67]	48	Median 40 (range, 15–70)	42
SC + UFC + dexa suppression	Atkinson et al. [21]	63	Mean 115.2 (range, 1–252)	71.4 (early) 56 (overall)
	Bakiri et al. [89]	50	Median 71.5 (range 21–219)	72
	Barbetta et al. [90]	68	Median 57.5 (range, 12–252)	68 (early) 79 (persistent)
	Jehle et al. [87]	193	Mean 57.6±42 (range, 8.4–148.8)	80.8 (overall)
	Alwani et al. [17]	79	Median 84 (range, 6–197)	65 (immediate)
	Shimon et al. [91]	74	Mean 50.4±34.8	78

Overnight dexamethasone (dexa), urine-free cortisol (UFC), mean serum cortisol (mSC), and morning SC

In a single-center study [71] in 52 patients with CD followed over a minimum of 6 years, early postoperative cortisol <2 µg/dL and ACTH<5 pg/mL was a sensitive predictor of remission. The positive predictive value (PPV) for remission with postoperative nadir cortisol <2 µg/dL and ACTH<5 pg/mL was 100% (p<0.005). The PPV for non-remission of ACTH>15 pg/mL was 87.5%. Interestingly, no patients with postoperative cortisol >10 µg/dL were found to have delayed remission. While this study found a lower cutoff value for ACTH and cortisol (<5 pg/mL and <2 µg/dL, respectively) than other studies to be highly predictive of remission, no level predicted the lack of recurrence. The addition of ACTH to cortisol measurements might increase accuracy of remission assessments [72].

Late-Night Salivary Cortisol

In another single-center study that included 164 surgical CD patients, LNSC [40] had a 94% sensitivity and 80% specificity for remission at a cut off of 1.9 nmol/L within 3 months of TSS. A nadir morning serum cortisol of <5 µg/dL and nadir 24 h UFC of <23 µg was used to define remission, in these patients. Recurrence was established with LNSC at a cutoff of 7.4 nmol/L (75% sensitivity and 95% specificity) and 1.6-fold above normal 24 h UFC (68% sensitivity and 100% specificity), respectively, at a median follow-up of 53.5 months.

Delayed Remission After Surgery

Hormonal assessment in the immediate postoperative period, in rare cases, may be misleading in a subset of patients after TSS for CD because of delayed remission. A retrospective review of 620 patients who underwent TSS for CD between 1982 and 2007 in two large centers [43] classified outcomes into three groups based on the postoperative pattern of cortisol testing: IC for immediate control, NC for no control, and DC for delayed control. The IC group had a 70.5% rate (437 of 620 patients) of hypocortisolism and/or cortisol normalization throughout the postoperative follow-up period. The NC group reported 23.9% (148 of the 620 patients) with persistent hypercortisolism, while the DC group reported 5.6% (35 of 620 patients) with early elevated or normal UFC levels and developed delayed and persistent cortisol decrease after an average of 38±50 days [43].

Degree of Tumor Invasiveness and Remission

Shin et al. studied 49 patients who underwent EEA resection at a single institution over an 11-year period. The endocrinologic remission rates were analyzed according to degree of invasiveness, by Knosp score [35]. The Knosp score (ranging from

0 to 4) is based on the tumors relationship, as seen on preoperative MRI, to the cavernous segment of the internal carotid artery (ICA) [73]. In this study, the initial remission rate (36 h to 2 weeks postoperatively) was 79.6 % and was 70 % in patients with a mean follow-up of 37.5 ± 4.6 months. An initial remission rate of 80 % was reported in MRI negative adenomas, 84.8 % among noninvasive/minimally invasive adenomas and 50 % among invasive adenomas.

This further highlights the challenges of treating patients with invasive tumors. Interestingly, preoperative UFC levels were not significantly different with respect to degree of tumor invasiveness and had no significant effect on remission rate in this series. However, a higher preoperative ACTH level was associated with a higher degree of invasiveness.

Timeline to HPA Recovery After Remission of CD

Hypothalamic–pituitary–adrenal axis recovery after remission of CD is variable, between 13 and 25 months [38, 44, 74–76]. In a study of 91 patients with CS, 54 with CD [77], CD patients were divided into three groups: group 1, patients with normal postoperative pituitary function and no recurrence; group 2, patients with later recurrence after successful surgery; and group 3, patients who displayed postoperative additional anterior and posterior pituitary insufficiencies, presumably because of a more radical surgical approach. Those cured were defined by the development of postoperative tertiary AI requiring glucocorticoid replacement therapy. Recurrence occurred between 2.4 and 14.4 years after surgery (mean, 7.2 ± 4.6 years). The three CD groups were not different with respect to age, preoperative BMI, male-to-female ratio, duration of symptoms, or other biochemical parameters [77]. The authors hypothesized that this stratification would enable them to identify if normal pituitary gland tissue damage, as a result of surgery, significantly influenced HPA recovery. Plasma cortisol and ACTH, UFC, and salivary cortisol were all studied. A subgroup analysis showed that the probability of recovery at 5 years was 71 % in group 1 and 100 % in group 2. Group 3 patients had the poorest rate of recovery. Only in group 1, the probability to recovery of adrenal function was associated with younger age independent of sex, BMI, duration of symptoms, basal cortisol, and basal ACTH levels. The mean age of patients experiencing recovery was 37 years of age at the time of surgery, compared with 48 in patients without recovery.

The long-term occurrence of hypocortisolism after TSS has been hypothesized to be associated with the number of Crook's cells present [76]. Similarly, Saeger et al. [78] associated Crook's cell count and severity of glucocorticoid excess in CD. Crooke's hyaline change was first described in 1935 in the normal anterior pituitary surrounding an ACTH-secreting adenoma. Non-neoplastic corticotrophs have increased eosinophilic cytoplasm filled with perinuclear cytokeratin while the adenoma itself does not. The cause of Crooke's hyaline change is uncertain, but it is related to increased glucocorticoid or cortisol levels [79].

Another factor at play may be the duration of exogenous glucorticoid received by the patient and its effects on the adrenal gland. Sacre et al. [80] analyzed AI rates following pharmacologic glucocorticoid treatment for various inflammatory disorders and found that cumulative dose and exposure time were independent predictors of AI. Berr et al. concluded that the recovery of corticotroph function is due to residual tumor cell clusters rather than by hypothalamic CRH-mediated stimulation on normal corticotroph cells [77]. After multivariate analysis, this study identified younger patient age as an independent significant factor influencing HPA recovery in patients with CD. The preoperative degree of hypercortisolism and postoperative glucocorticoid replacement doses did not seem to be relevant.

Cushing's disease also has accompanying disturbances in growth hormone (GH) and prolactin (PRL) secretion [81–84]. A small study compared eight adults (five females and three males) with CD in remission with eight healthy patients matched for gender, BMI, and age. Remission was established by the absence of signs and symptoms during long-term follow-up of 8.2 ± 1.7 years, normalized 24 h UFC, and suppression of morning plasma cortisol concentration below 0.10 µmol/L after the administration of 1 mg dexamethasone, orally, at 2300 h (at yearly visits in an outpatient clinic). Before TSS, ACTH and cortisol levels were found to have elevated basal rates, augmented secretory pulse amplitudes, blunted or absent diurnal variation characteristics, and a loss of orderly secretory patterns [85, 86] but the 24 h secretion properties of ACTH, cortisol are normalized after clinically successful TSS [84]. Physiological recovery was determined by total secretory activity (pulsatile and non-pulsatile), diurnal rhythmicity, and the orderliness of the release process. Further studies are needed to determine if all physiological characteristics of ACTH, cortisol, GH, and PRL secretion can be consistently normalized after TSS in CD.

How to Diagnose Early Recurrent Cushing's Disease

Recurrence Rates

Definitions of recurrence are poorly characterized, heterogeneous, and furthermore, infrequently reported in the literature. The general criteria used to define recurrence include a combination of a relapse of symptoms, clinical features, and/or biochemical confirmation.

Some studies defined recurrence just on the basis of questionnaire response and results of routine endocrine reevaluation locally, without independent repeated assay in the initial center [42, 43, 47, 48, 51, 52, 57–62, 70, 87, 88]. The most frequently utilized biochemical tests to detect recurrence are 24 h UFC and 1 mg DST. However, measuring LNSC has been shown to be sensitive and is becoming more commonly used [40–42]. In the literature, some studies used the same criteria to determine remission and recurrence while others used a separate criteria for each [23]. Urine-free cortisol testing was used either alone or more commonly in combination with serum cortisol and/or LDDST as an endpoint for establishing recurrence (Table 2) [42, 43, 47, 48, 51, 52, 57–62, 70, 87, 88].

Table 2 Recurrence: clinical and biochemical assessments

	Author	No. of patients (n)	Follow-up (months)	Recurrence rate (%)
Morning cortisol + UFC	Bou Khalil et al. [58]	127	Median 50.4 (range, 7–99)	20.8 (mild or overt)
mSC + UFC + dexa	Carrasco et al. [42]	68	Mean 51 ± 30 (range, 9–90)	14.3
	Jehle et al. [87]	193	Mean 57.6 ± 42 (range, 8.4–148.8)	13.5
	Kim et al. [47]	54	Median 50.7	32.4 (at 5 years) 54.6 (at 10 years)
UFC	Fomekong et al. [59]	40	Mean 95 ± 35	9.4 (overall) 14 (microadenoma) 0 (macroadenoma)
	Guilhaume et al. [88]	64	Range, 24–36	14.28
	Patil et al. [52]	215	Mean 45	17.4
UFC + dexa	Hassan-Smith et al. [60]	80	Median 55.2	11
	Pereira et al. [70]	78	Median 84 Median 174 (range, 120–288)	9 (n = 5 of 56) 16.7 (n = 4 of 24)
	Sonino et al. [48]	162	Median 84	Not stated
	Yap et al. [51]	89	Mean 36.3	11.5
mSC + UFC	Acebes et al. [57]	44	Mean 25 (range, 2–102)	7.7
	Rees et al. [61]	54	Median 17 (range, 6–50)	5.1
	Salenave et al. [62]	54	Mean 19.9 ± 22.7 (range, 1–89)	19.5
	Valassi et al. [43]	620	Median 66	13 (total)

Overnight dexamethasone (dexa), urine-free cortisol (UFC), mean serum cortisol (mSC), and morning serum cortisol

Recurrence is reportedly lowest when a combination of LDDST and UFC was used as biochemical endpoints regardless of whether serum cortisol and/or clinical parameters were also included in the assessment of recurrence [17, 21, 87, 89–91]. Overall, recurrence rates in all studies, regardless of methods used to determine recurrence, were slightly less in the group of patients with microadenomas vs. macroadenomas, 13.4 % vs. 17.6 %, respectively, but not statistically significant [23]. The duration of follow-up ranged from 13 to 96 months [23], but follow-up time did not predict rate of recurrence. The limitations in interpreting these resultant conclusions need to be taken into account given the small number of studies involved. If one further analyzes the data for the studies that reported rates of recurrence for both microadenomas and macroadenomas, the mean rate of recurrence for microadenomas was 10.9 % (four studies) and 23.6 % (two studies) for macroadenomas [55, 63, 92]. Interestingly, studies that only used biochemical tests to determine overall recurrence rates reported a relatively similar rate (15.7 %) vs. the ones that used both clinical and biochemical endpoints to determine rate of recurrence reported 14.4 % [23]. The overall calculated recurrence rate was 15.2 % and meantime to recurrence (in the 23 studies where was reported) was 50.8 months (range 3–158 months) [23].

Despite initial data suggesting otherwise, recurrence rates in patients with cortisol in the immediate postoperative period between 2 and 5 µg/dL appear to be no greater than those seen in patients with postoperative serum cortisol levels less than 2 µg/dL [5, 15, 18, 21, 45, 51, 56, 61, 69, 70].

Biochemical Testing Timeline

Different timelines of change from normal to abnormal in some biochemical tests are also interesting. A study that looked at sequential alterations over time after surgery in 101 patients [58] found that 21 (20.8 %) presented with recurrence, 'mild' or 'overt', during long-term follow-up (median 50.4 months, range 7–99). Interestingly, vasopressin analogs and CRH tests were eventually positive in 85 and 93 % of all patients who experienced disease recurrence. Recurrence occurred less frequently and later in patients with early AI compared with patients with normal cortisol after surgery. Increase in LNSC occurred in a mean time of 38.2 months, while UFC elevation was observed at 50.6 months; however, a positive response to vasopressin analogs or CRH preceded the increase in midnight cortisol or UFC in 71 % and 64 % of patients, respectively.

Combined Biochemical Testing

Coupled dexamethasone desmopressin test (CDDT) has been also suggested as good predictor of recurrence of CD after surgery [93]. In a small study (38 patients) followed for a median of 60 months, CDDT became positive in eight of ten patients with recurrence 6–60 months before classical markers of CD. Similar to other studies, AI did not ensure lack of recurrence: six patients with immediate postsurgical corticotroph deficiency presented with recurrence; however, all patients had abnormal CDDT positivity during the 3 years after surgery with recurrence 6–60 months after CDDT positivity. CDDT has been considered an early predictor of recurrence of CD and could be of particular interest in the first 3 years after surgery, by selecting patients at high risk of recurrence despite falsely reassuring classical hormonal markers [93]. However, a comparison with LNSC in predicting recurrence remains to be determined.

Degree of Tumor Invasiveness and Recurrence

The degree of tumor invasiveness has also been shown to play a role in potentially influencing recurrence rates [35].

Impact of Delayed Remission on Recurrence

In a large study [43] (described in more detail in the remission section), 35 of 620 patients (5.6 %) had delayed control defined as early elevated or normal UFC levels and developed a delayed and persistent cortisol decrease after an average of 38 ± 50 postoperative days. These patients with *delayed remission* vs. those with *immediate control* of CD after TSS seem to have significantly higher cumulative rate of recurrence at 4.5 years, 43 % vs. 14 %, respectively over a median of 66 months after TSS with a total recurrence rate of 13 % [43]. Criteria for recurrence in this particular study included at least two abnormal tests from the following four: elevated serum cortisol or 24-h UFC, abnormal ODST, here defined as cortisol >5 µg/dL (138 nmol/L), or abnormal serum cortisol during the combination of low-dose dexamethasone suppression test and ovine or human CRH stimulation test.

Assessment Criteria in Patients with Cushing's Disease Treated with Medical Therapy

The assessment of CD remission after a patient is started on medical therapy is very complex and remains controversial, overall [2, 11]. Therapies with agents acting at the pituitary level (cabergoline, pasireotide), adrenal steroidogenesis inhibitors, and a glucocorticoid receptor blocker (mifepristone) are reviewed below, with a focus on biochemical markers and clinical improvements; mechanism of action of each drug, study design, and adverse events have been previously and extensively reviewed [25–29].

Biochemical Testing

24-Hour Urine-Free Cortisol

A retrospective analysis of 137 patients with clinical conditions suggestive of hypercortisolism, 38 with confirmed CS diagnosis and 99 without, found that UFC revealed both a combined higher positive and a lower negative likelihood ratio for diagnosing CS among first-line tests (10.7 and 0.03, respectively) [94]. Computing a receiver operating characteristic (ROC)-contrast analysis to compare the power of each single test with that of the others, alone or combined (DST+LNSC, DST+UFC and LNSC+UFC), or with that of all the tests together (DST+LNSC+UFC), UFC assay was at least as good as all the other possible combinations. The different results noted compared with other studies could be related to the liquid chromatography–mass spectrometry/ mass spectrometry (LC-MS/MS) method used for UFC. In that particular study, LNSC was measured by radio-immunometric method and serum cortisol by chemiluminescence immunoassay [94].

The reliability and reproducibility of UFC are both very important [94, 95]. Newer methods such as LC-MS/MS have revealed that the analytical performance of UFC is better than urinary cortisol:cortisone ratio in detecting CS.

Intra-patient UFC variability at diagnosis is a well-known caveat; large studies have shown up to 50 % variability [96] and overall variability in mUFC increased as UFC levels increased. However, there were no correlations between UFC and clinical features of hypercortisolism. The assay used is even more important at potentially lower values of UFC when determining remission or recurrence. Most clinical studies looking at the effects of medical therapies have measured UFC during treatment; furthermore, new clinical guidelines [14] emphasize that despite some caveats, UFC is a good marker to monitor therapy response. One important exception represents treatment with a glucocorticoid receptor blocker, in which case UFC is not reliable and monitoring has to rely solely on clinical grounds and other biochemical assessments such as glucose for example.

While for diagnosis, at least two 24-h UFC are recommended [10], the number of UFCs needed to ensure correct assessment for remission is still unclear. The UFC variability with regards to medical treatment is largely unknown. A summary of studies using UFC as marker for biochemical response on medical therapy can be found in Table *3* [9, 93, 97–108].

Late-Night Salivary Cortisol

Most of the available data on the use of LNSC in patients with CS comes from screening studies [109]; however, data looking at salivary cortisol response to short-term medical therapy are emerging. Ease of use and patient preference represent a great advantage when periodic assessments are needed.

Table 3 Summary of biochemical and clinical markers identified as endpoints for response to medical therapy in Cushing's disease and Cushing's syndrome

Author	Study design	Follow-up (months)	Medication	Patients (n)	Definition of response			
					24 h UFC	Late-night SC	Overnight Dexa	Serum Cortisol
Boscaro and Arnaldi [9]	Prospective	15 days	Pasireotide	39	Normal			
Colao et al. [100]	Prospective	6	Pasireotide	162	Normal			
Fleseriu et al. [101][a]	Prospective	6	Mifepristone	50				
Bertagna et al. [98]	Prospective	12 weeks	Osilodrostat	12	Normal and/or 50 % decrease			
Castinetti et al. [93]	Retrospective	Mean 23 (range, 6–72)	Ketoconazole	38	Normal			
Castinetti et al. [99]	Retrospective	Mean 4.05±4.1	Ketoconazole	200	Normal			
Pivonello et al. [102]	Prospective	24	Cabergoline	27	Normal			
Verhelst et al. [107]	Retrospective		Metyrapone					Normal 'mean cortisol levels'
Baudry et al. [97]	Retrospective	Mean 97 (range, 6.3–192)	Mitotane	76	Normal			
Lila et al. [104]	Prospective	5–12	Cabergoline	20			Normal	Normal midnight
Vilar et al. [108]	Prospective		Cabergoline + Ketoconazole	Normal				
Feelders et al. [103]	Prospective	82 days	Pasireotide + Cabergoline + Ketoconazole	17	Normal			

van der Pas et al. [114]	Prospective	82 days	Pasireotide + Cabergoline + Ketoconazole	17	Recovery of circadian rhythm was defined by LNSC of less than 75 % of the 09.00 am value
Trementino et al. [105]	Prospective	1–9	Pasireotide treatment Acute pasireotide test	19	Normal	A fall of >27 % of LNSC during PST calculated by ROC curve; best parameter in predicting a positive response to Pasireotide (sensitivity 91 %; specificity 100 %; positive predictive value 100 %; negative predictive value 75 %)
Trementino et al. [105]	Retrospective	15 days	Pasireotide	7	Normal

Late-night salivary cortisol (LNSC); receiver operating characteristic (ROC); serum cortisol (SC); dexamethasone (Dexa); urine-free cortisol (UFC); PST (pasireotide suppression test)

[a]Clinical improvement as primary endpoint of the study

The routine immunoassay for salivary cortisol seems to have better diagnostic performance than LC/tandem MS, although measurement of normal salivary cortisone concentrations with the latter technique is very useful in identifying samples contaminated with topical hydrocortisone [110].

Is also well known that age and metabolic syndrome affect salivary cortisol rhythm [111]. In a study which included almost 1000 samples, gender, sampling time, smoking, and interestingly perceived social support were determinants of cortisol secretion [112].

In a small subset of patients treated with subcutaneous pasireotide 600 μg bid for 15 days, LNSC was reduced in six patients at day 15 [105]. For this study, all patients had elevated LNSC, which correlated significantly with UFC levels ($r=0.97$) at baseline. Late-night salivary cortisol decreases were observed from day 1 (−20 %) and persisted until day 15 (overall mean reduction from baseline −51 %), with the greatest decrease on day 5 (−58 %). At day 15, UFC levels were decreased in all patients and normalized in a patient that also restored salivary cortisol rhythm.

Furthermore, a small study of 19 patients with active CD followed for a median of 6 months (range 1–9 months) showed that a decrease in LNSC after one dose of pasireotide might predict response to treatment [106]. Late-night salivary cortisol, serum cortisol, and plasma ACTH were assessed before and after a single dose of 600 μg pasireotide. LNSC decreased in about 82 % of patients (14/17), achieving normalization in five. Short-term pasireotide treatment was associated with a normalization of 24 h UFC at last follow-up in about 68 % of patients. Interestingly, a decrease of >27 % in LNSC during acute pasireotide (calculated by ROC curve) was the best parameter in predicting a positive response to treatment with pasireotide (positive predictive value 100 %; negative predictive value 75 %).

Despite these encouraging results, the decrease in LNSC in patients treated with pasireotide in a larger study (12-month, multicenter, Phase III study with 93 patients who had LNSC measured) did not always correlate with decrease in UFC [113]. At baseline, the linear correlation was strong ($r=0.9$). LNSC was normalized at 6 months in 37.3 % patients with baseline abnormal LNSC, comprising 40.0 % and 33.3 % patients in the 600 and 900 μg groups, respectively. However, just 10/25 patients with normalized LNSC at 6 months also had normalized UFC; seven had partial UFC control. In both 600 and 900 μg groups, LNSC decreased in UFC controlled/partially controlled patients and increased in uncontrolled patients; however, numbers within each subgroup were low. An exploratory analysis showed weak linear correlation ($r=0.2$), but moderate correlation ($r=0.5$) on the log scale between LNSC and UFC when all time points were pooled.

The effect of triple combination therapy (pasireotide, cabergoline, and ketoconazole) [114] on HPA axis has been even less well studied. Circadian rhythm (CR) at baseline was abnormal in 12 patients, but preserved in 5 patients, though there was no difference in baseline UFC between these groups. While the complete biochemical response (defined by normal 24 h UFC) was 88 % in this study, a midnight decrease of serum and salivary cortisol levels to less than 75 % of morning values (CR recovery) was noted in 6 of the 12 patients with abnormal baseline CR (3 mono-, 1 duo-, and 2 triple-therapy). Serum cortisol levels at 10 pm and midnight salivary cortisol ($p<0.05$) at day 80 were significantly lower in patients in whom CR recovered. Interestingly, CR did not recover at 80 days, despite normalization of UFC in five of these patients.

The group of patients with recovered CR and not-recovered CR (defined by midnight cortisol decrease) had no significant differences at 80 days and furthermore, despite CR recovery, patients did not report more sleep improvement vs. those without CR recovery. Theoretically, it is possible that a longer duration of CR improvement might have an effect, but further investigation and data collection is needed.

This suggests that normalization of cortisol production by medical therapy allows for recovery of hypothalamic control of normal corticotroph cell function in patients with CD. It is unclear if the centrally acting agents, pasireotide and cabergoline, have an influence on CR.

In conclusion, salivary cortisol may be a simpler and more convenient biomarker than 24-h UFC. As discussed earlier in this chapter, salivary cortisol seems more accurate than 24 h UFC in detecting recurrence during long-term follow-up after surgery [40], but its role in assessing response to medical therapy or furthermore predicting long-term response remains to be determined.

Adrenocorticotropic Hormone

The effects of medical therapy on ACTH secretion differ depending on mechanism of action. Cabergoline and pasireotide decrease ACTH, while all the other drugs, either adrenal steroidogenesis inhibitors or a glucocorticoid receptor blocker will increase ACTH [27]. Notably, ketoconazole has been shown in some studies to also have an effect on ACTH secretion [115], but this remains controversial [22].

ACTH decreased significantly ($p=0.002$) in patients who responded to treatment with Cabergoline, while was essentially unchanged in non-responders [102].

In vitro studies [114] showed that after achieving normal cortisol with medical therapy, cortisol-mediated somatostatin receptor subtype 2 (sst_2) downregulation on corticotroph adenomas is reversible at the mRNA but not at the protein level. However, octreotide remained less potent than pasireotide and cabergoline with respect to in vitro inhibition of ACTH secretion.

In the phase III study of patients with CD treated with pasireotide, the mean percentage change in plasma ACTH level was -12.8% (95 % CI, -20.1 to -5.4) and -16.9% (95 % CI, -27.0 to -6.8) at months 6 and 12, respectively [100]. In this study, the reduction in UFC levels in response to pasireotide was accompanied by reductions in serum cortisol and plasma ACTH levels, as well as improvements in signs and symptoms of CD, but no direct correlations were analyzed.

In a 22-week, prospective, open-label, multicenter, Phase II study of osilodrostat (LCI 699), overall response rate defined by a mean of two 24 h UFC was 89.5 % (17/19). Mean baseline ACTH levels in the overall population were >ULN (20.2 pmol/L; normal range 1.8–9.2) and increased fourfold at week 22 after treatment, primarily driven by two patients' data [116].

In patients treated with mifepristone, ACTH will increase. A $\geq$2-fold increase in ACTH was observed in 72 % of patients treated for a median duration of almost a year [117]. The mean peak increase in ACTH was 2.76 ± 1.65-fold during the first 6 months of therapy in the main study, but remained stable during long-term treatment. ACTH increase was directly correlated with mifepristone dose and declined to near baseline levels after stopping the drug. Increases in ACTH seen with mifepristone therapy [117] do not seem to correlate with increased in tumor size.

ACTH might be a predictor of escape/recurrence of disease on medical therapy in some patients. Mean plasma ACTH started to increase and then was even higher than at baseline in patients who escaped treatment with cabergoline [102]. In most patients treated with ketoconazole in a recent large retrospective study, the increase of ACTH induced by long-term cortisol inhibition lead to cortisol escape in 15 % of the patients treated for more than 2 years. However, being a retrospective study, patients who were not controlled earlier were not excluded from the study [99].

On the other hand, a high plasma ACTH concentration at the time of treatment withdrawal with Mitotane seems to be associated with a lower probability of recurrence [97]. The authors hypothesized that a higher ACTH concentration reflects the extent of adrenal suppression in these patients; increased tumoral ACTH secretion secondary to reduced cortisol feedback on the tumor cells and, to a lesser extent, reactivation of the normal corticotroph cells with cortisol excess correction.

Clinical Improvements Associated with CD Remission

Improvement of clinical features should feature importantly and highly as a treatment goal. In a small prospective study, clinical features overall improved during treatment in responders to treatment with cabergoline [102]. Interestingly, BMI slightly increased initially, but significantly decreased after 3–6 months, while waist to hip ratio progressively decreased overtime. The prevalence of overweight or obesity decreased from 87.5 at 62.5 % after 2 years of treatment. Hypertension decreased from 50 % at baseline to 0 % after 24 months of treatment after a trial of stopping antihypertensive medications. Fasting serum glucose and insulin were also significantly decreased. As expected, the clinical picture slightly worsened in the patients who experienced treatment escape [102].

In the phase III pasireotide study, reductions in blood pressure were observed even without full UFC control and were greatest in patients who did not receive antihypertensive medications during the study. Significant reductions in total cholesterol and low-density lipoprotein (LDL) cholesterol were observed in patients who achieved UFC control. Reductions in BMI, weight, and waist circumference occurred during the study even without full UFC control [100, 118].

Mifepristone, studied in the SEISMIC study, induced improvement in global clinical response (GCR) in 87 %; 37 % of patients had positive GCR by week 10 that persisted through study end, whereas only 6.5 % of patients had a positive GCR during the study that was not maintained [101]. As a group, women tended to have a slower onset of positive GCR compared with men. Four features have been found to be significant predictors of a graded positive GCR: (1) weight loss, (2) 120-min serum glucose after 75 g glucose during the oral glucose tolerance test, (3) diastolic blood pressure, and (4) investigator-graded Cushingoid appearance. Assessment of multiple [119] clinical variables can be used by clinicians to assess mifepristone response and dosing in CS.

Mitotane has also been shown to induce statistically significant improvements in some metabolic parameters after 6 months of treatment, except systolic blood pressure and lipid profile. Both total cholesterol and LDL, and triglycerides increased [97].

In one of the subgroup of patients treated for more than 2 years with ketoconazole [99], the clinical improvement followed closely the biochemical response, but was not observed uniformly across all patients. UFC was normalized in 33 of 51 patients (64.7 %), and it had decreased by at least 50 % in 12 of 51 patients (23.5 %), but hypertension was improved in 15 of 27 patients (55.5 %), diabetes in 7 of 14 patients (50 %), and hypokalemia in 7 of 8 patients (87.5 %).

Predictors of Response

Data on which is the best predictor of response to medical therapy, which would amount to a giant step in the future of individualized patient-centered therapy, are lacking. Normalization of UFC was more likely to be achieved in patients with lower baseline levels than in patients with higher baseline levels in the pasireotide phase III trial [100]; however, pasireotide also decreased UFC levels in some patients with severe hypercortisolism.

For ketoconazole [99] there were no significant differences between responders and non-responders regarding age at diagnosis, previous treatments, and initial dose. Surprisingly, gender appeared to be a predictive factor despite the fact that the maximal dose was not statistically different between both groups (750 ± 236.7 vs. 716 ± 281.5 mg/day in males vs. females, respectively).

It has even been suggested, in a small study [120] that preoperative medical treatment (with ketoconazole or metyrapone) might be associated with low postoperative cortisol concentration and higher rates of long-term remission. However, further studies are needed to address the persistence of the drug response and the effects on the dynamics of the HPA axis.

A collaborative, multi-site, patient treatment registry in which standardized biochemical markers along with clinical parameters are used to determine time and rate of remission and time to and rate of recurrence would assist in standardizing study design and therefore analysis and guiding best practice treatments.

Conclusion

In summary, CD remains difficult to diagnose and treat. Biochemical testing and imaging are key for a definitive diagnosis. Treatment goals are reversal of clinical features and normalization of biochemical changes with long-term disease control. Surgical intervention is a first-line treatment in most cases; however, CD can persist and/or recur. There are no firm established criteria for remission and furthermore there are many more challenges in how to diagnose early recurrent CD. Predictors of response to medical therapy are elusive. Lifelong individualized follow-up and biochemical assessment for disease remission or recurrence, and management is required.

References

1. Cushing H. Medical classic. The functions of the pituitary body: Harvey Cushing. Am J Med Sci. 1981;281(2):70–8.
2. Pivonello R, De Leo M, Cozzolino A, Colao A. The treatment of Cushing's disease. Endocr Rev. 2015;36(4):385–486.
3. Plotz CM, Knowlton AI, Ragan C. The natural history of Cushing's syndrome. Am J Med. 1952;13(5):597–614.
4. Ragnarsson O, Johannsson G. Cushing's syndrome: a structured short- and long-term management plan for patients in remission. Eur J Endocrinol. 2013;169(5):R139–52.
5. Biller BMK, Grossman AB, Stewart PM, Melmed S, Bertagna X, Bertherat J, et al. Treatment of adrenocorticotropin-dependent Cushing's syndrome: a consensus statement. J Clin Endocrinol Metab. 2008;93(7):2454–62.
6. Dorn LD, Burgess ES, Dubbert B, Simpson SE, Friedman T, Kling M, et al. Psychopathology in patients with endogenous Cushing's syndrome: 'atypical' or melancholic features. Clin Endocrinol (Oxf). 1995;43(4):433–42.
7. Starkman MN, Giordani B, Berent S, Schork MA, Schteingart DE. Elevated cortisol levels in Cushing's disease are associated with cognitive decrements. Psychosom Med. 2001;63(6):985–93.
8. Blevins Jr LS, Verity DK, Allen G. Aggressive pituitary tumors. Oncology (Williston Park). 1998;12(9):1307–12. 15; discussion 15-8.
9. Boscaro M, Arnaldi G. Approach to the patient with possible Cushing's syndrome. J Clin Endocrinol Metab. 2009;94(9):3121–31.
10. Nieman LK, Biller BMK, Findling JW, Newell-Price J, Savage MO, Stewart PM, et al. The diagnosis of Cushing's syndrome: an endocrine society clinical practice guideline. J Clin Endocrinol Metab. 2008;93(5):1526–40.
11. Lacroix A, Feelders RA, Stratakis CA, Nieman LK. Cushing's syndrome. Lancet. 2015;386(9996):913–27.
12. Sun H, Yedinak C, Ozpinar A, Anderson J, Dogan A, Delashaw J, et al. Preoperative lateralization modalities for Cushing disease: is dynamic magnetic resonance imaging or cavernous sinus sampling more predictive of intraoperative findings? J Neurol Surg B Skull Base. 2015;76(3):218–24.
13. Swearingen B, Katznelson L, Miller K, Grinspoon S, Waltman A, Dorer DJ, et al. Diagnostic errors after inferior petrosal sinus sampling. J Clin Endocrinol Metab. 2004;89(8):3752–63.
14. Nieman LK, Biller BM, Findling JW, Murad MH, Newell-Price J, Savage MO, et al. Treatment of Cushing's syndrome: an endocrine society clinical practice guideline. J Clin Endocrinol Metab. 2015;100(8):2807–31.

15. Hammer GD, Tyrrell JB, Lamborn KR, Applebury CB, Hannegan ET, Bell S, et al. Transsphenoidal microsurgery for Cushing's disease: initial outcome and long-term results. J Clin Endocrinol Metab. 2004;89(12):6348–57.

16. Prevedello DMS, Pouratian N, Sherman JH, Jane JA, Lopes MB, Vance ML, et al. Analysis of 445 patients with Cushing's disease treated by transsphenoidal surgery. Neurosurgery. 2006;59(2):487.

17. Alwani RA, de Herder WW, van Aken MO, van den Berge JH, Delwel EJ, Dallenga AHG, et al. Biochemical predictors of outcome of pituitary surgery for Cushing's disease. Neuroendocrinology. 2010;91(2):169–78.

18. Esposito F, Dusick JR, Cohan P, Moftakhar P, McArthur D, Wang C, et al. Early morning cortisol levels as a predictor of remission after transsphenoidal surgery for Cushing's disease. J Clin Endocrinol Metab. 2006;91(1):7–13.

19. Rollin G, Ferreira NP, Czepielewski MA. Prospective evaluation of transsphenoidal pituitary surgery in 108 patients with Cushing's disease. Arquivos Brasileiros de Endocrinologia & Metabologia. 2007;51(8):1355–61.

20. Rollin GAFS, Ferreira NP, Junges M, Gross JL, Czepielewski MA. Dynamics of serum cortisol levels after transsphenoidal surgery in a cohort of patients with Cushing's disease. J Clin Endocrinol Metab. 2004;89(3):1131–9.

21. Atkinson AB, Kennedy A, Wiggam MI, McCance DR, Sheridan B. Long-term remission rates after pituitary surgery for Cushing's disease: the need for long-term surveillance. Clin Endocrinol. 2005;63(5):549–59.

22. Fleseriu M, Petersenn S. Medical management of Cushing's disease: what is the future? Pituitary. 2012;15(3):330–41.

23. Petersenn S, Beckers A, Ferone D, van der Lely A, Bollerslev J, Boscaro M, et al. Therapy of endocrine disease: outcomes in patients with Cushing's disease undergoing transsphenoidal surgery: systematic review assessing criteria used to define remission and recurrence. Eur J Endocrinol. 2015;172(6):R227–39.

24. Vance ML. Cushing's disease: radiation therapy. Pituitary. 2009;12(1):11–4.

25. Cuevas-Ramos D, Fleseriu M. Treatment of Cushing's disease: a mechanistic update. J Endocrinol. 2014;223(2):R19–39.

26. Fleseriu M. Medical treatment of Cushing disease: new targets, new hope. Endocrinol Metab Clin North Am. 2015;44(1):51–70.

27. Fleseriu M, Petersenn S. Medical therapy for Cushing's disease: adrenal steroidogenesis inhibitors and glucocorticoid receptor blockers. Pituitary. 2015;18(2):245–52.

28. Petersenn S, Fleseriu M. Pituitary-directed medical therapy in Cushing's disease. Pituitary. 2015;18(2):238–44.

29. Tritos NA, Biller BM. Cushing's disease. Handb Clin Neurol. 2014;124:221–34.

30. Berker M, Isikay I, Berker D, Bayraktar M, Gurlek A. Early promising results for the endoscopic surgical treatment of Cushing's disease. Neurosurg Rev. 2014;37:105–114.

31. Dehdashti AR, Gentili F. Current state of the art in the diagnosis and surgical treatment of Cushing disease: early experience with a purely endoscopic endonasal technique. Neurosurg Focus. 2007;23(3), E9.

32. Netea-Maier RT, van Lindert EJ, den Heijer M, van der Eerden A, Pieters GF, Sweep CG, et al. Transsphenoidal pituitary surgery via the endoscopic technique: results in 35 consecutive patients with Cushing's disease. Eur J Endocrinol. 2006;154(5):675–84.

33. Starke RM, Reames DL, Chen CJ, Laws ER, Jane Jr JA. Endoscopic transsphenoidal surgery for Cushing disease: techniques, outcomes, and predictors of remission. Neurosurgery. 2013;72(2):240–7. discussion 7.

34. Wagenmakers MA, Boogaarts HD, Roerink SH, Timmers HJ, Stikkelbroeck NM, Smit JW, et al. Endoscopic transsphenoidal pituitary surgery: a good and safe primary treatment option for Cushing's disease, even in case of macroadenomas or invasive adenomas. Eur J Endocrinol. 2013;169(3):329–37.

35. Shin SS, Gardner PA, Ng J, Faraji AH, Agarwal N, Chivukula S, et al. Endoscopic endonasal approach for ACTH- secreting pituitary adenomas: outcomes and analysis of remission rates and tumor biochemical activity with respect to tumor invasiveness. World Neurosurg. 2015.

36. Barkhoudarian G, Zada G, Laws ER. Endoscopic endonasal surgery for nonadenomatous sellar/parasellar lesions. World Neurosurg. 2014;82(6 Suppl):S138–46.
37. Blevins LS, Christy JH, Khajavi M, Tindall GT. Outcomes of therapy for Cushing's disease due to adrenocorticotropin-secreting pituitary macroadenomas 1. J Clin Endocrinol Metab. 1998;83(1):63–7.
38. Alexandraki KI, Kaltsas GA, Isidori AM, Storr HL, Afshar F, Sabin I, et al. Long-term remission and recurrence rates in Cushing's disease: predictive factors in a single-centre study. Eur J Endocrinol. 2013;168(4):639–48.
39. Dimopoulou C, Schopohl J, Rachinger W, Buchfelder M, Honegger J, Reincke M, et al. Long-term remission and recurrence rates after first and second transsphenoidal surgery for Cushing's disease: care reality in the Munich Metropolitan Region. Eur J Endocrinol. 2014;170(2):283–92.
40. Amlashi FG, Swearingen B, Faje AT, Nachtigall LB, Miller KK, Klibanski A, et al. Accuracy of late night salivary cortisol in evaluating postoperative remission and recurrence in Cushing's disease. J Clin Endocrinol Metab. 2015;100(10):3770–7. doi:10.1210/jc.2015-2107.
41. Bochicchio D, Losa M, Buchfelder M. Factors influencing the immediate and late outcome of Cushing's disease treated by transsphenoidal surgery: a retrospective study by the European Cushing's Disease Survey Group. J Clin Endocrinol Metab. 1995;80(11):3114–20.
42. Carrasco CA, Coste J, Guignat L, Groussin L, Dugué MA, Gaillard S, et al. Midnight salivary cortisol determination for assessing the outcome of transsphenoidal surgery in Cushing's disease. J Clin Endocrinol Metab. 2008;93(12):4728–34.
43. Valassi E, Biller BM, Swearingen B, Pecori Giraldi F, Losa M, Mortini P, et al. Delayed remission after transsphenoidal surgery in patients with Cushing's disease. J Clin Endocrinol Metab. 2010;95(2):601–10.
44. Ciric I, Zhao J-C, Du H, Findling JW, Molitch ME, Weiss RE, et al. Transsphenoidal surgery for Cushing disease. Neurosurgery. 2012;70(1):70–81.
45. Estrada J, García-Uría J, Lamas C, Alfaro J, Lucas T, Diez S, et al. The complete normalization of the adrenocortical function as the criterion of cure after transsphenoidal surgery for Cushing's disease. J Clin Endocrinol Metab. 2001;86(12):5695–9.
46. Flitsch J, Knappe UJ, Lüdecke DK. The use of postoperative ACTH levels as a marker for successful transsphenoidal microsurgery in Cushing's disease. Zentralbl Neurochir. 2003;64(1):6–11.
47. Kim JH, Shin CS, Paek SH, Jung HW, Kim SW, Kim SY. Recurrence of Cushing's disease after primary transsphenoidal surgery in a university hospital in Korea. Endocr J. 2012;59(10):881–8.
48. Sonino N, Zielezny M, Fava GA, Fallo F, Boscaro M. Risk factors and long-term outcome in pituitary-dependent Cushing's disease. J Clin Endocrinol Metab. 1996;81(7):2647–52.
49. Swearingen B, Biller BM, Barker 2nd FG, Katznelson L, Grinspoon S, Klibanski A, et al. Long-term mortality after transsphenoidal surgery for Cushing disease. Ann Intern Med. 1999;130(10):821–4.
50. Tahir AH. Recurrent Cushing's disease after transsphenoidal surgery. Arch Intern Med. 1992;152(5):977.
51. Yap LB, Turner HE, Adams CBT, Wass JAH. Undetectable postoperative cortisol does not always predict long-term remission in Cushing's disease: a single centre audit*. Clin Endocrinol. 2002;56(1):25–31.
52. Patil CG, Prevedello DM, Lad SP, Vance ML, Thorner MO, Katznelson L, et al. Late recurrences of Cushing's disease after initial successful transsphenoidal surgery. J Clin Endocrinol Metab. 2008;93(2):358–62.
53. Aghi MK, Petit J, Chapman P, Loeffler J, Klibanski A, Biller BM, et al. Management of recurrent and refractory Cushing's disease with reoperation and/or proton beam radiosurgery. Clin Neurosurg. 2008;55:141–4.
54. Tindall GT, Herring CJ, Clark RV, Adams DA, Watts NB. Cushing's disease: results of transsphenoidal microsurgery with emphasis on surgical failures. J Neurosurg. 1990;72(3):363–9.

55. Boggan JE, Tyrrell JB, Wilson CB. Transsphenoidal microsurgical management of Cushing's disease. J Neurosurg. 1983;59(2):195–200.
56. Chee GH, Mathias DB, James RA, Kendall-Taylor P. Transsphenoidal pituitary surgery in Cushing's disease: can we predict outcome? Clin Endocrinol. 2001;54(5):617–26.
57. Acebes JJ, Martino J, Masuet C, Montanya E, Soler J. Early post-operative ACTH and cortisol as predictors of remission in Cushing's disease. Acta Neurochir (Wien). 2007;149(5):471–7. discussion 7-9.
58. Bou Khalil R, Baudry C, Guignat L, Carrasco C, Guibourdenche J, Gaillard S, et al. Sequential hormonal changes in 21 patients with recurrent Cushing's disease after successful pituitary surgery. Eur J Endocrinol. 2011;165(5):729–37.
59. Fomekong E, Maiter D, Grandin C, Raftopoulos C. Outcome of transsphenoidal surgery for Cushing's disease: a high remission rate in ACTH-secreting macroadenomas. Clin Neurol Neurosurg. 2009;111(5):442–9.
60. Hassan-Smith ZK, Sherlock M, Reulen RC, Arlt W, Ayuk J, Toogood AA, et al. Outcome of Cushing's disease following transsphenoidal surgery in a single center over 20 years. J Clin Endocrinol Metab. 2012;97(4):1194–201.
61. Rees DA, Hanna FWF, Davies JS, Mills RG, Vafidis J, Scanlon MF. Long-term follow-up results of transsphenoidal surgery for Cushing's disease in a single centre using strict criteria for remission. Clin Endocrinol. 2002;56(4):541–51.
62. Salenave S, Gatta B, Pecheur S, San-Galli F, Visot A, Lasjaunias P, et al. Pituitary magnetic resonance imaging findings do not influence surgical outcome in adrenocorticotropin-secreting microadenomas. J Clin Endocrinol Metab. 2004;89(7):3371–6.
63. Storr HL, Alexandraki KI, Martin L, Isidori AM, Kaltsas GA, Monson JP, et al. Comparisons in the epidemiology, diagnostic features and cure rate by transsphenoidal surgery between paediatric and adult-onset Cushing's disease. Eur J Endocrinol. 2011;164(5):667–74.
64. Prevedello DM, Pouratian N, Sherman J, Jane Jr JA, Vance ML, Lopes MB, et al. Management of Cushing's disease: outcome in patients with microadenoma detected on pituitary magnetic resonance imaging. J Neurosurg. 2008;109(4):751–9.
65. Hofmann BM, Hlavac M, Martinez R, Buchfelder M, Muller OA, Fahlbusch R. Long-term results after microsurgery for Cushing disease: experience with 426 primary operations over 35 years. J Neurosurg. 2008;108(1):9–18.
66. Invitti C, Pecori Giraldi F, de Martin M, Cavagnini F. Diagnosis and management of Cushing's syndrome: results of an Italian multicentre study. Study Group of the Italian Society of Endocrinology on the Pathophysiology of the Hypothalamic-Pituitary-Adrenal Axis. J Clin Endocrinol Metab. 1999;84(2):440–8.
67. Trainer PJ, Lawrie HS, Verhelst J, Howlett TA, Lowe DG, Grossman AB, et al. Transsphenoidal resection in Cushing's disease: undetectable serum cortisol as the definition of successful treatment. Clin Endocrinol (Oxf). 1993;38(1):73–8.
68. Jagannathan J, Smith R, DeVroom HL, Vortmeyer AO, Stratakis CA, Nieman LK, et al. Outcome of using the histological pseudocapsule as a surgical capsule in Cushing disease. J Neurosurg. 2009;111(3):531–9.
69. Chen JCT, Amar AP, Choi S, Singer P, Couldwell WT, Weiss MH. Transsphenoidal microsurgical treatment of Cushing disease: postoperative assessment of surgical efficacy by application of an overnight low-dose dexamethasone suppression test. J Neurosurg. 2003;98(5):967–73.
70. Pereira AM, van Aken MO, van Dulken H, Schutte PJ, Biermasz NR, Smit JWA, et al. Long-term predictive value of postsurgical cortisol concentrations for cure and risk of recurrence in Cushing's disease. J Clin Endocrinol Metab. 2003;88(12):5858–64.
71. Hameed N, Yedinak CG, Brzana J, Gultekin SH, Coppa ND, Dogan A, et al. Remission rate after transsphenoidal surgery in patients with pathologically confirmed Cushing's disease, the role of cortisol, ACTH assessment and immediate reoperation: a large single center experience. Pituitary. 2013;16(4):452–8.
72. Salmon PM, Loftus PD, Dodd RL, Harsh G, Chu OS, Katznelson L. Utility of adrenocorticotropic hormone in assessing the response to transsphenoidal surgery for Cushing's disease. Endocr Pract. 2014;20(11):1159–64.

73. Knosp E, Steiner E, Kitz K, Matula C. Pituitary adenomas with invasion of the cavernous sinus space: a magnetic resonance imaging classification compared with surgical findings. Neurosurgery. 1993;33(4):610–7. discussion 7-8.

74. Aranda G, Ensenat J, Mora M, Puig-Domingo M, Martinez de Osaba MJ, Casals G, et al. Long-term remission and recurrence rate in a cohort of Cushing's disease: the need for long-term follow-up. Pituitary. 2015;18(1):142–9.

75. Costenaro F, Rodrigues TC, Rollin GA, Ferreira NP, Czepielewski MA. Evaluation of Cushing's disease remission after transsphenoidal surgery based on early serum cortisol dynamics. Clin Endocrinol (Oxf). 2014;80(3):411–8.

76. Flitsch J, Ludecke DK, Knappe UJ, Saeger W. Correlates of long-term hypocortisolism after transsphenoidal microsurgery for Cushing's disease. Exp Clin Endocrinol Diabetes. 1999;107(3):183–9.

77. Berr CM, Di Dalmazi G, Osswald A, Ritzel K, Bidlingmaier M, Geyer LL, et al. Time to recovery of adrenal function after curative surgery for Cushing's syndrome depends on etiology. J Clin Endocrinol Metab. 2015;100(4):1300–8.

78. Saeger W, Geisler F, Ludecke DK. Pituitary pathology in Cushing's disease. Pathol Res Pract. 1988;183(5):592–5.

79. Syro LV, Rotondo F, Ramirez A, Di Ieva A, Sav MA, Restrepo LM, et al. Progress in the diagnosis and classification of pituitary adenomas. Front Endocrinol (Lausanne). 2015;6:97.

80. Sacre K, Dehoux M, Chauveheid MP, Chauchard M, Lidove O, Roussel R, et al. Pituitary-adrenal function after prolonged glucocorticoid therapy for systemic inflammatory disorders: an observational study. J Clin Endocrinol Metab. 2013;98(8):3199–205.

81. Inzucchi SE, Robbins RJ. Clinical review 61: effects of growth hormone on human bone biology. J Clin Endocrinol Metab. 1994;79(3):691–4.

82. Ross EJ, Linch DC. Cushing's syndrome—killing disease: discriminatory value of signs and symptoms aiding early diagnosis. Lancet. 1982;2(8299):646–9.

83. Urbanic RC, George JM. Cushing's disease—18 years' experience. Medicine (Baltimore). 1981;60(1):14–24.

84. Veldman RG, Frolich M, Pincus SM, Veldhuis JD, Roelfsema F. Apparently complete restoration of normal daily adrenocorticotropin, cortisol, growth hormone, and prolactin secretory dynamics in adults with Cushing's disease after clinically successful transsphenoidal adenomectomy. J Clin Endocrinol Metab. 2000;85(11):4039–46.

85. van den Berg G, Frolich M, Veldhuis JD, Roelfsema F. Combined amplification of the pulsatile and basal modes of adrenocorticotropin and cortisol secretion in patients with Cushing's disease: evidence for decreased responsiveness of the adrenal glands. J Clin Endocrinol Metab. 1995;80(12):3750–7.

86. van den Berg G, Pincus SM, Veldhuis JD, Frolich M, Roelfsema F. Greater disorderliness of ACTH and cortisol release accompanies pituitary-dependent Cushing's disease. Eur J Endocrinol. 1997;136(4):394–400.

87. Jehle S, Walsh JE, Freda PU, Post KD. Selective use of bilateral inferior petrosal sinus sampling in patients with adrenocorticotropin-dependent Cushing's syndrome prior to transsphenoidal surgery. J Clin Endocrinol Metab. 2008;93(12):4624–32.

88. Guilhaume B, Bertagna X, Thomsen M, Bricaire C, Vila-Porcile E, Olivier L, et al. Transsphenoidal pituitary surgery for the treatment of Cushing's disease: results in 64 patients and long term follow-up studies. J Clin Endocrinol Metab. 1988;66(5):1056–64.

89. Bakiri F, Tatai S, Aouali R, Semrouni M, Derome P, Chitour F, et al. Treatment of Cushing's disease by transsphenoidal, pituitary microsurgery: prognosis factors and long-term follow-up. J Endocrinol Invest. 1996;19(9):572–80.

90. Barbetta L, Dall'Asta C, Tomei G, Locatelli M, Giovanelli M, Ambrosi B. Assessment of cure and recurrence after pituitary surgery for Cushing's disease. Acta Neurochir. 2001;143(5):477–82.

91. Shimon I, Ram Z, Cohen ZR, Hadani M. Transsphenoidal surgery for Cushing's disease: endocrinological follow-up monitoring of 82 patients. Neurosurgery. 2002;51(1):57–62.

92. Honegger J, Schmalisch K, Beuschlein F, Kaufmann S, Schnauder G, Naegele T, et al. Contemporary microsurgical concept for the treatment of Cushing's disease: endocrine outcome in 83 consecutive patients. Clin Endocrinol (Oxf). 2012;76(4):560–7.

93. Castinetti F, Martinie M, Morange I, Dufour H, Sturm N, Passagia J-G, et al. A combined dexamethasone desmopressin test as an early marker of postsurgical recurrence in Cushing's disease. J Clin Endocrinol Metab. 2009;94(6):1897–903.

94. Ceccato F, Barbot M, Zilio M, Frigo AC, Albiger N, Camozzi V, et al. Screening tests for Cushing's syndrome: urinary free cortisol role measured by LC-MS/MS. J Clin Endocrinol Metab. 2015;100(10):3856–61.

95. Elamin MB, Murad MH, Mullan R, Erickson D, Harris K, Nadeem S, et al. Accuracy of diagnostic tests for Cushing's syndrome: a systematic review and metaanalyses. J Clin Endocrinol Metab. 2008;93(5):1553–62.

96. Petersenn S, Newell-Price J, Findling JW, Gu F, Maldonado M, Sen K, et al. High variability in baseline urinary free cortisol values in patients with Cushing's disease. Clin Endocrinol. 2013;80(2):261–9.

97. Baudry C, Coste J, Bou Khalil R, Silvera S, Guignat L, Guibourdenche J, et al. Efficiency and tolerance of mitotane in Cushing's disease in 76 patients from a single center. Eur J Endocrinol. 2012;167(4):473–81.

98. Bertagna X, Pivonello R, Fleseriu M, Zhang Y, Robinson P, Taylor A, et al. LCI699, a potent 11beta-hydroxylase inhibitor, normalizes urinary cortisol in patients with Cushing's disease: results from a multicenter, proof-of-concept study. J Clin Endocrinol Metab. 2014;99(4):1375–83.

99. Castinetti F, Guignat L, Giraud P, Muller M, Kamenicky P, Drui D, et al. Ketoconazole in Cushing's disease: is it worth a try? J Clin Endocrinol Metab. 2014;99(5):1623–30.

100. Colao A, Petersenn S, Newell-Price J, Findling JW, Gu F, Maldonado M, et al. A 12-month phase 3 study of pasireotide in Cushing's disease. N Engl J Med. 2012;366(10):914–24.

101. Fleseriu M, Biller BM, Findling JW, Molitch ME, Schteingart DE, Gross C, et al. Mifepristone, a glucocorticoid receptor antagonist, produces clinical and metabolic benefits in patients with Cushing's syndrome. J Clin Endocrinol Metab. 2012;97(6):2039–49.

102. Pivonello R, De Martino MC, Cappabianca P, De Leo M, Faggiano A, Lombardi G, et al. The medical treatment of Cushing's disease: effectiveness of chronic treatment with the dopamine agonist cabergoline in patients unsuccessfully treated by surgery. J Clin Endocrinol Metab. 2009;94(1):223–30.

103. Feelders RA, de Bruin C, Pereira AM, Romijn JA, Netea-Maier RT, Hermus AR, et al. Pasireotide alone or with cabergoline and ketoconazole in Cushing's disease. N Engl J Med. 2010;362(19):1846–8.

104. Lila AR, Gopal RA, Acharya SV, George J, Sarathi V, Bandgar T, et al. Efficacy of cabergoline in uncured (persistent or recurrent) Cushing disease after pituitary surgical treatment with or without radiotherapy. Endocr Pract. 2010;16(6):968–76.

105. Trementino L, Cardinaletti M, Concettoni C, Marcelli G, Polenta B, Spinello M, et al. Salivary cortisol is a useful tool to assess the early response to pasireotide in patients with Cushing's disease. Pituitary. 2015;18(1):60–7.

106. Trementino L, Zilio M, Marcelli G, Michetti G, Barbot M, Ceccato F, et al. The role of an acute pasireotide suppression test in predicting response to treatment in patients with Cushing's disease: findings from a pilot study. Endocrine. 2014;50(1):154–61.

107. Verhelst JA, Trainer PJ, Howlett TA, Perry L, Rees LH, Grossman AB, et al. Short and long-term responses to metyrapone in the medical management of 91 patients with Cushing's syndrome. Clin Endocrinol (Oxf). 1991;35(2):169–78.

108. Vilar L, Naves LA, Azevedo MF, Arruda MJ, Arahata CM, Moura ESL, et al. Effectiveness of cabergoline in monotherapy and combined with ketoconazole in the management of Cushing's disease. Pituitary. 2010;13(2):123–9.

109. Ceccato F, Barbot M, Zilio M, Ferasin S, Occhi G, Daniele A, et al. Performance of salivary cortisol in the diagnosis of Cushing's syndrome, adrenal incidentaloma, and adrenal insufficiency. Eur J Endocrinol. 2013;169(1):31–6.

110. Raff H. Update on late-night salivary cortisol for the diagnosis of Cushing's syndrome: methodological considerations. Endocrine. 2013;44(2):346–9.
111. Ceccato F, Barbot M, Zilio M, Ferasin S, De Lazzari P, Lizzul L, et al. Age and the metabolic syndrome affect salivary cortisol rhythm: data from a community sample. HJ. 2015.
112. Lederbogen F, Kuhner C, Kirschbaum C, Meisinger C, Lammich J, Holle R, et al. Salivary cortisol in a middle-aged community sample: results from 990 men and women of the KORA-F3 Augsburg study. Eur J Endocrinol. 2010;163(3):443–51.
113. Biller BM, Petersenn S, Pivonello R, Findling JW, Fleseriu M, Trovato A, et al. Evaluation of late-night salivary cortisol during 12 months of Pasireotide treatment in patients with Cushing's disease. In: Endocrine Society's 96th Annual Meeting and Expo, June 15–18, 2013; San Francisco2013. p. MON 89.
114. van der Pas R, de Bruin C, Pereira AM, Romijn JA, Netea-Maier RT, Hermus AR, et al. Cortisol diurnal rhythm and quality of life after successful medical treatment of Cushing's disease. Pituitary. 2013;16(4):536–44.
115. Terzolo M, Panarelli M, Piovesan A, Torta M, Paccotti P, Angeli A. Ketoconazole treatment in Cushing's disease. Effect on the circadian profile of plasma ACTH and Cortisol. J Endocrinol Invest. 1988;11(10):717–21.
116. Fleseriu M, Pivonello R, Young J, Hamrahian AH, Molitch ME, Shimizu C, et al. Osilodrostat, a potent oral 11beta-hydroxylase inhibitor: 22-week, prospective. Pituitary: Phase II study in Cushing's disease; 2015.
117. Fleseriu M, Findling JW, Koch CA, Schlaffer SM, Buchfelder M, Gross C. Changes in plasma ACTH levels and corticotroph tumor size in patients with Cushing's disease during long-term treatment with the glucocorticoid receptor antagonist mifepristone. J Clin Endocrinol Metab. 2014;99(10):3718–27.
118. Pivonello R, Petersenn S, Newell-Price J, Findling JW, Gu F, Maldonado M, et al. Pasireotide treatment significantly improves clinical signs and symptoms in patients with Cushing's disease: results from a Phase III study. Clin Endocrinol (Oxf). 2014;81(3):408–17.
119. Katznelson L, Loriaux DL, Feldman D, Braunstein GD, Schteingart DE, Gross C. Global clinical response in Cushing's syndrome patients treated with mifepristone. Clin Endocrinol (Oxf). 2014;80(4):562–9.
120. van den Bosch OFC, Stades AME, Zelissen PMJ. Increased long-term remission after adequate medical cortisol suppression therapy as presurgical treatment in Cushing's disease. Clin Endocrinol. 2013;80(2):184–90.

Part III
Beyond Cushing's: Glucocorticoid Sensitivity, Regulation, and the Metabolic Syndrome

Primary Generalized Glucocorticoid Resistance or Chrousos Syndrome: Allostasis Through a Mutated Glucocorticoid Receptor

Nicolas C. Nicolaides, Agaristi Lamprokostopoulou, Amalia Sertedaki, George P. Chrousos, and Evangelia Charmandari

Abstract Primary generalized glucocorticoid resistance or Chrousos syndrome is a rare familial or sporadic condition, which affects almost all organs and is characterized by partial target tissue insensitivity to glucocorticoids. Patients with this condition may be asymptomatic or may present with clinical manifestations of mineralocorticoid and/or androgen excess. The molecular basis of Chrousos syndrome has been associated with point mutations, insertions or deletions in the *NR3C1* gene that expresses the human glucocorticoid receptor, a member of the steroid receptor family of the nuclear receptor superfamily of transcription factors. We and others have systematically investigated the molecular mechanisms of action

N.C. Nicolaides, M.D., Ph.D. (✉) • A. Lamprokostopoulou, M.Sc., Ph.D.
Division of Endocrinology and Metabolism, Center of Clinical, Experimental Surgery and Translational Research, Biomedical Research Foundation of the Academy of Athens,
4 Soranou tou Efessiou Street, Athens 11527, Greece
e-mail: nnicolaides@bioacademy.gr; alamprok@bioacademy.gr

A. Sertedaki, B.Sc., Ph.D.
Division of Endocrinology, Metabolism and Diabetes, First Department of Pediatrics,
University of Athens Medical School, "Aghia Sophia" Children's Hospital,
Thivon and Papadiamantopoulou Street, Athens 11527, Greece
e-mail: aserted@med.uoa.gr

G.P. Chrousos, M.D., M.A.C.P., M.A.C.E., F.R.C.P. (London)
Division of Endocrinology and Metabolism, Center of Clinical, Experimental Surgery and Translational Research, Biomedical Research Foundation of the Academy of Athens,
4 Soranou tou Efessiou Street, Athens 11527, Greece

Division of Endocrinology, Metabolism and Diabetes, First Department of Pediatrics,
University of Athens Medical School, "Aghia Sophia" Children's Hospital,
Thivon and Papadiamantopoulou Street, Athens 11527, Greece

UNESCO Chair on Adolescent Health Care, University of Athens Medical School, "Aghia Sophia" Children's Hospital, Thivon and Papadiamantopoulou Street, Athens 11527, Greece

National Institute of Child Health and Human Development, National Institutes of Health, Bethesda, MD 20892, USA

King Abdulaziz University, Jeddah, S. Arabia
e-mail: chrousge@med.uoa.gr; chrousog@mail.nih.gov

© Springer International Publishing Switzerland 2017
E.B. Geer (ed.), *The Hypothalamic-Pituitary-Adrenal Axis in Health and Disease*, DOI 10.1007/978-3-319-45950-9_13

of the mutant glucocorticoid receptors causing Chrousos syndrome by applying standard methods of molecular and structural biology. In this chapter, we discuss the clinical manifestations, pathophysiology, molecular pathogenesis, diagnostic approach, and therapeutic management of Chrousos syndrome.

Keywords Adrenal androgens • Chrousos syndrome • Dexamethasone suppression test • Glucocorticoid receptor • Glucocorticoid signaling • Glucocorticoids • Mineralocorticoids • *NR3C1* gene mutations • Primary generalized glucocorticoid resistance • Sequencing • Urinary free cortisol excretion

Abbreviations

ACTH	Adrenocorticotropic hormone
AP-1	Activator protein 1
AVP	Arginine vasopressin
CRH	corticotropin-releasing hormone
DBD	DNA-binding domain
DHEA	Dehydroepiandrosterone
DHEAS	DHEA-sulfate
GFP	Green fluorescent protein
GR	Glucocorticoid receptor
GREs	Glucocorticoid response elements
GRIP1	Glucocorticoid receptor-interacting protein 1
GST	Glutathione-S-transferase
HDL	High density lipoprotein
HPA axis	Hypothalamic–pituitary–adrenal axis
HSPs	Heat shock proteins
LBD	Ligand-binding domain
LDL	Low density lipoprotein
NF-kB	Nuclear factor kB
NTD	N-terminal domain
STAT	Signal transducer and activator of transcription
UFC	Urinary free cortisol

E. Charmandari, M.D., M.Sc., Ph.D., M.R.C.P.(UK)., C.C.S.T.(UK)
Division of Endocrinology and Metabolism, Center of Clinical, Experimental Surgery and Translational Research, Biomedical Research Foundation of the Academy of Athens, 4 Soranou tou Efessiou Street, Athens 11527, Greece

Division of Endocrinology, Metabolism and Diabetes, First Department of Pediatrics, University of Athens Medical School, "Aghia Sophia" Children's Hospital, Thivon and Papadiamantopoulou Street, Athens 11527, Greece
e-mail: evangelia.charmandari@googlemail.com

Introduction

Homeostasis (from the Greek *homoios*, or similar, and *stasis*, or position), a term proposed by Walter Bradford Cannon to describe all the physiologic processes that maintain the steady state of the organism, is tightly achieved through the coordinated functions of numerous systems [1–4]. All homeostatic systems operate through an inverted U-type activity-effect curve, which means that homeostasis is adequately maintained in the middle range of homeostatic activity. If any of the homeostatic systems has too much or too little activity, then homeostasis is turned to *allostasis* or *cacostasis*, causing several pathologic conditions [1–4]. One of the fundamental homeostatic systems that plays crucial role in the stress response is the glucocorticoid system, which mediates all the well-known genomic and nongenomic actions of glucocorticoid hormones (cortisol in human, corticosterone in most rodents) through a ubiquitously expressed protein, the glucocorticoid receptor [5]. Undoubtedly, any dysfunction of the glucocorticoid system contributes to allostasis. In terms of glucocorticoid secretion from the adrenal cortex, elevated concentrations of glucocorticoids cause the cardinal clinical manifestations of Cushing syndrome, whereas glucocorticoid deficiency is responsible for the life-threatening Addison's disease [6, 7]. In terms of the molecular mechanisms of glucocorticoid action at the tissue level, alterations in any step of the glucocorticoid signaling cascade may cause impaired tissue sensitivity to glucocorticoids, which may take the form of *glucocorticoid resistance* or *glucocorticoid hypersensitivity*, both with significant morbidity (Table 1) [8–12]. One such condition that we and others have thoroughly investigated at the clinical, hormonal, and molecular level is primary generalized glucocorticoid resistance or Chrousos syndrome [9–16].

Table 1 Expected clinical manifestations in tissue-specific glucocorticoid excess or hypersensitivity and deficiency or resistance

Target tissue	Glucocorticoid hypersensitivity = Glucocorticoid excess	Glucocorticoid resistance = Glucocorticoid deficiency
Central nervous system	Insomnia, anxiety, depression, defective cognition	Fatigue, somnolence, malaise, defective cognition
Liver	+ Gluconeogenesis, + lipogenesis	Hypoglycemia, resistance to diabetes mellitus
Fat	Accumulation of visceral fat (metabolic syndrome)	Loss of weight, resistance to weight gain
Blood vessels	Hypertension	Hypotension
Bone	Stunted growth, osteoporosis	
Inflammation/immunity	Immune suppression, anti-inflammation, vulnerability to certain infections and tumors	+ Inflammation, + autoimmunity, + allergy

Modified from Reference [9]

Primary Generalized Glucocorticoid Resistance or Chrousos Syndrome

Primary Generalized Glucocorticoid Resistance or Chrousos syndrome is a familial or sporadic allostatic condition, which is characterized by target tissue insensitivity to glucocorticoids in almost all organs [9–16]. Because of the generalized nature of glucocorticoid resistance, all the neuroanatomic structures participating in the formation of the glucocorticoid negative feedback loops display decreased response to glucocorticoids, leading to compensatory activation of the hypothalamic–pituitary–adrenal (HPA) axis. As a result, the increased secretion of corticotropin-releasing hormone (CRH) and arginine vasopressin (AVP) from the hypothalamus into the hypophysial portal system triggers the production and release of adrenocorticotropic hormone (ACTH) by the anterior pituitary gland. Hypersecretion of ACTH results in adrenal cortex hypertrophy and triggers the production of cortisol, adrenal androgens [androstenedione, dehydroepiandrosterone (DHEA), and DHEA-sulfate (DHEAS)], and steroid precursors with mineralocorticoid activity (deoxycorticosterone and corticosterone) [9–16].

The clinical spectrum of Chrousos syndrome is broad, ranging from completely asymptomatic cases to mild or even severe cases of mineralocorticoid and/or androgen excess. The increased concentrations of steroid precursors with mineralocorticoid activity may cause hypertension and/or hypokalemic alkalosis, while adrenal androgen excess may result in ambiguous genitalia in karyotypic females, precocious puberty, acne, hirsutism, male-pattern hair loss and hypofertility in both sexes, oligo-amenorrhea and menstrual irregularities in women, and oligospermia in men [9–16]. It is worth noting that clinical manifestations of glucocorticoid deficiency are rare and have been reported in adults with chronic fatigue [14, 17, 18], in a child with hypoglycemic generalized tonic–clonic seizures during an episode of febrile illness [19], and in a newborn with profound hypoglycemia, reported easy "fatigability" with feeding and growth hormone deficiency [20]. Interestingly, the increased concentrations of CRH may account for anxiety and depression in some patients with Chrousos syndrome [16].

The aforementioned clinical heterogeneity of Chrousos syndrome occurs because of differences in target tissues' sensitivity to glucocorticoids, mineralocorticoids, and adrenal androgens among patients [9–16]. In addition to their cognate receptors, other molecules participating in steroid signaling pathways, such as hormone inactivating or activating enzymes, immunophilins, and heat shock proteins, as well as genetic and epigenetic factors contribute substantially to tissue response to steroid hormones [13, 15, 16].

The Molecular Basis of Chrousos Syndrome

The molecular basis of Chrousos syndrome has been ascribed to point mutations, insertions or deletions in the *NR3C1* gene, which encodes the human glucocorticoid receptor (hGR) [9–16]. The *NR3C1* gene is located on the short arm of chromosome

5 and contains 10 exons. Exons 2–9 express all the protein isoforms, whereas exon 1 consists of several promoters, which enable the initiation of transcription in a promoter- or tissue-specific fashion [4, 5, 16, 21]. The alternative splicing of exon 9 gives rise to the two main protein isoforms, the hGRα and the hGRβ, which have distinct properties with respect to localization, ligand-binding ability, and transcriptional activity [22–26]. Moreover, the alternative splicing of the *NR3C1* gene generates three more receptor subtypes, the hGRγ, hGR-A, and hGR-P [23]. At the mRNA level, the alternative translation initiation of hGRα generates eight receptor isoforms α (hGRα-A, hGRα-B, hGRα-C1, hGRα-C2, hGRα-C3, hGRα-D1, hGRα-D2, and hGRα-D3) and possibly eight β isoforms as well, with distinct intracellular localization and transcriptional activity [27, 28].

The classic hGRα belongs to the steroid hormone receptor family of the nuclear receptor superfamily and functions as a ligand-induced transcription factor influencing the transcription rate of numerous genes [4, 5, 16]. At the protein level, the hGRα consists of four functional domains: (1) the N-terminal or immunogenic (NTD), which contains important amino acids that undergo several posttranslational modifications; (2) the DNA-binding domain (DBD), which consists of the characteristic and highly conserved motif of two zinc fingers, and enables the interaction between the receptor and its target DNA sequences in the glucocorticoid-responsive genes; (3) the hinge region, which provides the appropriate structural flexibility to the protein and allows the receptor to interact with different target genes; and (4) the ligand-binding domain (LBD), which is responsible for the binding of the receptor to glucocorticoids and contains sequences important for the translocation of the protein from the cytoplasm to the nucleus following activation, as well as amino acids that mediate the interaction of the receptor with coactivators in a ligand-dependent fashion [4, 5, 16].

At the target cell, the glucocorticoid signaling pathway is activated upon the binding of the receptor to synthetic and/or natural glucocorticoids, which causes the appropriate conformational changes to the protein, enabling the receptor to dissociate from chaperon heat shock proteins (HSPs) and immunophilins, and to translocate into the nucleus [4, 5, 16]. Within the nucleus, the activated receptor forms homo- or heterodimers and binds to the specific glucocorticoid response elements (GREs) within the promoter sequences of target genes, thereby inducing or repressing the transcription of the latter. Furthermore, the ligand-bound hGRα can modulate gene expression independently of DNA binding by physically interacting with other fundamental transcription factors, such as the activator protein-1 (AP-1), nuclear factor-kB (NF-kB), and signal transducers and activators of transcription (STATs) [4, 5, 16].

Patients with Chrousos syndrome usually harbor a point mutation, insertion or deletion in the *NR3C1* gene, which generally results in a defective glucocorticoid receptor and impaired glucocorticoid signal transduction, leading to reduced tissue sensitivity to glucocorticoids. The majority of the reported mutations are located in the LBD (Fig. 1), leading to a broad spectrum of clinical manifestations [17, 19, 20, 29–44]. The first identified *NR3C1* gene mutation was an adenine to thymine substitution at nucleotide position 1922, which resulted in substitution of aspartic acid

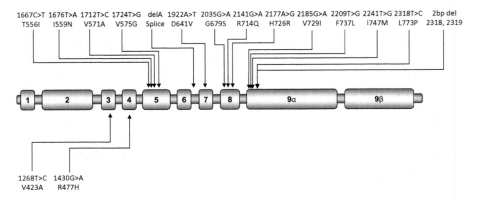

Fig. 1 Schematic representation of the known mutations of the *NR3C1* gene causing Chrousos syndrome. Mutations in the upper panel are located in the LBD of the receptor, while the V423A and R477H mutations are located in the DBD of the receptor

to valine at amino acid residue 641 at the LBD [17]. Within the last four decades, the tremendous progress of molecular and structural biology has provided us with the appropriate methods to study in depth the molecular mechanisms of action of the mutant glucocorticoid receptors.

From the Bedside to the Bench: Molecular and Structural Biology of Chrousos Syndrome

We and others have thoroughly investigated the molecular mechanisms of action of the defective natural hGRs [17, 19, 20, 29–44]. We systematically investigated: (1) the transcriptional activity of the mutant receptors through reporter assays; (2) the protein expression via Western blotting; (3) the ability of the mutant receptors to exert a dominant negative effect upon the hGRα-mediated transcriptional activity using reporter assays; (4) the ability of the mutant receptors to transrepress the NF-kB signaling pathway using reporter assays; (5) the affinity of the mutant receptors for the ligand through dexamethasone-binding assays; (6) the subcellular localization of the mutant receptors and the time required to translocate from the cytoplasm to nucleus following exposure to the ligand using green fluorescent protein (GFP)-fused plasmids; (7) the ability of the mutant receptors to bind to GREs via in vitro binding assays; (8) the interaction of the mutant receptors with the glucocorticoid receptor-interacting protein 1 (GRIP1) coactivator using Glutathione-S-Transferase (GST) pull-down assays; and (9) the conformational change of the mutant receptor that causes glucocorticoid resistance by structural biology studies. The molecular defects of the mutant receptors that have been identified in patients with Chrousos syndrome are presented in Table 2 [17, 19, 20, 29–44].

Table 2 Mutations of the human glucocorticoid receptor gene causing Chrousos syndrome

| Author (Reference) | Mutation position | | Molecular mechanisms | Genotype | Phenotype |
	cDNA	Amino acid			
Chrousos et al. [17]	1922 (A→T)	641 (D→V)	Transactivation ↓	Homozygous	Hypertension
Hurley et al. [30]			Affinity for ligand ↓ (×3)		Hypokalemic alkalosis
Charmandari et al. [37]			Nuclear translocation: 22 min		
			Abnormal interaction with GRIP1		
Karl et al. [31]		4 bp deletion in exon-intron 6	hGRα number: 50 % of control	Heterozygous	Hirsutism
			Inactivation of the affected allele		Male-pattern hair loss
					Menstrual irregularities
Malchoff et al. [32]	2185 (G→A)	729 (V→I)	Transactivation ↓	Homozygous	Precocious puberty
Charmandari et al. [37]			Affinity for ligand ↓ (×2)		Hyperandrogenism
			Nuclear translocation: 120 min		
			Abnormal interaction with GRIP1		
Karl et al. [29]	1676 (T→A)	559 (I→N)	Transactivation ↓	Heterozygous	Hypertension
Kino et al. [33]			Decrease in hGR binding sites		Oligospermia
Charmandari et al. [37]			Transdominance (+)		Infertility
			Nuclear translocation: 180		
			Abnormal interaction with GRIP1		
Ruiz et al. [34]	1430 (G→A)	477 (R→H)	Transactivation ↓	Heterozygous	Hirsutism
Charmandari et al. [39]			No DNA binding		Fatigue
			Nuclear translocation: 20 min		Hypertension
Ruiz et al. [34]	2035 (G→A)	679 (G→S)	Transactivation ↓	Heterozygous	Hirsutism
Charmandari et al. [39]			Affinity for ligand ↓ (×2)		Fatigue
			Nuclear translocation: 30 min		Hypertension
			Abnormal interaction with GRIP1		

(continued)

Table 2 (continued)

Author (Reference)	Mutation position		Genotype	Molecular mechanisms	Phenotype
	cDNA	Amino acid			
Mendonca et al. [35]	1712 (T→C)	571 (V→A)	Homozygous	Transactivation ↓	Ambiguous genitalia
Charmandari et al. [37]				Affinity for ligand ↓ (×6)	Hypertension
				Nuclear translocation: 25 min	Hypokalemia
				Abnormal interaction with GRIP1	Hyperandrogenism
Vottero et al. [36]	2241 (T→G)	747 (I→M)	Heterozygous	Transactivation ↓	Cystic acne
Charmandari et al. [37]				Transdominance (+)	Hirsutism
				Affinity for ligand ↓ (×2)	Oligo-amenorrhea
				Nuclear translocation ↓	
				Abnormal interaction with GRIP1	
Charmandari et al. [38]	2318 (T→C)	773 (L→P)	Heterozygous	Transactivation ↓	Fatigue
				Transdominance (+)	Anxiety
				Affinity for ligand ↓ (×2.6)	Acne
				Nuclear translocation: 30 min	Hirsutism
				Abnormal interaction with GRIP1	Hypertension
Charmandari et al. [40]	2209 (T→C)	737 (F→L)	Heterozygous	Transactivation ↓	Hypertension
				Transdominance (+)	Hypokalemia
				Affinity for ligand ↓ (×1.5)	
				Nuclear translocation: 180 min	
McMahon et al. [20]	2 bp deletion at nt 2318-9	773	Homozygous	Transactivation ↓	Hypoglycemia
				Affinity for ligand: absent	Fatigability with feeding
				No suppression of IL-6	Hypertension

Reference	Nucleotide	Amino acid	Molecular findings	Zygosity	Clinical features
Nader et al. [19]	2141 (G→A)	714 (R→Q)	Transactivation ↓ Transdominance (+) Affinity for ligand ↓ (×2) Nuclear translocation ↓ Abnormal interaction with GRIP1	Heterozygous	Hypoglycemia Hypokalemia Hypertension Mild clitoromegaly Advanced bone age Precocious pubarche
Zhu Hui-juan et al. [41]	1667 (G→T)	556 (T→I)	Not studied yet	Heterozygous	Adrenal incidentaloma
Roberts et al. [42]	1268 (T→C)	423 (V→A)	Transactivation ↓ Affinity for ligand: N No DNA binding Nuclear translocation: 35 min Interaction with GRIP1: N	Heterozygous	Fatigue Anxiety Hypertension
Nicolaides et al. [43]	1724 (T→G)	575 (V→G)	Transactivation ↓ Transrepression ↑ Affinity for ligand ↓ (×2) Nuclear translocation ↓ Abnormal interaction with GRIP1	Heterozygous	Melanoma Asymptomatic daughters
Nicolaides et al. [44]	2177 (A→G)	726 (H→R)	Transactivation ↓ Transrepression ↓ Affinity for ligand ↓ (×2) Nuclear translocation ↓ Abnormal interaction with GRIP1	Heterozygous	Hirsutism, Acne Alopecia, Anxiety Fatigue Irregular menstrual cycles

Modified from Reference [16]

Numbers in the parentheses following authors' names indicate the corresponding references

We have recently identified a novel point mutation in the *NR3C1* gene associated with Chrousos syndrome in a patient that presented with hirsutism, acne, alopecia, anxiety, fatigue, and irregular menstrual cycles, but no clinical manifestations suggestive of Cushing's syndrome [44]. The patient harbored a novel A>G transition at nucleotide position 2177, which resulted in histidine (H) to arginine (R) substitution at amino acid position 726 of the receptor [44]. Following identification, we applied the abovementioned methods in an attempt to investigate how the mutant receptor hGRαH726R caused glucocorticoid resistance. Compared with the wild-type receptor, the hGRαH726R displayed reduced ability to transactivate target genes and to transrepress the NF-kB signaling pathway, had 55 % lower affinity for the ligand and a fourfold delay in cytoplasmic-to-nuclear translocation, and interacted with the GRIP1 coactivator mostly through its activation function-1 domain [44] (Fig. 2).

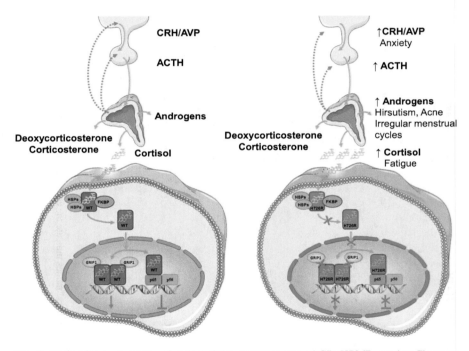

Fig. 2 Molecular mechanisms of action of the mutant receptor hGRαH726R causing Chrousos syndrome. Both the wild-type hGRα and the mutant receptor hGRαH726R reside in the cytoplasm in the absence of ligand by forming a heterocomplex with heat shock proteins (HSPs) and FKBP51 (FKBP). Upon ligand binding, the wild-type hGRα dissociates from the heterocomplex and translocates into the nucleus, while this process of the mutant hGRαH726R is significantly delayed due to decreased ligand binding and/or impaired nuclear translocation. The wild-type hGRα induces or represses the transcriptional activity of glucocorticoid target genes by attracting to GREs several coactivators including the glucocorticoid receptor-interacting protein 1 (GRIP1), or by interacting with other transcription factors, such as the NF-kB. On the other hand, the mutant receptor hGRαH726R has impaired interaction with the GRIP1, and displays reduced ability to transactivate glucocorticoid-responsive genes and to transrepress the NF-kB signaling pathway. FKBP: immunophilins; GRIP1: glucocorticoid receptor-interacting protein 1; H726R: human glucocorticoid receptor H726R; HSP: heat shock proteins; p65: transcription factor p65; p50: transcription factor p50; WT: wild-type human glucocorticoid receptor

Finally, structural biology studies showed that the H726R mutation revealed a significant structural shift in the rigidity of helix 10 of the receptor, which caused reduced flexibility and decreased affinity of the mutant receptor for the ligand [44] (Table 2).

Diagnostic Approach

The diagnostic approach to subjects suspected to have Chrousos syndrome consists of a detailed personal and family history [9–11, 13–16]. Particular emphasis should be given to any symptoms indicating alterations in HPA axis activity. Therefore, headaches, seizures, or visual impairment should be carefully evaluated. Moreover, the regularity of menstrual cycles in women should be documented. Furthermore, growth, development, and sexual maturation should be evaluated in detail in children suspected to have Chrousos syndrome. On clinical examination, physicians should pay particular attention to signs suggestive of mineralocorticoid and/or androgen excess [9–11, 13–16].

The endocrinologic evaluation of patients suspected to have Chrousos syndrome consists of measurement of the 08:00 h concentrations of serum cortisol, plasma ACTH, plasma renin activity (recumbent), serum aldosterone, androgens (testosterone, androstenedione, DHEA, DHEAS), total cholesterol, HDL, LDL, triglycerides, and fasting glucose and insulin [9–11, 13–16]. Affected subjects have increased morning serum cortisol concentrations and elevated 24-h urinary free cortisol (UFC) excretion without any symptoms or signs of hypercortisolism; therefore, the 24-h UFC excretion should be determined on 2 or 3 consecutive days to enable accurate diagnosis of the syndrome [9–11, 13–16]. Interestingly, there is a high variation in the increased 24-h UFC excretion and the elevated serum cortisol concentrations among patients with Chrousos syndrome due to the different degree of impairment of glucocorticoid signal transduction. More specifically, serum cortisol concentrations and 24-h UFC excretion may be, respectively, up to 7- and 50-fold higher compared with the highest value of their normal range. On the other hand, morning plasma ACTH concentrations may be normal or high, whereas the circadian pattern of both ACTH and cortisol secretion and their responsiveness to stressors are maintained, albeit at higher concentrations than normal [9–11, 13–16].

To evaluate the responsiveness of the HPA axis to exogenously administered glucocorticoids, subjects suspected to have Chrousos syndrome should undergo a dexamethasone suppression test [9–11, 13–16]. Dexamethasone should be given *per os* at midnight every other day at progressively increasing doses of 0.3, 0.6, 1.0, 1.5, 2.0, 2.5, and 3.0 mg, and serum cortisol concentrations should be determined the following morning. To avoid any nonadherence to the treatment, or to exclude the possibility of increased metabolic clearance or reduced absorption of the administered medication, dexamethasone concentrations should also be measured at the same time [16]. Patients with Chrousos syndrome generally display resistance of the HPA axis to dexamethasone suppression with high variation that depends on the severity of the pathologic condition. Therefore, dexamethasone should be given to subjects suspected to have Chrousos syndrome in a dose up to 7.5-fold higher compared with that required to achieve suppression of serum cortisol concentrations by 50 % in normal subjects [16].

There are two in vitro methods that allow us to confirm the diagnosis of Chrousos syndrome: dexamethasone-binding assays and thymidine incorporation assays, both on peripheral leukocytes obtained by the patient and a matched-control subject [9–11, 13–16]. In dexamethasone-binding assays, the defective glucocorticoid receptor has lower affinity for the ligand compared to that of the control subject. In thymidine incorporation assays, the patient shows resistance to dexamethasone-induced suppression of phytohemagglutinin-stimulated thymidine incorporation, compared with the control subject. Finally, to identify any mutations, if present, the coding region of the *NR3C1* gene, including the junctions between introns and exons, must be sequenced [9–11, 13–16].

Therapeutic Management

Patients with Chrousos syndrome should be treated with high doses of mineralocorticoid-sparing synthetic glucocorticoids, such as dexamethasone (1–3 mg given once daily) to reduce the excess secretion of ACTH, which triggers the high production of mineralocorticoids and/or adrenal androgens [9–11, 13–16]. It is particularly important that the dose of dexamethasone be carefully titrated based on the severity of clinical manifestations and biochemical profile of the patients, given that the HPA axis should be adequately suppressed to avoid the development of ACTH-secreting adenomas secondary to long-standing ACTH hypersecretion, as this was the case with the patient carrying the hGRαI559N mutation [29]. Treatment with high doses of mineralocorticoid-sparing synthetic glucocorticoids ameliorates the clinical manifestations of the condition and normalizes the concentrations of plasma ACTH and serum androgens [9–11, 13–16].

Concluding Remarks and Future Directions

Many clinical cases of Chrousos syndrome remain unrecognized for a long time, because of the variable clinical manifestations of the syndrome and the difficulty in establishing the diagnosis. Therefore, we recommend determination of the 24-h UFC excretion followed by sequencing of the *NR3C1* gene in patients with hyperandrogenism and/or hypertension of unknown origin. Once the diagnosis is established, patients should be treated with high doses of dexamethasone that should be carefully titrated to adequately suppress the excess ACTH secretion and to effectively achieve the minimum glucocorticoid side effects.

Although most cases of Chrousos syndrome have been attributed to point mutations, insertions or deletions in the *NR3C1* gene, sequencing analysis does not always reveal these defects in the gene encoding the human glucocorticoid receptor, suggesting that other molecules (e.g., HSPs, immunophilins) might contribute to the impaired glucocorticoid signal transduction. In the era of next-generation sequencing, when Chrousos syndrome is suspected, we suggest the sequencing of at least a panel of genes that express proteins participating in the glucocorticoid signaling system. Undoubtedly, the application of whole-exome sequencing will uncover numerous other unknown genes expressing hGR protein partners or cofactors.

References

1. Chrousos GP. Stress and disorders of the stress system. Nat Rev Endocrinol. 2009;5:374–81.
2. Chrousos GP, Gold PW. The concepts of stress and stress system disorders: overview of physical and behavioral homeostasis. JAMA. 1992;267:1244–52.
3. Kontopoulou TD, Marketos SG. Homeostasis. The ancient Greek origin of a modern scientific principle. Hormones. 2002;1:124–5.
4. Nicolaides NC, Kyratzi E, Lamprokostopoulou A, Chrousos GP, Charmandari E. Stress, the stress system and the role of glucocorticoids. Neuroimmunomodulation. 2015;22:6–19.
5. Nicolaides NC, Galata Z, Kino T, Chrousos GP, Charmandari E. The human glucocorticoid receptor: Molecular basis of biologic function. Steroids. 2010;75:1–12.
6. Nicolaides NC, Charmandari E, Chrousos GP. The hypothalamic-pituitary-adrenal axis in human health and disease. In: Cokkinos DV, editor. Introduction in translational cardiovascular research. Cham: Springer; 2015. p. 91–107.
7. Charmandari E, Nicolaides NC, Chrousos GP. Adrenal insufficiency. Lancet. 2014;383: 2152–67.
8. Quax RA, Manenschijn L, Koper JW, Hazes JM, Lamberts SW, van Rossum EF, Feelders RA. Glucocorticoid sensitivity in health and disease. Nat Rev Endocrinol. 2013;9:670–86.
9. Charmandari E. Primary generalized glucocorticoid resistance and hypersensitivity. Horm Res Paediatr. 2011;76:145–55.
10. Charmandari E. Primary generalized glucocorticoid resistance and hypersensitivity: the end-organ involvement in the stress response. Sci Signal. 2012;5:pt5.
11. Charmandari E, Kino T, Chrousos GP. Primary generalized familial and sporadic glucocorticoid resistance (Chrousos syndrome) and hypersensitivity. Endocr Dev. 2013;24:67–85.
12. Nicolaides NC, Charmandari E, Chrousos GP, Kino T. Recent advances in the molecular mechanisms determining tissue sensitivity to glucocorticoids: novel mutations, circadian rhythm and ligand-induced repression of the human glucocorticoid receptor. BMC Endocr Disord. 2014;14:71.
13. Charmandari E, Kino T. Chrousos syndrome: a seminal report, a phylogenetic enigma and the clinical implications of glucocorticoid signaling changes. Eur J Clin Invest. 2010;40:932–42.
14. Charmandari E, Kino T, Ichijo T, Chrousos GP. Generalized glucocorticoid resistance: clinical aspects, molecular mechanisms, and implications of a rare genetic disorder. J Clin Endocrinol Metab. 2008;93:1563–72.
15. Chrousos G. Q&A: primary generalized glucocorticoid resistance. BMC Med. 2011;9:27.
16. Nicolaides NC, Charmandari E. Chrousos syndrome: from molecular pathogenesis to therapeutic management. Eur J Clin Invest. 2015;45:504–14.
17. Chrousos GP, Vingerhoeds A, Brandon D, Eil C, Pugeat M, DeVroede M, Loriaux DL, Lipsett MB. Primary cortisol resistance in man. A glucocorticoid receptor-mediated disease. J Clin Invest. 1982;69:1261–9.
18. Chrousos GP, Detera-Wadleigh SD, Karl M. Syndromes of glucocorticoid resistance. Ann Intern Med. 1993;119:1113–24.
19. Nader N, Bachrach BE, Hurt DE, Gajula S, Pittman A, Lescher R, Kino T. A novel point mutation in the helix 10 of the human glucocorticoid receptor causes generalized glucocorticoid resistance by disrupting the structure of the ligand-binding domain. J Clin Endocrinol Metab. 2010;95:2281–5.

20. McMahon SK, Pretorius CJ, Ungerer JP, Salmon NJ, Conwell LS, Pearen MA, Batch JA. Neonatal complete generalized glucocorticoid resistance and growth hormone deficiency caused by a novel homozygous mutation in Helix 12 of the ligand binding domain of the glucocorticoid receptor gene (NR3C1). J Clin Endocrinol Metab. 2010;95:297–302.
21. Zhou J, Cidlowski JA. The human glucocorticoid receptor: one gene, multiple proteins and diverse responses. Steroids. 2005;70:407–17.
22. Bamberger CM, Bamberger AM, de Castro M, Chrousos GP. Glucocorticoid receptor β, a potential endogenous inhibitor of glucocorticoid action in humans. J Clin Invest. 1995;95: 2435–41.
23. Charmandari E, Chrousos GP, Ichijo T, Bhattacharyya N, Vottero A, Souvatzoglou E, Kino T. The human glucocorticoid receptor (hGR) β isoform suppresses the transcriptional activity of hGRα by interfering with formation of active coactivator complexes. Mol Endocrinol. 2005;19:52–64.
24. Yudt MR, Jewell CM, Bienstock RJ, Cidlowski JA. Molecular origins for the dominant negative function of human glucocorticoid receptor β. Mol Cell Biol. 2003;23:4319–30.
25. Kino T, Manoli I, Kelkar S, Wang Y, Su YA, Chrousos GP. Glucocorticoid receptor (GR) beta has intrinsic, GRalpha-independent transcriptional activity. Biochem Biophys Res Commun. 2009;381:671–5.
26. Kino T, Su YA, Chrousos GP. Human glucocorticoid receptor isoform beta: recent understanding of its potential implications in physiology and pathophysiology. Cell Mol Life Sci. 2009;66:3435–48.
27. Oakley RH, Cidlowski JA. Cellular processing of the glucocorticoid receptor gene and protein: new mechanisms for generating tissue-specific actions of glucocorticoids. J Biol Chem. 2011;286:3177–84.
28. Lu NZ, Cidlowski JA. Translational regulatory mechanisms generate N-terminal glucocorticoid receptor isoforms with unique transcriptional target genes. Mol Cell. 2005;18:331–42.
29. Karl M, Lamberts SW, Koper JW, Katz DA, Huizenga NE, Kino T, Haddad BR, Hughes MR, Chrousos GP. Cushing's disease preceded by generalized glucocorticoid resistance: clinical consequences of a novel, dominant-negative glucocorticoid receptor mutation. Proc Assoc Am Physicians. 1996;108:296–307.
30. Hurley DM, Accili D, Stratakis CA, Karl M, Vamvakopoulos N, Rorer E, Constantine K, Taylor SI, Chrousos GP. Point mutation causing a single amino acid substitution in the hormone binding domain of the glucocorticoid receptor in familial glucocorticoid resistance. J Clin Invest. 1991;87:680–6.
31. Karl M, Lamberts SW, Detera-Wadleigh SD, Encio IJ, Stratakis CA, Hurley DM, Accili D, Chrousos GP. Familial glucocorticoid resistance caused by a splice site deletion in the human glucocorticoid receptor gene. J Clin Endocrinol Metab. 1993;76:683–9.
32. Malchoff DM, Brufsky A, Reardon G, McDermott P, Javier EC, Bergh CH, Rowe D, Malchoff CD. A mutation of the glucocorticoid receptor in primary cortisol resistance. J Clin Invest. 1993;91:1918–25.
33. Kino T, Stauber RH, Resau JH, Pavlakis GN, Chrousos GP. Pathologic human GR mutant has a transdominant negative effect on the wild-type GR by inhibiting its translocation into the nucleus: importance of the ligand-binding domain for intracellular GR trafficking. J Clin Endocrinol Metab. 2001;86:5600–8.
34. Ruiz M, Lind U, Gafvels M, Eggertsen G, Carlstedt-Duke J, Nilsson L, Holtmann M, Stierna P, Wikstrom AC, Werner S. Characterization of two novel mutations in the glucocorticoid receptor gene in patients with primary cortisol resistance. Clin Endocrinol (Oxf). 2001;55:363–71.
35. Mendonca BB, Leite MV, de Castro M, Kino T, Elias LL, Bachega TA, Arnhold IJ, Chrousos GP, Latronico AC. Female pseudohermaphroditism caused by a novel homozygous missense mutation of the GR gene. J Clin Endocrinol Metab. 2002;87:1805–9.
36. Vottero A, Kino T, Combe H, Lecomte P, Chrousos GP. A novel, C-terminal dominant negative mutation of the GR causes familial glucocorticoid resistance through abnormal interactions with p160 steroid receptor coactivators. J Clin Endocrinol Metab. 2002;87:2658–67.

37. Charmandari E, Kino T, Vottero A, Souvatzoglou E, Bhattacharyya N, Chrousos GP. Natural glucocorticoid receptor mutants causing generalized glucocorticoid resistance: Molecular genotype, genetic transmission and clinical phenotype. J Clin Endocrinol Metab. 2004;89:1939–49.

38. Charmandari E, Raji A, Kino T, Ichijo T, Tiulpakov A, Zachman K, Chrousos GP. A novel point mutation in the ligand-binding domain (LBD) of the human glucocorticoid receptor (hGR) causing generalized glucocorticoid resistance: the importance of the C terminus of hGR LBD in conferring transactivational activity. J Clin Endocrinol Metab. 2005;90:3696–705.

39. Charmandari E, Kino T, Ichijo T, Zachman K, Alatsatianos A, Chrousos GP. Functional characterization of the natural human glucocorticoid receptor (hGR) mutants hGRαR477H and hGRαG679S associated with generalized glucocorticoid resistance. J Clin Endocrinol Metab. 2006;91:1535–43.

40. Charmandari E, Kino T, Ichijo T, Jubiz W, Mejia L, Zachman K, Chrousos GP. A novel point mutation in helix 11 of the ligand-binding domain of the human glucocorticoid receptor gene causing generalized glucocorticoid resistance. J Clin Endocrinol Metab. 2007;92:3986–90.

41. Zhu HJ, Dai YF, Wang O, Li M, Lu L, Zhao WG, Xing XP, Pan H, Li NS, Gong FY. Generalized glucocorticoid resistance accompanied with an adrenocortical adenoma and caused by a novel point mutation of human glucocorticoid receptor gene. Chin Med J (Engl). 2011;124:551–5.

42. Roberts ML, Kino T, Nicolaides NC, Hurt DE, Katsantoni E, Sertedaki A, Komianou F, Kassiou K, Chrousos GP, Charmandari E. A novel point mutation in the DNA-binding domain (DBD) of the human glucocorticoid receptor causes primary generalized glucocorticoid resistance by disrupting the hydrophobic structure of its DBD. J Clin Endocrinol Metab. 2013;98:E790–5.

43. Nicolaides NC, Roberts ML, Kino T, Braatvedt G, Hurt DE, Katsantoni E, Sertedaki A, Chrousos GP, Charmandari E. A novel point mutation of the human glucocorticoid receptor gene causes primary generalized glucocorticoid resistance through impaired interaction with the LXXLL motif of the p160 coactivators: dissociation of the transactivating and transrepressive activities. J Clin Endocrinol Metab. 2014;99:E902–7.

44. Nicolaides NC, Geer EB, Vlachakis D, Roberts ML, Psarra AM, Moutsatsou P, Sertedaki A, Kossida S, Charmandari E. A novel mutation of the hGR gene causing chrousos syndrome. Eur J Clin Invest. 2015;45:782–91.

Cortisol Metabolism as a Regulator of the Tissue-Specific Glucocorticoid Action

Emilia Sbardella and Jeremy W. Tomlinson

Abstract Glucocorticoids have a diverse array of functions affecting almost all tissues in the body. While circulating cortisol levels are under the control of the hypothalamo–pituitary–adrenal axis, within individual organs and tissues, a series of enzymes is able to metabolize, either inactivating or reactivating glucocorticoids to control their availability to bind and activate the glucocorticoid receptor. The most studied of these enzymes are the 11β-hydroxysteroid dehydrogenases (type 1 and type 2) and the A-ring reductases (5α-reductase type 1 and 2 and 5β-reductase). 11β-Hydroxysteroid dehydrogenase type 1 regenerates active glucocorticoid (cortisol) from inactive cortisone and thus amplifies local glucocorticoid action. In contrast, 11β-hydroxysteroid dehydrogenase type 2 and the A-ring reductases clear and inactivate glucocorticoids. All have tissue-specific patterns of expression and regulation and have been implicated in the pathogenesis of many diseases that are discussed as part of this chapter. In addition, 11β-hydroxysteroid dehydrogenases type 1 represents a novel therapeutic target and selective inhibitors that decease tissue-specific glucocorticoid levels have reached phase II clinical trials. The prereceptor regulation of glucocorticoid action is therefore not only of fundamental physiological and pathological importance, but continues to represent an area of intense scientific and therapeutic interest.

Keywords 11β-Hydroxysteroid dehydrogenases • A-Ring reductases • 5α-Reductase • 5β-Reductase • Cortisol • Cortisone • Obesity • Adipose • Liver

E. Sbardella
Department of Experimental Medicine, Sapienza University of Rome, Rome, Italy

Oxford Centre for Diabetes, Endocrinology and Metabolism, NIHR Oxford Biomedical Research Centre, Churchill Hospital, University of Oxford, Oxford OX3 7LE, UK

J.W. Tomlinson, Ph.D., F.R.C.P. (✉)
Oxford Centre for Diabetes, Endocrinology and Metabolism, NIHR Oxford Biomedical Research Centre, Churchill Hospital, University of Oxford, Oxford OX3 7LE, UK
e-mail: jeremy.tomlinson@ocdem.ox.ac.uk

© Springer International Publishing Switzerland 2017   271
E.B. Geer (ed.), *The Hypothalamic-Pituitary-Adrenal Axis in Health and Disease*, DOI 10.1007/978-3-319-45950-9_14

Introduction

Glucocorticoids (GC) have a diverse array of functions in almost all tissues of the body and are crucial regulators of fundamental physiological processes that include glucose and amino acid metabolism, inflammation, and immunity [1]. Classical GC action is dependent upon binding of ligand to the glucocorticoid receptor (GR), dissociation from its associated heat shock protein and other chaperones, translocation from the cytosol to the nucleus, dimerization, and subsequent regulation of gene transcription (Fig. 1). Since their discovery in the 1940s by Kendall and Hench, GCs are now one of the most commonly prescribed class of therapeutic agents for conditions including rheumatoid arthritis, and asthma and are a fundamental component of antirejection medication regimes in organ transplant recipients.

Circulating GC levels are tightly controlled by the hypothalamo–pituitary–adrenal (HPA) axis, which regulates secretion from the adrenal glands via a classical negative feedback loop. Healthy adults secrete 10–15 mg cortisol/day [2] and the majority is bound to cortisol-binding globulin (CBG). Estimates suggest that only 5 % of circulating cortisol is "free" and biologically active [3, 4]. The half-life of free cortisol is brief (only a few minutes) whereas protein-bound cortisol has a much longer half-life between 70 and 120 min [4–6]. Importantly the biological availability of GCs represents a balance between synthesis/secretion and metabolism/clearance.

Within GC target tissues, there is an added layer of complexity to the regulation of GC action. Cortisol delivered from the circulation into cells can be subjected to a series of metabolic pathways which are able to modify the access of the active ligand, cortisol, to the GR, the so-called prereceptor regulation (Fig. 1).

Once inside the cell, a series of enzymes are able to metabolize cortisol and these include the 11β-hydroxysteroid dehydrogenases (11β-HSD1 and 2) and the A-ring reductases (5α-reductase type 1 [5αR1] and 2 [5αR2] and 5β-reductase). All have tissue-specific patterns of expression and all have been implicated in the pathogenesis of various conditions (Fig. 2). Within this chapter we will describe the enzymes involved and summarize on a tissue-by-tissue basis the contribution of each enzyme system to the regulation of GC actions.

11β-Hydroxysteroid Dehydrogenase

11β-Hydroxysteroid Dehydrogenase Type 1

GCs were identified more than 60 years ago and were heralded as a potentially curative treatment for many diseases [7]. Kendall et al. published the discovery of what they believed to be a treatment that could reverse rheumatoid arthritis in the 1950s [8]. They identified Compound E, now recognized to be cortisone, an inactive GC metabolite that requires reactivation to cortisol (Compound F), to allow it to bind and activate the GR. It is now recognized that the enzyme responsible for the

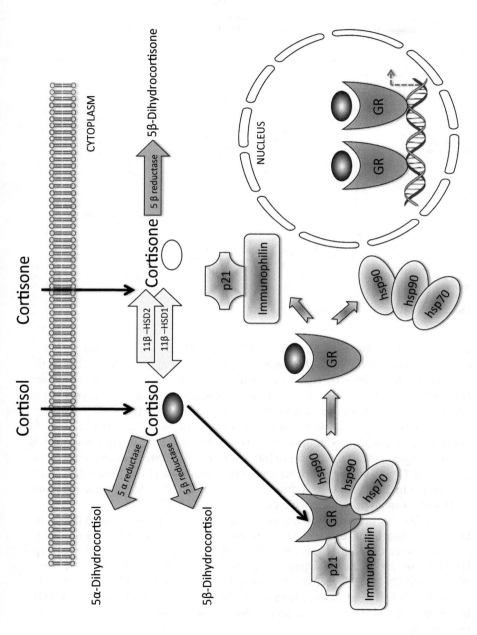

Fig. 1 Pre-receptor regulation of glucocorticoid availability governs access to bind and activate the glucocorticoid receptor

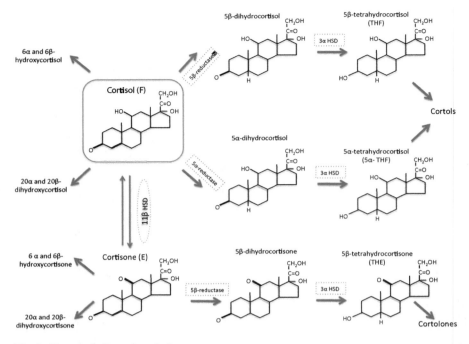

Fig. 2 The metabolism of cortisol

conversion of cortisone to cortisol is *11β-HSD1*. The hydroxyl group at C11 is crucially important for cortisol to be active [9, 10]. *11β-HSD1* is a member of the short-chain dehydrogenase/reductase (SDR) superfamily of enzymes which are NADP(H)-dependent enzymes, which have a fundamental role in the regulation of hormone signaling with in excess of 3000 family members [11].

11β-HSD1 was purified and cloned from rodent tissues in the 1980s [12, 13]. In humans the gene that encodes the protein, HSD11B1, is located on chromosome 1, is 30 kb in length, and has 6 exons and 5 introns. *11β-HSD1* comprises 292 amino acids and shares 77 % homology with rat amino acid sequence [14]. Human *11β-HSD1* was cloned in 2002, exists as a dimer, and is bound to the endoplasmic reticulum (ER) with its catalytic domain within the ER lumen [15–17].

11β-HSD1 is a bidirectional enzyme, which in vivo acts primarily as an oxoreductase, converting inactive cortisone (11-dehydrocorticosterone in rodents) to active cortisol. The catalytic directionality of the enzyme is based on the position of 11β-HSD1 within the ER lumen where it colocalizes with hexose-6-phosphate dehydrogenase (H6PD). H6PD generates the reduced cosubstrate, NADPH. Thus, the ratio of NADPH/NADP confers directionality to 11β-HSD1 [18, 19]. Purified 11β-HSD1, in the absence of H6PD, behaves principally as a dehydrogenase, oxidizing cortisol to cortisone.

The ontogeny of 11β-HSD1 has been studied mainly in animal models and is predominantly expressed in the postnatal period. While 11β-HSD1 is detectable in many tissues, there is a lack of activity in early gestation with reductase activity only becoming apparent after delivery and rising steadily throughout infancy [20–22]. In humans, 11β-HSD1 ontogeny is less well characterized with few published studies. Cortisone therapy is ineffective in treating congenital adrenal hyperplasia in early infancy most likely due to absent or significantly reduced liver 11β-HSD1 [23]. Both reductase and dehydrogenase activity have been demonstrated in fetal lung tissue [24]. 11β-HSD1 activity remains similar throughout childhood in both boys and girls [25]. At puberty, there is a reduction in 11β-HSD1 activity in women which continues into adult life. In adults, there is a well-described dimorphism in cortisol metabolism between men and women with an apparent reduction in 11β-HSD1 activity in women [26, 27] although this is not consistent across all studies [28].

11β-HSD1 is expressed in many tissues including liver, adipose tissue, gonads, GI tract, kidney, eye, anterior pituitary, leukocytes, and bone [20]. Expression is highest in liver, brain, gonads, and adipose tissue. Many factors regulate expression and activity of 11β-HSD1. In most studies, GCs, proinflammatory cytokines (TNFα, IL-1β) peroxisome proliferator-activated receptor γ agonists, and CCAAT/enhancer-binding proteins (CEBPs) increase expression and/or activity. In contrast, growth hormone (GH) and liver X receptor (LXR) agonists decrease expression [20]. Recently, salicylates have been shown to downregulate 11β-HSD1 expression in adipose tissue and improve insulin sensitivity [29]. The effects of sex steroids, insulin, and other hormones are variable across tissues and between species. Estradiol has been shown to decrease 11β-HSD1 expression in rat liver and kidney, but testosterone was without effect [30].

Genetic defects in both HSDB1 and H6PD have been described. Cortisone reductase deficiency (CRD) is caused by HSDB1 gene defects and apparent cortisone reductase deficiency (ACRD) by mutations in H6PD. Both cause a reduction in tissue 11β-HSD1 activity with low urinary cortisol metabolites, significantly elevated cortisone metabolites with a consequent compensatory increased HPA activity leading to hyperandrogenism, premature adrenarche, and PCOS in women, and precocious puberty in males [19, 31–35].

11β-Hydroxysteroid Dehydrogenase Type 2

In 1993, an enzyme with exclusive 11β-HSD dehydrogenase activity was identified from both human placenta and rat kidney [36], and in 1994, Krozoski et al. isolated human 11β-HSD from human kidney that was identical to the dehydrogenase enzyme found in the placenta [37]. This second enzyme was found to be distinct from 11β-HSD1 and was called 11β-HSD2 and is also a member of the SDR family.

The human HSD11B2 gene is located on chromosome 16, has 5 exons and is only 6 kb in length [38]. Human 11β-HSD2 contains 405 amino acids with a molecular weight of 44 kDa. It is also anchored to the ER and loses its dehydrogenase activity once dissociated from tissue membranes [37, 39]. 11β-HSD2 acts exclusively as a dehydrogenase across all species and has a Km for cortisol of 50–60 and 10–13 nM for cortisone [40]. Mutations in HSD11B2 lead to the syndrome of apparent mineralocorticoid excess (AME), a hereditary cause of life-threatening hypertension and hypokalemia, suppressed renin activity, and a metabolic alkalosis [41–45]. The underpinning mechanism relies upon the fact that cortisol is able to activate both the GR and mineralocorticoid receptor (MR) with equal affinity. Circulating cortisol concentrations far exceed those of aldosterone, the natural ligand for the MR, and therefore to prevent cortisol activating the MR in mineralocorticoid target tissues, 11β-HSD2 inactivates cortisol (to cortisone) locally. The condition is characterized by an increased urinary ratio of cortisol to cortisone metabolites. It can be treated with the synthetic GC, dexamethasone, which lacks mineralocorticoid activity, but is able to suppress endogenous cortisol production. Functional inhibition of 11β-HSD2 activity within the kidney is also the mechanism underpinning liquorice-induced hypertension [46].

11β-HSD2 is therefore expressed in aldosterone sensitive tissues, mainly in the distal nephron, colonic epithelium, salivary and sweat glands, and in the fetus and placenta during gestation. During gestation in humans and mammals, high levels of expression within the placenta protect the developing fetal tissues against excess GC exposure. Expression within the placenta steadily rises throughout gestation and declines two weeks prior to labor [22]. Altered or disrupted 11β-HSD2 activity, with subsequent excess intrauterine exposure to GC, has been implicated in "programming" effects upon the developing fetus leading to low birth weight and lifelong physiological consequences such as increased cardiovascular, metabolic, and psychiatric complications [47].

Unlike 11β-HSD1, there are considerable data published on the epigenetic influence on 11β-HSD2 activity in humans and in rodent models [48]. HSD11B2 is susceptible to epigenetic influence, with methylation of the promoter region of particular interest. Increased methylation of this region has been inversely associated with 11β-HSD2 expression and has been linked with the development of hypertension, intrauterine growth retardation, reduced birth weight, and neurobehavioral movement disorders [49]. In rodent models, intrauterine growth retardation has been associated with increased methylation of HSD11B2 gene promoter with subsequent repression of 11β-HSD2 expression in adult kidneys [50].

Factors that increase 11β-HSD1 expression tend to reduce 11β-HSD2 and include pro-inflammatory cytokines such as TNFα [51]. Estrogen increases 11β-HSD2 expression [30, 52]. Vasopressin has been shown to stimulate 11β-HSD2 [53]. Glucocorticoids downregulate 11β-HSD2 in fetal placenta and lung cells, but not fetal kidney [54, 55]. Hypoxia has also been shown to reduce 11β-HSD2 expression [56] whereas in colonic epithelium, aldosterone increases 11β-HSD2 expression [57].

A-Ring-Reductases

The A-ring reductases are important regulators of GC availability. 5αR1 and 2 have an important dual role in the prereceptor regulation of steroid hormone availability. They inactivate cortisol to dihydrocortisol which is then subsequently converted to tetrahydrocortisol through the activity of 3α-hydroxysteroid dehydrogenase (3α-HSD) and are therefore of crucial importance in local GC clearance. In addition to this role, 5αRs are fundamentally important in the reduction of testosterone to the more potent androgen, dihydrotestosterone (DHT), and therefore sit at an importance interface that sets the balance at a cellular level between GC and androgen action. Apart from GC and androgens, 5αRs can also metabolize other steroid substrates including progesterone and mineralocorticoids. 5αRs are microsomal enzymes and are NADPH dependent. While three isoforms of 5αRs have been identified to date [58, 59], with different biochemical properties and sensitivity to substrates, it is only type 1 and type 2 that appear to have a role in the regulation of steroid hormone availability.

5α-Reductase Type 1

The gene encoding 5αR1 (SRD5A1) lies on chromosomes 5 and has 5 exons and 4 introns. It consists of 259 amino acids with a molecular weight of 29 kDa. It is expressed in both human and mouse liver and also in skin (nongenital) and adipose tissue [58]. Although testosterone is the most widely recognized substrate of this enzyme, progesterone has a lower K_m and therefore enzymatically may be the preferred substrate [60, 61]. To date, no mutations have been identified in SRD5A1.

Dutasteride is a dual inhibitor of both isoforms, 5αR1 and 5αR2, reducing circulating DHT by nearly 95 % compared to the baseline. MK-386 was reported to be a selective inhibitor of 5αR1 with 90 % efficiency but this compound is neither commercially available nor used in clinical practice [62].

5α-Reductase Type 2

The gene encoding 5αR2 (SRD5A2) lies on chromosomes 2 and has 5 exons and 4 introns. It has 254 amino acids with a molecular weight of 28 kDa and shares less than 50 % homology with 5αR1 [58, 63]. 5αR2 is expressed in human liver but not in mouse liver. 5αR2 is predominantly expressed in androgen-target organs such as prostate, epididymis, and seminal vesicles [58].

Finasteride is a selective 5αR2 inhibitor, while 5αR1 has a low sensitivity to this inhibitor. In comparison with 5αR1, 5αR2 has much higher affinity for androgen substrates such as testosterone. Many mutations and polymorphisms have been identified throughout the coding and noncoding regions of SRD5A2 [64]. Since 5αR2 converts testosterone to a more potent androgen (DHT), mutations in this enzyme lead to 46XY DSD (disorder of sex development) with consequent lack of virilization and poor development of the external genitalia. However, excessive androgen generation through the activity of 5αR2 has been implicated in conditions including polycystic ovary syndrome, breast cancer, and prostate cancer, as well as male pattern baldness [65–67].

5β-Reductase

The gene encoding 5β-reductase (or AKR1D1) is located on chromosome 7. 5β-Reductase is also highly expressed in hepatocytes and its crystal structure has been determined [68]. AKR1D1 is able to metabolize both cortisol and cortisone, and following 3α-HSD activity, it generates 5β-tetrahydrocortisol (5β-THF) and 5β-tetrahydrocortisone (5βTHE). While 5αRs reduce testosterone to the more potent 5α-DHT, 5β reductase generates 5β-DHT, which is inactive and thus limits androgen action locally. 5β-Reductase has a significant role in clearing the majority of all C-19-C21 steroids and therefore disruption of its activity has the potential to impact upon clearance of GCs, mineralocorticoids, and sex steroids. It also has an important role in bile acid production. Mutations in the gene encoding 5β-reductase lead to bile acid deficiency and form neonatal cholestatic liver disease which can progress to liver failure [69]. However, spontaneous recovery and survival into adult hood is reported [70].

Tissue-Specific Cortisol Metabolism

Adipose Tissue

In metabolic disease, alterations in adipose 11β-HSD1 in rodent models are well described. Activity is increased in visceral adipose tissue of obese, compared to lean, Zucker, and Wistar/obese (WNIN/ob) rats, and diabetic (db/db) mice [71, 72]. Additionally, in obese WNIN/ob and db/db but not Zucker diabetic fatty (ZDF) animals, 11β-HSD1 activity was increased in the subcutaneous depot [73, 74]. Interestingly, in Wistar rats short-term, but not long-term, high fat diet decreased 11β-HSD1 activity in subcutaneous and omental depots [75] suggesting an adaptive mechanism to protect against the short-term effects of high fat feeding.

11β-HSD1 knockout mice have an improved metabolic phenotype in comparison with wild-type littermates. They resist diet-induced obesity, have a more metabolically safe adipose distribution, gaining fat in the epididymal rather than the visceral depot, display improved glucose tolerance and insulin sensitivity, and have decreased circulating plasma fatty acids. Isolated adipocytes have increased insulin sensitivity [76, 77]. Transgenic mice overexpressing 11β-HSD1 specifically in adipocytes have a 15–30 % increase in adipose corticosterone (the predominant active GC in rodents) concentration and have increased food intake, and a small increase in subcutaneous and a dramatic increase in visceral adipose tissue mass [78]. These animals were also hypertensive, hyperglycemic, hyperinsulinemic, and glucose intolerant, with raised serum fatty acids and triglycerides [78, 79]. In a comparative study, a mouse overexpressing 11β-HSD2 in adipose tissue developed adipose tissue-specific GC deficiency. These mice had reduced fat mass and were resistant to weight gain on a high fat diet. Unexpectedly, the reduction in fat mass was predominantly due to a decrease in the subcutaneous depot, with a less dramatic upon visceral adipose. Globally, mice had improved glucose tolerance and insulin sensitivity; however, food intake was decreased and energy expenditure increased [80].

High levels of 11β-HSD1, but not 11β-HSD2, are expressed in human adipose tissue [81] where it functions largely as an oxoreductase, generating active GC and being induced by GCs and pro-inflammatory cytokines [82–84]. Whole tissue subcutaneous and omental adipose tissue depot expression levels are similar; however, H6PD and GR are more highly expressed in omental adipose tissue [85]. 11β-HSD1 expression is higher in omental compared to subcutaneous adipose stromal cells (contrasting with whole tissue expression data) and increases across adipocyte differentiation [86]. 11β-HSD1 inhibition blocks cortisone-induced differentiation [86, 87] and regulates GC-induced lipid accumulation [88].

Human expression studies have mainly focused on subcutaneous adipose tissue, and the majority of studies have shown that 11β-HSD1 expression and activity correlate positively with BMI and insulin resistance [89–96]. A few studies have examined omental adipose, and overall data suggest increased expression in obesity [85, 97–99]; however, this is not consistent across all studies [90, 100]. Stable isotope techniques have been used to demonstrate functional activity of 11β-HSD1 and while it is clear that adipose tissue is able to generate significant amounts of active GC, there is little evidence to suggest that intra-abdominal adipose actively 'exports' this to distant tissues [101]; however, active GC generated within subcutaneous adipose tissue can be exported to distant organs [101]. Importantly, in both intra-abdominal and subcutaneous depots there is shuttling between active and inactive GCs [102], thus altering the amount of locally derived GC, which in turn can have a potent impact upon adipose tissue biology.

11β-HSD2 expression has been described in human adipocytes although its true functional role has not been determined [103]. At a functional level, studies utilizing stable isotopes of cortisol that are able to distinguish oxoreductase versus dehydrogenase activity suggest that exclusive activity is the generation of cortisol within adipose tissue, as a result of 11β-HSD1 activity [104].

5αR2 and 5β-reductase are not expressed in human adipose tissue; however, 5αR1 is expressed at reasonably high levels [105] and has functional activity in rodents and humans [106, 107]. Its true role in the regulation of adipose tissue biology is still emerging.

Liver

11β-HSD1 is expressed in rodent and human liver at high levels [14]. In rodent studies, hepatic 11β-HSD1 expression is decreased in some murine models of obesity [71, 73]. However, in the diabetic db/db mouse, hepatic 11β-HSD1 and GR expression are increased [108]. Global 11β-HSD1KO mice are protected from diet-induced hepatic steatosis [109] and, when fed a high fat diet, fasting glucose levels are significantly lower compared to controls [76]. In order to explore the role of hepatic 11β-HSD1 in global metabolic homeostasis, mouse models with liver-specific overexpression and knockdown have been developed. Transgenic mice overexpressing 11β-HSD1 under the hepatocyte-specific apoE promoter are hypertensive, dyslipidemic, and develop hepatic steatosis due to increased triglyceride accumulation and impaired lipid clearance. Interestingly, they do not develop steatohepatitis (NASH) and have only modest levels of insulin resistance when compared to adipose tissue-specific 11β-HSD1 overexpression [110]. Liver-specific 11β-HSD1KO mice have a mild metabolic phenotype, with a slight improvement in glucose tolerance (without significant improvement in insulin sensitivity) and no changes in hepatic lipid accumulation [111]. These data highlight the importance of extrahepatic 11β-HSD1 in regulating global and hepatic homeostasis.

In the human liver, 11β-HSD1 is localized centripetally with maximum expression around the central vein [112] and activity is exclusively oxoreductase [112, 113] generating active GC. In obese patients, the expression of GR, 11β-HSD1, and H6PD were all increased in the livers of patients with metabolic disease and were associated with disease severity [114]. However, in patients with proven non-alcoholic fatty liver disease (NAFLD), the expression of these genes was not altered [115, 116]. It is possible that 11β-HSD1 is differently regulated across the progression from steatosis to NASH. In patients with steatosis, total cortisol metabolites are increased, consistent with increased cortisol production yet hepatic 11β-HSD1 activity is decreased. However, in patients with NASH, activity was increased compared to controls and this might reflect the progression to a more inflammatory phenotype rather than simple lipid accumulation [117]. In patients with simple obesity, heaptic 11ß-HSD1 activity (as measured by cortisol generation form oral cortisone) is reduced [94, 118] as this is likely to largely (although not-exclusively [111]) reflect hepatic activity. However, stable isotope techniques have demonstrated preserved, rather than decreased, activity in patients with obesity and coexistent type 2 diabetes [119]. 11β-HSD2 is not expressed in the human liver.

There is an emerging role for the A-ring reductases in the prereceptor regulation of GC availability to modulate hepatic function. Rodent expression profiles differ from the human situation in that 5αR1 and not 5αR2 is expressed in rodent liver (both are expressed in humans). Rodent models have demonstrated that 5αR1 deletion is associated with increased hepatic steatosis as well as increased risk of progression to fibrosis and scarring in models of liver injury [120, 121]. As expected the changes were not seen in 5αR2 knockout models consistent with the lack of expression of 5αR2 in the normal rodent liver. 5β-Reductase is expressed in the rodent liver, but with the exception of its role in bile acid synthesis its contribution to other conditions using rodent models has not been explored.

Clinical studies have consistently demonstrated an association between worsening metabolic phenotype and increased 5αR activity as assessed most commonly by urinary steroid hormone metabolites [122–126]. In addition, patients with polycystic ovarian syndrome, which in itself is associated with insulin resistance and an adverse metabolic phenotype, have increased 5αR activity [125, 126]. Importantly, following aggressive weight loss in clinical studies 5αR activity decreases [124]. 5β-Reductase activity increases with hepatic lipid accumulation [127], but data on its role to regulate other aspects of metabolic pehnotype have not been explored.

Pancreatic Islet of Langerhans

There is continued debate about the localization and functional role of 11β-HSD1 in the pancreatic islet; studies have demonstrated colocalizations to the β-cell [169], while others have shown colocalization with glucagon in the periphery of murine and human islets, but not with insulin or stomatostatin, suggesting α- and not β-cell expression [128]. Several studies have demonstrated that pharmacological inhibition of 11β-HSD1 can regulate insulin secretion both in vitro [128–131] and in rodent models in vivo. Expression is increased in islets from obese ob/ob mice [129] and diabetic ZDF fa/fa rats, where 11β-HSD1 activity increased in proportion to hyperglycemia [132]. Prevention of hyperglycemia and hyperlipidemia by troglitazone, a PPAR gamma agonist, blocked the increase in 11β-HSD1; however, expression in isolated prediabetic islets was not altered by incubation with high glucose or oleate/palmitate, indicating that this was not a nutritional effect [132].

In a transgenic rodent model with β-cell-specific overexpression of 11β-HSD1, β-cell function was compromised with suppression of glucose-stimulated insulin secretion, but interestingly, in hemizygous mice fed there was reversal of β-cell failure on a high fat diet. This was thought to be due to an increased number and function of small islets, enhanced insulin secretion, and enhanced β-cell differentiation and survival. However, global 11β-HSD1 knockout mice have impaired β-cell function, with decreased glucose-stimulated insulin secretion [133]. Overall there remain many unanswered questions as to the role of 11β-HSD1 in the pancreatic islet and its true function is yet to be determined. 11β-HSD2 is expressed in whole islets although detailed localization and functional assessments have not been performed [134]. There are little if any data that have been published on the expression or activity of the A-ring reductase in the pancreatic islets. However, there does appear to be functional 5αR activity in fetal and pancreatic carcinoma tissue [135].

Skeletal Muscle

The role of 11β-HSD1 in skeletal muscle has not been examined in detail and the relative amount and activity in comparison with liver and adipose tissue is low, but oxoreductase activity has been demonstrated in human muscle explants, human primary cultures, murine explants, and transformed cell lines [136, 137]. Importantly, there are indications that skeletal muscle 11β-HSD1 activity may have a role in metabolic disease. Activity is increased in the gastrocnemius muscle of a rodent model of type 2 diabetes [138] and 11β-HSD1 inhibition increased skeletal muscle insulin receptor substrate 1 (IRS-1) mRNA expression and decreased expression of genes involved in lipid metabolism (lipolysis, lipogenesis, and lipid oxidation) [137]. Similar findings have been identified using human cell culture models [139], and in translational clinical studies, expression is increased in myotubes from obese type 2 diabetics, when compared to BMI-matched controls [140]. Increased expression is also associated with decreased grip strength with age [141, 142].

11β-HSD2, 5αR2, and 5β-reductase are not expressed to any significant level in skeletal muscle. 5αR1 however is expressed although its precise role is yet to be defined; however, inhibition of both 5αR1 and 2 using dutasteride was associated with decreased glucose disposal and this has been suggested to reflect a specific role of 5αR1 within skeletal muscle [105].

Cardiovascular System

11β-HSD1 and 2 are expressed in blood vessel walls and heart; however, oxoreductase directionality (11β-HSD1) predominates in vascular smooth muscle [143, 144]. 11β-HSD1 inhibition in apoE knockout mice achieved significant reduction atherosclerotic load suggesting a role in plaque formation [145]. Carbenoxolone (a nonspecific 11β-HSD1 and 2 inhibitor) treatment has been shown to reduce atherosclerosis in mice [146]. 11β-HSD1 in blood vessel epithelial cells may play a role in maintaining an antiangiogenic tone in vivo. In obesity, rapidly expanding adipose tissue becomes hypoxic, and this may drive inflammation, fibrosis, and insulin resistance. 11β-HSD1 knockout mice have enhanced vascularization and oxygenation of adipose tissue depots paralleled by increased expression of potent angiogenic factors including VEGF, apelin, and angiopoetin-like protein 4 [147]. Furthermore, 7 days after coronary artery ligation, 11ß-HSD1 knockout mice show increased vascularization in the infarcted myocardium, associated with partial protection against myocardial dysfunction [148].

11β-HSD2 is expressed in vascular endothelium [143]. 11β-HSD2 knockout mice develop endothelial dysfunction [149]. Lack of 11β-HSD2 and MR activation is implicated in generation of severe atherosclerosis in mouse models [150].

There is evidence linking 11β-HSD1 activity with atherosclerosis, and mediastinal adipose tissue 11β-HSD1 expression has been associated with coronary atherosclerosis [151]. The same authors demonstrated increased 11β-HSD1 expression in aortas of obese patients with the metabolic syndrome [152].

5αR1 is expressed in the vascular endothelium and smooth muscle. Most studies have evaluated its role in the context of functional inhibition or in the context of androgen administration. In rodent models, 5αR inhibition is associated with some endothelial damage and dysfunction [153], but currently data are lacking as to the contribution that cortisol clearance makes to these observations.

Central Nervous System

11β-HSD1 is widely distributed in the adult brain, while 11β-HSD2 is only expressed at low levels. 11β-HSD1 is most highly expressed in the hippocampus, cortex, cerebellum, and anterior pituitary although expression is also found in the hypothalamus, amygdala, and brain stem. Additionally, expression and activity have been demonstrated in the choroid plexus and arachnoid granulation tissue of the brain ventricular system [154], as well as in the ciliary epithelium and trabecular meshwork of the eye [155]. Although 11β-HSD1 appears to act predominantly as an oxoreductase in the central nervous system (CNS) [156], its cofactor generating enzyme, H6PDH, does not universally colocalize with 11β-HSD1 and this has raised the suggestion that provision of NADPH to 11β-HSD1 in the CNS may not be exclusively related to H6PD [157].

A role for 11β-HSD1 in mediating memory loss and hippocampal atrophy is supported by data demonstrating that inhibition of 11β-HSD1 in cultured hippocampal cells reduced GC-induced neurotoxicity [156]. In aged mice and humans, 11β-HSD inhibition improves cognitive function, with similar results in aged 11β-HSD1KO mice [158–160]. However, in a recent study, selective 11β-HSD1 inhibition did not improve cognitive function in patients with Alzheimer's disease [161].

In the ocular ciliary epithelium, 11β-HSD1 regulates aqueous humor production through increased local cortisol generation [155]. In a proof-of-principle study, the nonspecific 11β-HSD inhibitor, carbenoxolone, decreased intraocular pressure in patients with ocular hypertension [162].

Although the causal link is yet to be established, idiopathic intracranial hypertension (IHH) is associated with GC excess and also with simple obesity. In obese patients, dysregulation of 11β-HSD1 in the choroid plexus and arachnoid granulation tissue may be important in disease development. In obese subjects with IIH, global 11β-HSD1 activity decreases with weight loss and those with the greatest decrease in activity have the largest fall in intracranial pressure. In this study, weight loss was correlated inversely with CSF cortisone levels, suggesting decreased local 11β-HSD1 activity [163]. While the published data do suggest a role for 11β-HSD1 in the pathogenesis of IIH, proof-of-concept studies need to be undertaken using selective inhibitors in this group of patients.

11β-HSD2 is expressed at low levels in adult human brain. However, 11β-HSD2 is highly expressed in fetal (rat) brain [164] and has an important role in brain development. In normal anterior pituitary tissue, 11β-HSD2 mRNA is detected, but immunofluorescence has not been able to convincingly demonstrate protein expression. Interestingly, 11β-HSD2 expression is increased in ACTH-secreting corticotroph adenomas and therefore the consequent enhanced local inactivation of cortisol may explain, at last in part, their lack of response to circulating cortisol excess with resultant autonomous ACTH secretion [165].

While there is no doubt that the 5α-reduced steroids can impact brain function [166], it remains unclear as to how much of this impact is reliant upon their actions upon GCs. Similarly, 5β-reductase is expressed widely within the brain and reports have suggested that it is important for the local regulation of neuroactive steroid availability as well as potentially regulating extrahepatic bile acid synthesis that may function as neuroregulatory signaling molecules [167].

Inflammation and Immunity

GCs in pharmacological doses are immunosuppressive and produce powerful anti-inflammatory effects [168]. They achieve this by altering gene transcription and altering pro- and anti-inflammatory mediators including cytokines and signaling pathways. 11β-HSD1 is believed to play a key role in local inflammation and immune response to stimuli and allergens [169].

11β-HSD1 expression increases during monocyte to macrophage differentiation. In these cells expression is unaffected by pro-inflammatory cytokines, but is increased by IL-4, IL-13 and LPS [170]. Expression also increases when monocytes differentiate to dendritic cells under the influence of gm-CSF and IL-4. Activity is further increased by innate immune stimuli acting via toll-like receptors (TLRs), but is rapidly decreased by binding of the CD40 receptor, an adaptive immune stimulus [171]. Additionally, expression has been detected in murine CD4 and CD8 positive lymphocytes, B cells, and dendritic cells. Expression in CD4 positive lymphocytes increases with cellular activation or polarization into Th1 or Th2 cellular subsets [172].

11β-HSD1 knockout mice have defects in macrophage phagocytosis of apoptotic neutrophils during peritoneal inflammation [173]. In addition, they also display enhanced endotoxemia in response to LPS injection [174]. Furthermore, in models of joint inflammation, peritonitis, and lung inflammation, the inflammatory response was greater, and resolution slower, in 11β-HSD1 knockout mice [175], an observation that has raised concerns yabout the clinical use of selective 11β-HSD1 inhibitors.

A limited number of studies have examined 11β-HSD1 in inflammation in humans in vivo. 11β-HSD1 activity is increased in patients with rheumatoid arthritis [176] and mRNA and protein levels are higher in biopsies of colonic tissue obtained from patients with colitis compared to control patient samples [177]. Additionally, acute exacerbations of inflammatory bowel disease are associated with a significant increase in systemic 11β-HSD1 activity, most likely originating from the inflamed bowel [178]. Interestingly, patients in remission also have high systemic activity, suggesting that local GC production within inflamed tissues might be sufficient to suppress the clinical features of inflammation. Taken together, these studies implicate 11β-HSD1 in having a role to limit the acute inflammatory response.

The response of the 5αR isoforms to inflammation is not fully defined. In a single study in patients with inflammatory arthritis, 5αR activity increased with anti-TNFα treatment although this was paralleled by a decrease in 11β-HSD1 [179]. Any potential role for 5β-reductase has noty been examined.

Bone and Joint

Osteosarcoma cells only express 11β-HSD2 and this contrasts with primary osteoblasts which exclusively express 11β-HSD1 [180, 181]. Ex vivo assays using bone chips have shown bidirectional interconversion of cortisone and cortisol, with kinetics suggesting 11β-HSD1 rather than 11β-HSD2 activity [182]. Although expression appears primarily localized to osteoblasts, some expression is also seen in osteoclasts in human adult bone. Expression of 11ß-HSD1 is regualted across osteoblast differentiation and cortisone treatment of cells in culture enahnces cellular differentiation [183]. Activity of 11β-HSD1 in osteoblasts is increased by both proinflammatory cytokines and glucocorticoids [184] in a synergistic fashion [185], mediated via the nuclear factor-kB (NF-kB) and p38 mitogen-activated protein kinases (MAPK) pathways.

In global 11β-HSD1 knockout mice there are no changes in commonly measured parameters of bone mass and geometry [186]; however, increased circulating levels of corticosterone in this model limit the significance of these findings. Additionally, the phenotype has only been examined in young mice and it is likely that any bone phenotype would be most evident in older animals. Interestingly, targeted overexpression of 11β-HSD2 within osteoblasts, resulting in cell specific GC deficiency, causes subtle abnormalities of skeletal structure including reduced vertebral size and density and reduced cortical width [187].

The presence of 11β-HSD1 in bone raises the possibility that its activity may predict clinical susceptibility to GC-induced osteoporosis. In healthy subjects, the ratio of urinary cortisol to cortisone metabolites predicts the response of bone formation markers to prednisolone treatment [188].

The role of 5α-reductase isoforms has begun to be explored in rodent models although not in humans at present. 5αR1 knockout mice have a sexually dimorphic phenotype with decreased bone mineral content and bone density in male mice with increased bone mass in female mice. The authors postulate that this reflects local changes in androgen availability, but the contribution of alteration in tissue-specific GC concentrations was not assessed [189].

Skin and Salivary Glands

Cortisol metabolism within skin is rapidly becoming an area of interest and investigation. Skin has been shown to be an active site of cortisol production and metabolism [190, 191]. Excess skin exposure to GCs causes skin changes similar to the natural aging process including reduced elasticity, reduced collagen and fibroblast numbers, thinning of dermis, and epidermis and a reduction in the repair capacity of skin. Increased exposure to GCs has been postulated as a factor in age-related changes, inflammatory, and autoimmunity changes seen in skin [192]. It has been postulated that skin changes seen over time are in part a result of 11β-HSD1 activity [190].

Both 11β-HSD1 and 2 are expressed in skin [190, 191, 193]. 11β-HSD2 is expressed in association with the mineralocorticoid receptor on sweat glands; however, its role (if any) within the dermis and epidermis is debated. In wound healing, 11β-HSD2 expression has been shown to be induced 48 h after tissue injury with subsequent return to basal levels at 96 h [191]. This has been postulated to be a mechanism to reduce local cortisol excess following inflammation.

11β-HSD1 is widely expressed in human and mouse dermis and epidermis [190, 193, 194]. Upon differentiation of keratinocytes 11β-HSD1 expression increases [193], somewhat akin to the changes seen with preadipocyte differentiation [195]. Interestingly despite reducing levels of expression of 11β-HSD1 in elderly subjects, a paradoxical rise in 11β-HSD1 activity is seen with increasing age in both humans and mice [190]. This gives credence to the concept of age related skin atrophic changes being in part due to increased cortisol exposure secondary to increased 11β-HSD1 activity.

11β-HSD1 has been shown to have a pivotal role in skin repair following injury and tissue remodeling [193, 196, 197]. In mice, 11β-HSD1 contributes to impaired wound healing. Blocking 11β-HSD1 enhances wound healing in mice and prevents age-induced skin changes [196]. These data suggest that local cortisol, generated by 11β-HSD1, is critically important in wound healing and in aging skin changes. Inhibitors of 11β-HSD1 (topical or oral) may therefore have therapeutic potential.

As mineralocorticoid target tissues, both skin and salivary glands express 11β-HSD2. In the skin expression is mainly restricted to sweat glands [198]. 11β-HSD2 is expressed in both parotid and submandibular glands [198, 199] and measuring salivary cortisone has been postulated a potential biomarker of serum-free cortisol [200]. In addition, reduced activity of 11β-HSD2 in sweat glands has also been linked with essential hypertension [201].

The role of the A-ring reductases in skin has been extensively examined in the context of androgen generation and in particular its relationship to the development of hirsutism and the potential for local generation of DHT. Their role in cortisol metabolism within the skin has not been determined.

Kidney

11β-HSD2 is the predominant isoform in the human kidney, although 11β-HSD1 is expressed in the rodent kidney. The role of 11β-HSD2 in the kidney is to protect the MR from excess exposure to GC. 11β-HSD2 is widely expressed in distal nephrons [39]. Although the inherent enzyme ability of 11β-HSD2 to clear cortisol (converting it to cortisone) should not be enough, given concentrations and binding affinities, in reality it protects the mineralocorticoid receptor from GC exposure [202]. Lack of 11β-HSD2 in kidney leads to life-threatening hypertension and hypokalemia. Reduced 11β-HSD2 activity, as measured by urinary steroid metabolites ratios, has been associated with essential hypertension in aging populations [203] as well as in those with underlying renal impairment [204, 205].

5αR1 is expressed and active in the kidney [206], but as with many tissues already described, there are not data in the published domain that have examined its functional significance with regards to cortisol metabolism in the kidney. 5αR2 and 5β-reductase are not expressed [206].

Colon

11β-HSD2 is expressed in colonic epithelium [207]. Expression is increased by aldosterone in rats [57]. In Inflammatory bowel disease 11β-HSD2 expression is downregulated in both humans and rats [177]. This is accompanied by an increase in 11β-HSD1 expression and so is presumed to be an attempt to locally control GC exposure to inflamed tissue. This has been discussed in the section on immunity and inflammation above. Zhang et al. showed that inhibiting 11β-HSD2 reduces colon carcinogenesis by inhibiting COX 2 pathways. The reduction in 11β-HSD2 blocked colorectal adenocarcinoma angiogenesis and metastasis [208]. There are currently no data with regards to the role of the A-ring reductase and GCs within the colon.

Pharmacological Targeting of Prereceptor GC Metabolism

11β-HSD1 Inhibition

While the clinical consequences of 11β-HSD2 inhibition (as exemplified by liquorice consumption) are detrimental, over the last 10–15 years, there has been a significant drive to develop selective 11β-HSD1 inhibitors based upon the premise that decreasing tissue-specific cortisol availability, notably in the context of metabolic disease, is likely to have a beneficial impact.

Carbenoxolone is a nonselective 11β-HSD inhibitor. In healthy individuals, it improves whole body insulin sensitivity [209], and in patients with type 2 diabetes, it decreases glucose production rates, principally through a reduction in glycogenolysis with no apparent effect on gluconeogenesis. In addition, it decreases total circulating cholesterol levels [210]. Its beneficial effects are modest, and while this is most likely to reflect its nonselective action, questions have arisen as to its ability to access key metabolic target tissues, including adipose, although studies have shown therapeutic levels within adipose interstitial fluid [96, 211]. Carbenoxolone has also been shown to impact upon bone biology in vivo. In a proof-of-principle study there were no changes in bone formation markers, but bone resorption decreased significantly [182].

Several phase II studies have now been published that have examined selective 11β-HSD1 inhibitors. INCB013739, when administered to patients with type 2 diabetes twice daily for 2 weeks, completely abolished all conversion of oral cortisone to cortisol. Metabolically hepatic glucose production rates decreased, without alteration in glucose disposal. Interestingly, the decrease in fasting glucose was most marked in the most hyperglycemic patients. In addition, total and LDL cholesterol decreased with no change in HDL-cholesterol or triglyceride levels. In a double-blind placebo-controlled study, patients with type 2 diabetes with inadequate glycemic control on metformin therapy (HbA1c 7–11 %) were randomized to receive 5, 15, 50, 100, or 200 mg INCB13739 in addition to metformin for 12 weeks. Weight, glycemic control, and lipid profile all improved although the effects were relatively modest with reductions in HbA1c of approximately 0.5 % and a small reduction in HOMA-IR consistent with insulin sensitization. As expected, treatment with this class of agent activates the HPA axis (as a consequence of decreased cortisol half-life) with consequent elevation of adrenal androgen secretion. There were no changes in HDL or free fatty acids and blood pressure was not affected [212].

Data have also been published on additional compounds; MK0916 was given to patients with type 2 diabetes. While it was well tolerated, MK0916 had only very modest effects on metabolic parameters. There was a decrease in weight and waist hip ratio in the 6-mg group and in this group there was also a small reduction in HbA1c (0.3 %); however, no change was seen in fasting plasma glucose, 2 h postprandial glucose, or fasting or postprandial serum insulin [213]. A further compound, MK0736, has also been tested in obese and overweight hypertensive patients. Both doses of the compound tested decreased blood pressure. Again, consistent with other studies all active treatments caused a small but significant decrease in weight [214].

PF-915275 is an effective 11β-HSD1 inhibitor as measured by changes in urinary steroid metabolite ratios and prednisone to prednisolone conversion, but to date there are no data on the impact of this compound on metabolic phenotype [215]. Most recently, RO5093151 has been trialed in the context of hepatic steatosis. The drug appeared safe and well tolerated and did reduce hepatic steatosis as measured by magnetic resonance spectroscopy although in absolute terms the reduction was once again modest, but the duration of the study was only 12 weeks [216].

5α-Reductase Inhibition

Clinical studies have highlighted the potential role for 5α-reductase in the regulation of metabolic phenotype, although there is still debate as to whether the abnormalities observed represent the cause or consequence of disease. As described above, cross-sectional and longitudinal studies have demonstrated increased 5α-reductase activity with insulin resistance and increasing adiposity and reductions following weight loss [122–124]. A recently published study has examined the metabolic impact of selective 5αR inhibition in humans [105]. Following a 3-month treatment period, the authors observed inhibition of glucose disposal under hyperinsulinemic conditions with dutasteride treatment (nonselective 5αR1 and 2 inhibitor) but not finasteride (selective 5αR2 inhibitor), which may reflect the impact of inhibition of 5αR1 activity within skeletal muscle. The long-term clinical consequences of these observations remain to be determined as well as the identification of the mechanisms responsible, in particular their dependence upon either GC and/or androgen metabolism.

Conclusion

GCs have multiple actions across almost all tissues in the body and the regulation of their action is complex. Prereceptor GC metabolism, either regeneration of active cortisol through the activity of 11β-HSD1 or clearance via 11β-HSD2, and the A-ring reductase are potently able to impact upon GR activation. The consequences of their activity not only are dependent upon the precise pattern of expression within specific tissues but also may reflect the broad range of substrates (including GCs) that they are able to metabolize. Dysregulated expression has been implicated in the pathogenesis of many diseases, and the fundamental importance of the prereceptor GC concept is highlighted by patients with genetic defects that are potentially life-threatening. Pharmacological intervention, specifically targeting 11β-HSD1, has progressed all the way through to phase II clinical trials and while the outcomes with respect to metabolic disease have been positive, the magnitude of the response has perhaps been less than had been anticipated. This may reflect targeting of therapy to specific tissues but also the fact that only the 'regenerated' part of GC has been blocked. In terms of the future, the role of 11β-HSD1 in the skin and its involvement in wound healing make it an attractive therapeutic prospect. In addition, there is emerging evidence that 11β-HSD1 may have a role in the regulation of tissue-specific exposure to exogenously administered GCs, raising the possibility that 11β-HSD1 inhibitors could have utility in reducing the adverse effects of prescribed GCs [217, 218].

References

1. Munck A, Naray-Fejes-Toth A. The ups and downs of glucocorticoid physiology. Permissive and suppressive effects revisited. Mol Cell Endocrinol. 1992;90(1):C1–4.
2. Esteban NV, Loughlin T, Yergey AL, Zawadzki JK, Booth JD, Winterer JC, et al. Daily cortisol production rate in man determined by stable isotope dilution/mass spectrometry. J Clin Endocrinol Metab. 1991;72(1):39–45.
3. Siiteri PK, Murai JT, Hammond GL, Nisker JA, Raymoure WJ, Kuhn RW. The serum transport of steroid hormones. Recent Prog Horm Res. 1982;38:457–510.
4. Keenan DM, Roelfsema F, Veldhuis JD. Endogenous ACTH concentration-dependent drive of pulsatile cortisol secretion in the human. Am J Physiol Endocrinol Metab. 2004;287(4):E652–61.
5. Dorin RI, Qiao Z, Qualls CR, Urban 3rd FK. Estimation of maximal cortisol secretion rate in healthy humans. J Clin Endocrinol Metab. 2012;97(4):1285–93.
6. Toothaker RD, Welling PG. Effect of dose size on the pharmacokinetics of intravenous hydrocortisone during endogenous hydrocortisone suppression. J Pharmacokinet Biopharm. 1982;10(2):147–56.

7. Kendall EC, Charles S's S. Cortisone. 1971.
8. Hench PS, Kendall EC, et al. The effect of a hormone of the adrenal cortex (17-hydroxy-11-dehydrocorticosterone; compound E) and of pituitary adrenocorticotropic hormone on rheumatoid arthritis. Proc Staff Meet Mayo Clin. 1949;24(8):181–97.
9. Burton RB, Keutmann EH, Waterhouse C, Schuler EA. The conversion of cortisone acetate to other alphaketolic steroids. J Clin Endocrinol Metab. 1953;13(1):48–63.
10. Amelung D, Hubener HJ, Rocka L, Meyerheim G. Conversion of cortisone to compound F. J Clin Endocrinol Metabol. 1953;13:1125.
11. Kavanagh KL, Jornvall H, Persson B, Oppermann U. Medium- and short-chain dehydrogenase/reductase gene and protein families: the SDR superfamily: functional and structural diversity within a family of metabolic and regulatory enzymes. Cell Mol Life Sci. 2008;65(24):3895–906.
12. Agarwal AK, Monder C, Eckstein B, White PC. Cloning and expression of rat cDNA encoding corticosteroid 11 beta-dehydrogenase. J Biol Chem. 1989;264(32):18939–43.
13. Lakshmi V, Monder C. Purification and characterization of the corticosteroid 11 beta-dehydrogenase component of the rat liver 11 beta-hydroxysteroid dehydrogenase complex. Endocrinology. 1988;123(5):2390–8.
14. Tannin GM, Agarwal AK, Monder C, New MI, White PC. The human gene for 11β-hydroxysteroid dehydrogenase. Structure, tissue distribution, and chromosomal localization. J Biol Chem. 1991;266:16653–8.
15. Nobel CS, Dunas F, Abrahmsen LB. Purification of full-length recombinant human and rat type 1 11beta-hydroxysteroid dehydrogenases with retained oxidoreductase activities. Protein Expr Purif. 2002;26(3):349–56.
16. Maser E, Volker B, Friebertshauser J. 11 Beta-hydroxysteroid dehydrogenase type 1 from human liver: dimerization and enzyme cooperativity support its postulated role as glucocorticoid reductase. Biochemistry. 2002;41(7):2459–65.
17. Odermatt A, Arnold P, Stauffer A, Frey BM, Frey FJ. The N-terminal anchor sequences of 11beta-hydroxysteroid dehydrogenases determine their orientation in the endoplasmic reticulum membrane. J Biol Chem. 1999;274(40):28762–70.
18. Bujalska IJ, Draper N, Michailidou Z, Tomlinson JW, White PC, Chapman KE, et al. Hexose-6-phosphate dehydrogenase confers oxo-reductase activity upon 11 beta-hydroxysteroid dehydrogenase type 1. J Mol Endocrinol. 2005;34(3):675–84.
19. Draper N, Walker EA, Bujalska IJ, Tomlinson JW, Chalder SM, Arlt W, et al. Mutations in the genes encoding 11beta-hydroxysteroid dehydrogenase type 1 and hexose-6-phosphate dehydrogenase interact to cause cortisone reductase deficiency. Nat Genet. 2003;34(4):434–9.
20. Tomlinson JW, Walker EA, Bujalska IJ, Draper N, Lavery GG, Cooper MS, et al. 11beta-hydroxysteroid dehydrogenase type 1: a tissue-specific regulator of glucocorticoid response. Endocr Rev. 2004;25(5):831–66.
21. Rogers SL, Hughes BA, Jones CA, Freedman L, Smart K, Taylor N, et al. Diminished 11beta-hydroxysteroid dehydrogenase type 2 activity is associated with decreased weight and weight gain across the first year of life. J Clin Endocrinol Metab. 2014;99(5):E821–31.
22. Murphy VE, Clifton VL. Alterations in human placental 11beta-hydroxysteroid dehydrogenase type 1 and 2 with gestational age and labour. Placenta. 2003;24(7):739–44.
23. Jinno K, Sakura N, Nomura S, Fujitaka M, Ueda K, Kihara M. Failure of cortisone acetate therapy in 21-hydroxylase deficiency in early infancy. Pediatr Int. 2001;43(5):478–82.
24. Abramovitz M, Branchaud CL, Murphy BE. Cortisol-cortisone interconversion in human fetal lung: contrasting results using explant and monolayer cultures suggest that 11 beta-hydroxysteroid dehydrogenase (EC 1.1.1.146) comprises two enzymes. J Clin Endocrinol Metab. 1982;54(3):563–8.
25. Dimitriou T, Maser-Gluth C, Remer T. Adrenocortical activity in healthy children is associated with fat mass. Am J Clin Nutr. 2003;77(3):731–6.
26. Toogood AA, Taylor NF, Shalet SM, Monson JP. Sexual dimorphism of cortisol metabolism is maintained in elderly subjects and is not oestrogen dependent. Clin Endocrinol (Oxf). 2000;52(1):61–6.

27. Vierhapper H, Heinze G, Nowotny P. Sex-specific difference in the interconversion of cortisol and cortisone in men and women. Obesity. 2007;15(4):820–4.
28. Finken MJ, Andrews RC, Andrew R, Walker BR. Cortisol metabolism in healthy young adults: sexual dimorphism in activities of A-ring reductases, but not 11beta-hydroxysteroid dehydrogenases. J Clin Endocrinol Metab. 1999;84(9):3316–21.
29. Nixon M, Wake DJ, Livingstone DE, Stimson RH, Esteves CL, Seckl JR, et al. Salicylate downregulates 11beta-HSD1 expression in adipose tissue in obese mice and in humans, mediating insulin sensitization. Diabetes. 2012;61(4):790–6.
30. Gomez-Sanchez EP, Ganjam V, Chen YJ, Liu Y, Zhou MY, Toroslu C, et al. Regulation of 11 beta-hydroxysteroid dehydrogenase enzymes in the rat kidney by estradiol. Am J Physiol Endocrinol Metab. 2003;285(2):E272–9.
31. Biason-Lauber A, Suter SL, Shackleton CH, Zachmann M. Apparent cortisone reductase deficiency: a rare cause of hyperandrogenemia and hypercortisolism. Horm Res. 2000;53(5):260–6.
32. Lavery GG, Walker EA, Tiganescu A, Ride JP, Shackleton CH, Tomlinson JW, et al. Steroid Biomarkers and Genetic Studies Reveal Inactivating Mutations in Hexose-6-Phosphate Dehydrogenase in Patients with Cortisone Reductase Deficiency. J Clin Endocrinol Metab. 2008;93(10):3827–32.
33. Lawson AJ, Walker EA, Lavery GG, Bujalska IJ, Hughes B, Arlt W, et al. Cortisone-reductase deficiency associated with heterozygous mutations in 11beta-hydroxysteroid dehydrogenase type 1. Proc Natl Acad Sci U S A. 2011;108(10):4111–6.
34. Jamieson A, Wallace AM, Andrew R, Nunez BS, Walker BR, Fraser R, et al. Apparent cortisone reductase deficiency: a functional defect in 11beta-hydroxysteroid dehydrogenase type 1. J Clin Endocrinol Metab. 1999;84(10):3570–4.
35. Nordenstrom A, Marcus C, Axelson M, Wedell A, Ritzen EM. Failure of cortisone acetate treatment in congenital adrenal hyperplasia because of defective 11beta-hydroxysteroid dehydrogenase reductase activity. J Clin Endocrinol Metab. 1999;84(4):1210–3.
36. Brown RW, Chapman KE, Edwards CRW, Seckl JR. Human placental 11β-hydroxysteroid dehydrogenase: Evidence for and partial purification of a distinct NAD-dependent isoform. Endocrinology. 1993;132:2614–21.
37. Albiston AL, Obeyesekere VR, Smith RE, Krozowski ZS. Cloning and tissue distribution of the human 11 beta-hydroxysteroid dehydrogenase type 2 enzyme. Mol Cell Endocrinol. 1994;105(2):R11–7.
38. Agarwal AK, Rogerson FM, Mune T, White PC. Gene structure and chromosomal localization of the human HSD11K gene encoding the kidney (type 2) isozyme of 11 beta-hydroxysteroid dehydrogenase. Genomics. 1995;29(1):195–9.
39. Brown RW, Chapman KE, Kotelevtsev Y, Yau JL, Lindsay RS, Brett L, et al. Cloning and production of antisera to human placental 11 beta-hydroxysteroid dehydrogenase type 2. Biochem J. 1996;313(Pt 3):1007–17.
40. Stewart PM, Murry BA, Mason JI. Human kidney 11 beta-hydroxysteroid dehydrogenase is a high affinity nicotinamide adenine dinucleotide-dependent enzyme and differs from the cloned type I isoform. J Clin Endocrinol Metab. 1994;79(2):480–4.
41. Ulick S, Levine LS, Gunczler P, Zanconato G, Ramirex LC, Rauh W, et al. A syndrome of apparent mineralocorticoid excess associated with defects in the peripheral metabolism of cortisol. J Clin Endocrinol Metab. 1979;49:757–64.
42. Dave-Sharma S, Wilson RC, Harbison MD, Newfield R, Azar MR, Krozowski ZS, et al. Examination of genotype and phenotype relationships in 14 patients with apparent mineralocorticoid excess. J Clin Endocrinol Metab. 1998;83(7):2244–54.
43. Wilson RC, Harbison MD, Krozowski ZS, Funder JW, Shackleton CHL, Hanauske-Abel HM, et al. Several homozygous mutations in the gene for 11β-hydroxysteroid dehydrogenase type 2 in patients with apparent mineralocorticoid excess. J Clin Endocrinol Metab. 1995;80:3145–50.
44. Wilson RC, Krozowski ZS, Li K, Obeyesekere VR, Razzaghy-Azar M, Harbison MD, et al. A mutation in the HSD11B2 gene in a family with apparent mineralocorticoid excess. J Clin Endocrinol Metabol. 1995;80:2263–6.

45. Mune T, Rogerson FM, Nikkil H, Agarwal AK, White PC. Human hypertension caused by mutations in the kidney isozyme of 11β-hydroxysteroid dehydrogenase. Nat Genet. 1995;10:394–9.
46. Stewart PM, Wallace AM, Valentino R, Burt D, Shackleton CHL, Edwards CRW. Mineralocorticoid activity of liquorice: 11β-hydroxysteroid dehydrogenase deficiency comes of age. Lancet. 1987;ii:821–4.
47. Seckl JR, Holmes MC. Mechanisms of disease: glucocorticoids, their placental metabolism and fetal 'programming' of adult pathophysiology. Nat Clin Pract Endocrinol Metab. 2007;3(6):479–88.
48. Alikhani-Koopaei R, Fouladkou F, Frey FJ, Frey BM. Epigenetic regulation of 11 beta-hydroxysteroid dehydrogenase type 2 expression. J Clin Invest. 2004;114(8):1146–57.
49. Marsit CJ, Maccani MA, Padbury JF, Lester BM. Placental 11-beta hydroxysteroid dehydrogenase methylation is associated with newborn growth and a measure of neurobehavioral outcome. PLoS One. 2012;7(3), e33794.
50. Baserga M, Kaur R, Hale MA, Bares A, Yu X, Callaway CW, et al. Fetal growth restriction alters transcription factor binding and epigenetic mechanisms of renal 11beta-hydroxysteroid dehydrogenase type 2 in a sex-specific manner. Am J Physiol Regul Integr Comp Physiol. 2010;299(1):R334–42.
51. Kostadinova RM, Nawrocki AR, Frey FJ, Frey BM. Tumor necrosis factor alpha and phorbol 12-myristate-13-acetate down-regulate human 11beta-hydroxysteroid dehydrogenase type 2 through p50/p50 NF-kappaB homodimers and Egr-1. FASEB J. 2005;19(6):650–2.
52. Low SC, Assaad SN, Rajan V, Chapman KE, Edwards CR, Seckl JR. Regulation of 11 beta-hydroxysteroid dehydrogenase by sex steroids in vivo: further evidence for the existence of a second dehydrogenase in rat kidney. J Endocrinol. 1993;139(1):27–35.
53. Rubis B, Krozowski Z, Trzeciak WH. Arginine vasopressin stimulates 11beta-hydroxysteroid dehydrogenase type 2 expression in the mineralocorticosteroid target cells. Mol Cell Endocrinol. 2006;256(1-2):17–22.
54. Clarke KA, Ward JW, Forhead AJ, Giussani DA, Fowden AL. Regulation of 11 beta-hydroxysteroid dehydrogenase type 2 activity in ovine placenta by fetal cortisol. J Endocrinol. 2002;172(3):527–34.
55. Suzuki S, Koyama K, Darnel A, Ishibashi H, Kobayashi S, Kubo H, et al. Dexamethasone upregulates 11beta-hydroxysteroid dehydrogenase type 2 in BEAS-2B cells. Am J Respir Crit Care Med. 2003;167(9):1244–9.
56. Heiniger CD, Kostadinova RM, Rochat MK, Serra A, Ferrari P, Dick B, et al. Hypoxia causes down-regulation of 11 beta-hydroxysteroid dehydrogenase type 2 by induction of Egr-1. FASEB J. 2003;17(8):917–9.
57. Fukushima K, Funayama Y, Yonezawa H, Takahashi K, Haneda S, Suzuki T, et al. Aldosterone enhances 11beta-hydroxysteroid dehydrogenase type 2 expression in colonic epithelial cells in vivo. Scand J Gastroenterol. 2005;40(7):850–7.
58. Russell DW, Wilson JD. Steroid 5 alpha-reductase: two genes/two enzymes. Annu Rev Biochem. 1994;63:25–61.
59. Uemura M, Tamura K, Chung S, Honma S, Okuyama A, Nakamura Y, et al. Novel 5 alpha-steroid reductase (SRD5A3, type-3) is overexpressed in hormone-refractory prostate cancer. Cancer Sci. 2008;99(1):81–6.
60. Andersson S, Russell DW. Structural and biochemical properties of cloned and expressed human and rat steroid 5 alpha-reductases. Proc Natl Acad Sci U S A. 1990;87(10):3640–4.
61. Normington K, Russell DW. Tissue distribution and kinetic characteristics of rat steroid 5 alpha-reductase isozymes. Evidence for distinct physiological functions. J Biol Chem. 1992;267(27):19548–54.
62. Schwartz JI, Tanaka WK, Wang DZ, Ebel DL, Geissler LA, Dallob A, et al. MK-386, an inhibitor of 5alpha-reductase type 1, reduces dihydrotestosterone concentrations in serum and sebum without affecting dihydrotestosterone concentrations in semen. J Clin Endocrinol Metab. 1997;82(5):1373–7.

63. Labrie F, Sugimoto Y, Luu-The V, Simard J, Lachance Y, Bachvarov D, et al. Structure of human type II 5 alpha-reductase gene. Endocrinology. 1992;131(3):1571–3.
64. Samtani R, Bajpai M, Ghosh PK, Saraswathy KN. SRD5A2 gene mutations—a population-based review. Pediatr Endocrinol Rev. 2010;8(1):34–40.
65. Ellis JA, Stebbing M, Harrap SB. Genetic analysis of male pattern baldness and the 5alpha-reductase genes. J Invest Dermatol. 1998;110(6):849–53.
66. Jakimiuk AJ, Weitsman SR, Magoffin DA. 5alpha-reductase activity in women with polycystic ovary syndrome. J Clin Endocrinol Metab. 1999;84(7):2414–8.
67. Labrie F, Dupont A, Simard J, Luu-The V, Belanger A. Intracrinology: the basis for the rational design of endocrine therapy at all stages of prostate cancer. Eur Urol. 1993;24 Suppl 2:94–105.
68. Faucher F, Cantin L, Luu-The V, Labrie F, Breton R. The crystal structure of human Delta4-3-ketosteroid 5beta-reductase defines the functional role of the residues of the catalytic tetrad in the steroid double bond reduction mechanism. Biochemistry. 2008;47(32):8261–70.
69. Lemonde HA, Custard EJ, Bouquet J, Duran M, Overmars H, Scambler PJ, et al. Mutations in SRD5B1 (AKR1D1), the gene encoding delta(4)-3-oxosteroid 5beta-reductase, in hepatitis and liver failure in infancy. Gut. 2003;52(10):1494–9.
70. Palermo M, Marazzi MG, Hughes BA, Stewart PM, Clayton PT, Shackleton CH. Human Delta4-3-oxosteroid 5beta-reductase (AKR1D1) deficiency and steroid metabolism. Steroids. 2008;73(4):417–23.
71. Livingstone DE, Jones GC, Smith K, Jamieson PM, Andrew R, Kenyon CJ, et al. Understanding the role of glucocorticoids in obesity: tissue-specific alterations of corticosterone metabolism in obese Zucker rats. Endocrinology. 2000;141(2):560–3.
72. Livingstone DE, Kenyon CJ, Walker BR. Mechanisms of dysregulation of 11beta-hydroxysteroid dehydrogenase type 1 in obese Zucker rats. J Endocrinol. 2000;167(3):533–9.
73. Prasad SS, Prashanth A, Kumar CP, Reddy SJ, Giridharan NV, Vajreswari A. A novel genetically-obese rat model with elevated 11 beta-hydroxysteroid dehydrogenase type 1 activity in subcutaneous adipose tissue. Lipids Health Dis. 2010;9:132.
74. Nakano S, Inada Y, Masuzaki H, Tanaka T, Yasue S, Ishii T, et al. Bezafibrate regulates the expression and enzyme activity of 11beta-hydroxysteroid dehydrogenase type 1 in murine adipose tissue and 3T3-L1 adipocytes. Am J Physiol Endocrinol Metab. 2007;292(4):E1213–22.
75. Drake AJ, Livingstone DE, Andrew R, Seckl JR, Morton NM, Walker BR. Reduced adipose glucocorticoid reactivation and increased hepatic glucocorticoid clearance as an early adaptation to high-fat feeding in Wistar rats. Endocrinology. 2005;146(2):913–9.
76. Kotelevtsev Y, Holmes MC, Burchell A, Houston PM, Schmoll D, Jamieson P, et al. 11beta-hydroxysteroid dehydrogenase type 1 knockout mice show attenuated glucocorticoid-inducible responses and resist hyperglycemia on obesity or stress. Proc Natl Acad Sci U S A. 1997;94(26):14924–9.
77. Morton NM, Paterson JM, Masuzaki H, Holmes MC, Staels B, Fievet C, et al. Novel adipose tissue-mediated resistance to diet-induced visceral obesity in 11 beta-hydroxysteroid dehydrogenase type 1-deficient mice. Diabetes. 2004;53(4):931–8.
78. Masuzaki H, Paterson J, Shinyama H, Morton NM, Mullins JJ, Seckl JR, et al. A transgenic model of visceral obesity and the metabolic syndrome. Science. 2001;294(5549):2166–70.
79. Masuzaki H, Yamamoto H, Kenyon CJ, Elmquist JK, Morton NM, Paterson JM, et al. Transgenic amplification of glucocorticoid action in adipose tissue causes high blood pressure in mice. J Clin Invest. 2003;112(1):83–90.
80. Kershaw EE, Morton NM, Dhillon H, Ramage L, Seckl JR, Flier JS. Adipocyte-specific glucocorticoid inactivation protects against diet-induced obesity. Diabetes. 2005;54(4):1023–31.
81. Bujalska IJ, Kumar S, Stewart PM. Does central obesity reflect "Cushing's disease of the omentum"? Lancet. 1997;349:1210–3.

82. Tomlinson JW, Moore J, Cooper MS, Bujalska I, Shahmanesh M, Burt C, et al. Regulation of expression of 11beta-hydroxysteroid dehydrogenase type 1 in adipose tissue: tissue-specific induction by cytokines. Endocrinology. 2001;142(5):1982–9.

83. Handoko K, Yang K, Strutt B, Khalil W, Killinger D. Insulin attenuates the stimulatory effects of tumor necrosis factor alpha on 11beta-hydroxysteroid dehydrogenase 1 in human adipose stromal cells. J Steroid Biochem Mol Biol. 2000;72(3-4):163–8.

84. Friedberg M, Zoumakis E, Hiroi N, Bader T, Chrousos GP, Hochberg Z. Modulation of 11 beta-hydroxysteroid dehydrogenase type 1 in mature human subcutaneous adipocytes by hypothalamic messengers. J Clin Endocrinol Metab. 2003;88(1):385–93.

85. Veilleux A, Rheaume C, Daris M, Luu-The V, Tchernof A. Omental adipose tissue type 1 11 beta-hydroxysteroid dehydrogenase oxoreductase activity, body fat distribution, and metabolic alterations in women. J Clin Endocrinol Metab. 2009;94(9):3550–7.

86. Bujalska IJ, Kumar S, Hewison M, Stewart PM. Differentiation of adipose stromal cells: the roles of glucocorticoids and 11beta-hydroxysteroid dehydrogenase. Endocrinology. 1999;140(7):3188–96.

87. Bujalska IJ, Gathercole LL, Tomlinson JW, Darimont C, Ermolieff J, Fanjul AN, et al. A novel selective 11beta-hydroxysteroid dehydrogenase type 1 inhibitor prevents human adipogenesis. J Endocrinol. 2008;197(2):297–307.

88. Gathercole LL, Morgan SA, Bujalska IJ, Hauton D, Stewart PM, Tomlinson JW. Regulation of lipogenesis by glucocorticoids and insulin in human adipose tissue. PLoS One. 2011;6(10), e26223.

89. Engeli S, Bohnke J, Feldpausch M, Gorzelniak K, Heintze U, Janke J, et al. Regulation of 11beta-HSD genes in human adipose tissue: influence of central obesity and weight loss. Obes Res. 2004;12(1):9–17.

90. Goedecke JH, Wake DJ, Levitt NS, Lambert EV, Collins MR, Morton NM, et al. Glucocorticoid metabolism within superficial subcutaneous rather than visceral adipose tissue is associated with features of the metabolic syndrome in South African women. Clin Endocrinol (Oxf). 2006;65(1):81–7.

91. Lindsay RS, Wake DJ, Nair S, Bunt J, Livingstone DE, Permana PA, et al. Subcutaneous adipose 11beta-hydroxysteroid dehydrogenase type 1 activity and messenger ribonucleic Acid levels are associated with adiposity and insulinemia in Pima Indians and Caucasians. J Clin Endocrinol Metab. 2003;88(6):2738–44.

92. Kannisto K, Pietilainen KH, Ehrenborg E, Rissanen A, Kaprio J, Hamsten A, et al. Overexpression of 11beta-hydroxysteroid dehydrogenase-1 in adipose tissue is associated with acquired obesity and features of insulin resistance: studies in young adult monozygotic twins. J Clin Endocrinol Metab. 2004;89(9):4414–21.

93. Paulmyer-Lacroix O, Boullu S, Oliver C, Alessi MC, Grino M. Expression of the mRNA Coding for 11beta-Hydroxysteroid Dehydrogenase Type 1 in Adipose Tissue from Obese Patients: An in Situ Hybridization Study. J Clin Endocrinol Metab. 2002;87(6):2701–5.

94. Rask E, Olsson T, Soderberg S, Andrew R, Livingstone DE, Johnson O, et al. Tissue-specific dysregulation of cortisol metabolism in human obesity. J Clin Endocrinol Metab. 2001;86(3):1418–21.

95. Rask E, Walker BR, Soderberg S, Livingstone DE, Eliasson M, Johnson O, et al. Tissue-specific changes in peripheral cortisol metabolism in obese women: increased adipose 11beta-hydroxysteroid dehydrogenase type 1 activity. J Clin Endocrinol Metab. 2002;87(7):3330–6.

96. Sandeep TC, Andrew R, Homer NZ, Andrews RC, Smith K, Walker BR. Increased in vivo regeneration of cortisol in adipose tissue in human obesity and effects of the 11beta-hydroxysteroid dehydrogenase type 1 inhibitor carbenoxolone. Diabetes. 2005;54(3):872–9.

97. Desbriere R, Vuaroqueaux V, Achard V, Boullu-Ciocca S, Labuhn M, Dutour A, et al. 11beta-hydroxysteroid dehydrogenase type 1 mRNA is increased in both visceral and subcutaneous adipose tissue of obese patients. Obesity (Silver Spring). 2006;14(5):794–8.

98. Michailidou Z, Jensen MD, Dumesic DA, Chapman KE, Seckl JR, Walker BR, et al. Omental 11beta-hydroxysteroid dehydrogenase 1 correlates with fat cell size independently of obesity. Obesity (Silver Spring). 2007;15(5):1155–63.

99. Paulsen SK, Pedersen SB, Fisker S, Richelsen B. 11Beta-HSD type 1 expression in human adipose tissue: impact of gender, obesity, and fat localization. Obesity (Silver Spring). 2007;15(8):1954–60.
100. Tomlinson JW, Sinha B, Bujalska I, Hewison M, Stewart PM. Expression of 11beta-hydroxysteroid dehydrogenase type 1 in adipose tissue is not increased in human obesity. J Clin Endocrinol Metab. 2002;87(12):5630–5.
101. Stimson RH, Andersson J, Andrew R, Redhead DN, Karpe F, Hayes PC, et al. Cortisol release from adipose tissue by 11beta-hydroxysteroid dehydrogenase type 1 in humans. Diabetes. 2009;58(1):46–53.
102. Hughes KA, Manolopoulos KN, Iqbal J, Cruden NL, Stimson RH, Reynolds RM, et al. Recycling between cortisol and cortisone in human splanchnic, subcutaneous adipose, and skeletal muscle tissues in vivo. Diabetes. 2012;61(6):1357–64.
103. Lee MJ, Fried SK, Mundt SS, Wang Y, Sullivan S, Stefanni A, et al. Depot-specific regulation of the conversion of cortisone to cortisol in human adipose tissue. Obesity. 2008;16(6):1178–85.
104. Dube S, Norby BJ, Pattan V, Carter RE, Basu A, Basu R. 11beta-hydroxysteroid dehydroge-nase types 1 and 2 activity in subcutaneous adipose tissue in humans: implications in obesity and diabetes. J Clin Endocrinol Metab. 2015;100(1):E70–6.
105. Upreti R, Hughes KA, Livingstone DE, Gray CD, Minns FC, Macfarlane DP, et al. 5alpha-reductase type 1 modulates insulin sensitivity in men. J Clin Endocrinol Metab. 2014;99(8):E1397–406.
106. Zyirek M, Flood C, Longcope C. 5 Alpha-reductase activity in rat adipose tissue. Proc Soc Exp Biol Med. 1987;186(2):134–8.
107. Perel E, Daniilescu D, Kindler S, Kharlip L, Killinger DW. The formation of 5 alpha-reduced androgens in stromal cells from human breast adipose tissue. J Clin Endocrinol Metab. 1986;62(2):314–8.
108. Liu Y, Nakagawa Y, Wang Y, Sakurai R, Tripathi PV, Lutfy K, et al. Increased glucocorti-coid receptor and 11{beta}-hydroxysteroid dehydrogenase type 1 expression in hepato-cytes may contribute to the phenotype of type 2 diabetes in db/db mice. Diabetes. 2005;54(1):32–40.
109. Morton NM, Holmes MC, Fievet C, Staels B, Tailleux A, Mullins JJ, et al. Improved lipid and lipoprotein profile, hepatic insulin sensitivity, and glucose tolerance in 11beta-hydroxysteroid dehydrogenase type 1 null mice. J Biol Chem. 2001;276(44):41293–300.
110. Paterson JM, Morton NM, Fievet C, Kenyon CJ, Holmes MC, Staels B, et al. Metabolic syn-drome without obesity: Hepatic overexpression of 11beta-hydroxysteroid dehydrogenase type 1 in transgenic mice. Proc Natl Acad Sci U S A. 2004;101(18):7088–93.
111. Lavery GG, Zielinska AE, Gathercole LL, Hughes B, Semjonous N, Guest P, et al. Lack of significant metabolic abnormalities in mice with liver-specific disruption of 11β-hydroxysteroid dehydrogenase Type 1. Endocrinology. 2012;153(7):3236–48.
112. Ricketts ML, Verhaeg JM, Bujalska I, Howie AJ, Rainey WE, Stewart PM. Immunohistochemical localization of type 1 11beta-hydroxysteroid dehydrogenase in human tissues. J Clin Endocrinol Metab. 1998;83(4):1325–35.
113. Jamieson PM, Chapman KE, Edwards CR, Seckl JR. 11 beta-hydroxysteroid dehydrogenase is an exclusive 11 beta- reductase in primary cultures of rat hepatocytes: effect of physico-chemical and hormonal manipulations. Endocrinology. 1995;136(11):4754–61.
114. Torrecilla E, Fernandez-Vazquez G, Vicent D, Sanchez-Franco F, Barabash A, Cabrerizo L, et al. Liver upregulation of genes involved in cortisol production and action is associated with metabolic syndrome in morbidly obese patients. Obes Surg. 2012;22(3):478–86.
115. Candia R, Riquelme A, Baudrand R, Carvajal CA, Morales M, Solis N, et al. Overexpression of 11beta-hydroxysteroid dehydrogenase type 1 in visceral adipose tissue and portal hyper-cortisolism in non-alcoholic fatty liver disease. Liver Int. 2012;32(3):392–9.
116. Konopelska S, Kienitz T, Hughes B, Pirlich M, Bauditz J, Lochs H, et al. Hepatic 11beta-HSD1 mRNA expression in fatty liver and nonalcoholic steatohepatitis. Clin Endocrinol (Oxf). 2009;70(4):554–60.

117. Ahmed A, Rabbitt E, Brady T, Brown C, Guest P, Bujalska IJ, et al. A switch in hepatic cortisol metabolism across the spectrum of non alcoholic fatty liver disease. PLoS One. 2012;7(2), e29531.
118. Stewart PM, Boulton A, Kumar S, Clark PM, Shackleton CH. Cortisol metabolism in human obesity: impaired cortisone → cortisol conversion in subjects with central adiposity. J Clin Endocrinol Metab. 1999;84(3):1022–7.
119. Stimson RH, Andrew R, McAvoy NC, Tripathi D, Hayes PC, Walker BR. Increased whole-body and sustained liver cortisol regeneration by 11beta-hydroxysteroid dehydrogenase type 1 in obese men with type 2 diabetes provides a target for enzyme inhibition. Diabetes. 2011;60(3):720–5.
120. Dowman JK, Hopkins LJ, Reynolds GM, Armstrong MJ, Nasiri M, Nikolaou N, et al. Loss of 5alpha-reductase type 1 accelerates the development of hepatic steatosis but protects against hepatocellular carcinoma in male mice. Endocrinology. 2013;154(12):4536–47.
121. Livingstone DE, Barat P, Di Rollo EM, Rees GA, Weldin BA, Rog-Zielinska EA, et al. 5alpha-Reductase type 1 deficiency or inhibition predisposes to insulin resistance, hepatic steatosis and liver fibrosis in rodents. Diabetes. 2015;64(2):447–58.
122. Crowley RK, Hughes B, Gray J, McCarthy T, Hughes S, Shackleton CH, et al. Longitudinal changes in glucocorticoid metabolism are associated with later development of adverse metabolic phenotype. Eur J Endocrinol. 2014;171(4):433–42.
123. Tomlinson JW, Finney J, Gay C, Hughes BA, Hughes SV, Stewart PM. Impaired glucose tolerance and insulin resistance are associated with increased adipose 11β-hydroxysteroid dehydrogenase type 1 expression and elevated hepatic 5α-reductase activity. Diabetes. 2008;57(10):2652–60.
124. Tomlinson JW, Finney J, Hughes BA, Hughes SV, Stewart PM. Reduced glucocorticoid production rate, decreased 5α-reductase activity and adipose tissue insulin sensitization following weight loss. Diabetes. 2008;57(6):1536–43.
125. Tsilchorozidou T, Honour JW, Conway GS. Altered cortisol metabolism in polycystic ovary syndrome: insulin enhances 5alpha-reduction but not the elevated adrenal steroid production rates. J Clin Endocrinol Metab. 2003;88(12):5907–13.
126. Vassiliadi DA, Barber TM, Hughes BA, McCarthy MI, Wass JA, Franks S, et al. Increased 5{alpha}-Reductase Activity and Adrenocortical Drive in Women with Polycystic Ovary Syndrome. J Clin Endocrinol Metab. 2009;94(9):3558–66.
127. Westerbacka J, Yki-Jarvinen H, Vehkavaara S, Hakkinen AM, Andrew R, Wake DJ, et al. Body fat distribution and cortisol metabolism in healthy men: enhanced 5beta-reductase and lower cortisol/cortisone metabolite ratios in men with fatty liver. J Clin Endocrinol Metab. 2003;88(10):4924–31.
128. Swali A, Walker EA, Lavery GG, Tomlinson JW, Stewart PM. 11β-Hydroxysteroid dehydrogenase type 1 regulates insulin and glucagon secretion in pancreatic islets. Diabetologia. 2008;51(11):2003–11.
129. Ortsater H, Alberts P, Warpman U, Engblom LO, Abrahmsen L, Bergsten P. Regulation of 11beta-hydroxysteroid dehydrogenase type 1 and glucose-stimulated insulin secretion in pancreatic islets of Langerhans. Diabetes Metab Res Rev. 2005;21(4):359–66.
130. Hult M, Ortsater H, Schuster G, Graedler F, Beckers J, Adamski J, et al. Short-term glucocorticoid treatment increases insulin secretion in islets derived from lean mice through multiple pathways and mechanisms. Mol Cell Endocrinol. 2009;301(1-2):109–16.
131. Davani B, Khan A, Hult M, Martensson E, Okret S, Efendic S, et al. Type 1 11beta-hydroxysteroid dehydrogenase mediates glucocorticoid activation and insulin release in pancreatic islets. J Biol Chem. 2000;275(45):34841–4.
132. Duplomb L, Lee Y, Wang MY, Park BH, Takaishi K, Agarwal AK, et al. Increased expression and activity of 11beta-HSD-1 in diabetic islets and prevention with troglitazone. Biochem Biophys Res Commun. 2004;313(3):594–9.
133. Turban S, Liu X, Ramage L, Webster SP, Walker BR, Dunbar DR, et al. Optimal elevation of beta-cell 11beta-hydroxysteroid dehydrogenase type 1 is a compensatory mechanism that prevents high-fat diet-induced beta-cell failure. Diabetes. 2012;61(3):642–52.

134. Schmid J, Ludwig B, Schally AV, Steffen A, Ziegler CG, Block NL, et al. Modulation of pancreatic islets-stress axis by hypothalamic releasing hormones and 11beta-hydroxysteroid dehydrogenase. Proc Natl Acad Sci U S A. 2011;108(33):13722–7.
135. Iqbal MJ, Greenway B, Wilkinson ML, Johnson PJ, Williams R. Sex-steroid enzymes, aromatase and 5 alpha-reductase in the pancreas: a comparison of normal adult, foetal and malignant tissue. Clin Sci. 1983;65(1):71–5.
136. Dimitriadis G, Leighton B, Parry-Billings M, Sasson S, Young M, Krause U, et al. Effects of glucocorticoid excess on the sensitivity of glucose transport and metabolism to insulin in rat skeletal muscle. Biochem J. 1997;321(Pt 3):707–12.
137. Morgan SA, Sherlock M, Gathercole LL, Lavery GG, Lenaghan C, Bujalska IJ, et al. 11β-hydroxysteroid dehydrogenase type 1 regulates glucocorticoid-induced insulin resistance in skeletal muscle. Diabetes. 2009;58(11):2506–15.
138. Zhang M, Lv XY, Li J, Xu ZG, Chen L. Alteration of 11beta-hydroxysteroid dehydrogenase type 1 in skeletal muscle in a rat model of type 2 diabetes. Mol Cell Biochem. 2009;324(1-2):147–55.
139. Morgan SA, Gathercole LL, Simonet C, Hassan-Smith ZK, Bujalska I, Guest P, et al. Regulation of lipid metabolism by glucocorticoids and 11beta-HSD1 in skeletal muscle. Endocrinology. 2013;154(7):2374–84.
140. Abdallah BM, Beck-Nielsen H, Gaster M. Increased expression of 11beta-hydroxysteroid dehydrogenase type 1 in type 2 diabetic myotubes. Eur J Clin Invest. 2005;35(10):627–34.
141. Kilgour AH, Gallagher IJ, MacLullich AM, Andrew R, Gray CD, Hyde P, et al. Increased skeletal muscle 11betaHSD1 mRNA is associated with lower muscle strength in ageing. PLoS One. 2013;8(12), e84057.
142. Hassan-Smith ZK, Morgan SA, Sherlock M, Hughes B, Taylor AE, Lavery GG, et al. Gender-Specific Differences in Skeletal Muscle 11beta-HSD1 Expression Across Healthy Aging. J Clin Endocrinol Metab. 2015;100(7):2673–81.
143. Brem AS, Bina RB, King TC, Morris DJ. Localization of 2 11beta-OH steroid dehydrogenase isoforms in aortic endothelial cells. Hypertension. 1998;31(1 Pt 2):459–62.
144. Walker BR, Yau JL, Brett LP, Seckl JR, Monder C, Williams BC, et al. 11β-hydroxysteroid dehydrogenase in vascular smooth muscle and heart: implications for cardiovascular responses to glucocorticoids. Endocrinology. 1991;129:3305–12.
145. Hermanowski-Vosatka A, Balkovec JM, Cheng K, Chen HY, Hernandez M, Koo GC, et al. 11beta-HSD1 inhibition ameliorates metabolic syndrome and prevents progression of atherosclerosis in mice. J Exp Med. 2005;202(4):517–27.
146. Nuotio-Antar AM, Hachey DL, Hasty AH. Carbenoxolone treatment attenuates symptoms of metabolic syndrome and atherogenesis in obese, hyperlipidemic mice. Am J Physiol Endocrinol Metab. 2007;293(6):E1517–28.
147. Michailidou Z, Turban S, Miller E, Zou X, Schrader J, Ratcliffe PJ, et al. Increased angiogenesis protects against adipose hypoxia and fibrosis in metabolic disease-resistant 11beta-hydroxysteroid dehydrogenase type 1 (HSD1)-deficient mice. J Biol Chem. 2012;287(6):4188–97.
148. Small GR, Hadoke PW, Sharif I, Dover AR, Armour D, Kenyon CJ, et al. Preventing local regeneration of glucocorticoids by 11beta-hydroxysteroid dehydrogenase type 1 enhances angiogenesis. Proc Natl Acad Sci U S A. 2005;102(34):12165–70.
149. Hadoke PW, Christy C, Kotelevtsev YV, Williams BC, Kenyon CJ, Seckl JR, et al. Endothelial cell dysfunction in mice after transgenic knockout of type 2, but not type 1, 11beta-hydroxysteroid dehydrogenase. Circulation. 2001;104(23):2832–7.
150. Deuchar GA, McLean D, Hadoke PW, Brownstein DG, Webb DJ, Mullins JJ, et al. 11Beta-hydroxysteroid dehydrogenase type 2 deficiency accelerates atherogenesis and causes proinflammatory changes in the endothelium in apoe-/- mice. Endocrinology. 2011;152(1):236–46.
151. Atalar F, Gormez S, Caynak B, Akan G, Tanriverdi G, Bilgic-Gazioglu S, et al. The role of mediastinal adipose tissue 11beta-hydroxysteroid dehydrogenase type 1 and glucocorticoid expression in the development of coronary atherosclerosis in obese patients with ischemic heart disease. Cardiovasc Diabetol. 2012;11:115.

152. Atalar F, Vural B, Ciftci C, Demirkan A, Akan G, Susleyici-Duman B, et al. 11Beta-hydroxysteroid dehydrogenase type 1 gene expression is increased in ascending aorta tissue of metabolic syndrome patients with coronary artery disease. Genet Mol Res. 2012;11(3):3122–32.
153. Campelo AE, Cutini PH, Massheimer VL. Cellular actions of testosterone in vascular cells: mechanism independent of aromatization to estradiol. Steroids. 2012;77(11):1033–40.
154. Wyrwoll CS, Holmes MC, Seckl JR. 11beta-hydroxysteroid dehydrogenases and the brain: from zero to hero, a decade of progress. Front Neuroendocrinol. 2011;32(3):265–86.
155. Rauz S, Walker EA, Shackleton CH, Hewison M, Murray PI, Stewart PM. Expression and putative role of 11 beta-hydroxysteroid dehydrogenase isozymes within the human eye. Invest Ophthalmol Vis Sci. 2001;42(9):2037–42.
156. Rajan V, Edwards CR, Seckl JR. 11 beta-Hydroxysteroid dehydrogenase in cultured hippocampal cells reactivates inert 11-dehydrocorticosterone, potentiating neurotoxicity. J Neurosci. 1996;16(1):65–70.
157. Gomez-Sanchez EP, Romero DG, de Rodriguez AF, Warden MP, Krozowski Z, Gomez-Sanchez CE. Hexose-6-phosphate dehydrogenase and 11beta-hydroxysteroid dehydrogenase-1 tissue distribution in the rat. Endocrinology. 2008;149(2):525–33.
158. Sooy K, Webster SP, Noble J, Binnie M, Walker BR, Seckl JR, et al. Partial deficiency or short-term inhibition of 11beta-hydroxysteroid dehydrogenase type 1 improves cognitive function in aging mice. J Neurosci. 2010;30(41):13867–72.
159. Sandeep TC, Yau JL, MacLullich AM, Noble J, Deary IJ, Walker BR, et al. 11Beta-hydroxysteroid dehydrogenase inhibition improves cognitive function in healthy elderly men and type 2 diabetics. Proc Natl Acad Sci U S A. 2004;101(17):6734–9.
160. Yau JL, Noble J, Kenyon CJ, Hibberd C, Kotelevtsev Y, Mullins JJ, et al. Lack of tissue glucocorticoid reactivation in 11beta -hydroxysteroid dehydrogenase type 1 knockout mice ameliorates age-related learning impairments. Proc Natl Acad Sci U S A. 2001;98(8):4716–21.
161. Marek GJ, Katz DA, Meier A, Greco N, Zhang W, Liu W, et al. Efficacy and safety evaluation of HSD-1 inhibitor ABT-384 in Alzheimer's disease. Alzheimers Dement. 2014;10(5 Suppl):S364–73.
162. Rauz S, Cheung CM, Wood PJ, Coca-Prados M, Walker EA, Murray PI, et al. Inhibition of 11beta-hydroxysteroid dehydrogenase type 1 lowers intraocular pressure in patients with ocular hypertension. QJM. 2003;96(7):481–90.
163. Sinclair AJ, Walker EA, Burdon MA, van Beek AP, Kema IP, Hughes BA, et al. Cerebrospinal fluid corticosteroid levels and cortisol metabolism in patients with idiopathic intracranial hypertension: a link between 11beta-HSD1 and intracranial pressure regulation? J Clin Endocrinol Metab. 2010;95(12):5348–56.
164. Diaz R, Brown RW, Seckl JR. Distinct ontogeny of glucocorticoid and mineralocorticoid receptor and 11beta-hydroxysteroid dehydrogenase types I and II mRNAs in the fetal rat brain suggest a complex control of glucocorticoid actions. J Neurosci. 1998;18(7):2570–80.
165. Korbonits M, Bujalska I, Shimojo M, Nobes J, Jordan S, Grossman AB, et al. Expression of 11 beta-hydroxysteroid dehydrogenase isoenzymes in the human pituitary: induction of the type 2 enzyme in corticotropinomas and other pituitary tumors. J Clin Endocrinol Metab. 2001;86(6):2728–33.
166. Brunton PJ, Donadio MV, Yao ST, Greenwood M, Seckl JR, Murphy D, et al. 5Alpha-reduced neurosteroids sex-dependently reverse central prenatal programming of neuroendocrine stress responses in rats. J Neurosci. 2015;35(2):666–77.
167. Schubring SR, Fleischer W, Lin JS, Haas HL, Sergeeva OA. The bile steroid chenodeoxycholate is a potent antagonist at NMDA and GABA(A) receptors. Neurosci Lett. 2012;506(2):322–6.
168. Rhen T, Cidlowski JA. Antiinflammatory action of glucocorticoids—new mechanisms for old drugs. N Engl J Med. 2005;353(16):1711–23.
169. Chapman KE, Coutinho AE, Gray M, Gilmour JS, Savill JS, Seckl JR. The role and regulation of 11beta-hydroxysteroid dehydrogenase type 1 in the inflammatory response. Mol Cell Endocrinol. 2009;301(1-2):123–31.

170. Thieringer R, Le Grand CB, Carbin L, Cai TQ, Wong B, Wright SD, et al. 11 Beta-hydroxysteroid dehydrogenase type 1 is induced in human monocytes upon differentiation to macrophages. J Immunol. 2001;167(1):30–5.

171. Freeman L, Hewison M, Hughes SV, Evans KN, Hardie D, Means TK, et al. Expression of 11beta-hydroxysteroid dehydrogenase type 1 permits regulation of glucocorticoid bioavailability by human dendritic cells. Blood. 2005;106(6):2042–9.

172. Zhang TY, Ding X, Daynes RA. The expression of 11 beta-hydroxysteroid dehydrogenase type I by lymphocytes provides a novel means for intracrine regulation of glucocorticoid activities. J Immunol. 2005;174(2):879–89.

173. Gilmour JS, Coutinho AE, Cailhier JF, Man TY, Clay M, Thomas G, et al. Local amplification of glucocorticoids by 11 beta-hydroxysteroid dehydrogenase type 1 promotes macrophage phagocytosis of apoptotic leukocytes. J Immunol. 2006;176(12):7605–11.

174. Zhang TY, Daynes RA. Macrophages from 11beta-hydroxysteroid dehydrogenase type 1-deficient mice exhibit an increased sensitivity to lipopolysaccharide stimulation due to TGF-beta-mediated up-regulation of SHIP1 expression. J Immunol. 2007;179(9):6325–35.

175. Coutinho AE, Gray M, Brownstein DG, Salter DM, Sawatzky DA, Clay S, et al. 11beta-Hydroxysteroid dehydrogenase type 1, but not type 2, deficiency worsens acute inflammation and experimental arthritis in mice. Endocrinology. 2012;153(1):234–40.

176. Hardy R, Rabbitt EH, Filer A, Emery P, Hewison M, Stewart PM, et al. Local and systemic glucocorticoid metabolism in inflammatory arthritis. Ann Rheum Dis. 2008;67(9):1204–10.

177. Zbankova S, Bryndova J, Leden P, Kment M, Svec A, Pacha J. 11beta-hydroxysteroid dehydrogenase 1 and 2 expression in colon from patients with ulcerative colitis. J Gastroenterol Hepatol. 2007;22(7):1019–23.

178. Cooper MS, Kriel H, Sayers A, Fraser WD, Williams AM, Stewart PM, et al. Can 11beta-hydroxysteroid dehydrogenase activity predict the sensitivity of bone to therapeutic glucocorticoids in inflammatory bowel disease? Calcif Tissue Int. 2011;89(3):246–51.

179. Nanus DE, Filer AD, Hughes B, Fisher BA, Taylor PC, Stewart PM, et al. TNFalpha regulates cortisol metabolism in vivo in patients with inflammatory arthritis. Ann Rheum Dis. 2015;74(2):464–9.

180. Bland R, Worker CA, Noble BS, Eyre LJ, Bujalska IJ, Sheppard MC, et al. Characterization of 11beta-hydroxysteroid dehydrogenase activity and corticosteroid receptor expression in human osteosarcoma cell lines. J Endocrinol. 1999;161(3):455–64.

181. Eyre LJ, Rabbitt EH, Bland R, Hughes SV, Cooper MS, Sheppard MC, et al. Expression of 11 beta-hydroxysteroid dehydrogenase in rat osteoblastic cells: pre-receptor regulation of glucocorticoid responses in bone. J Cell Biochem. 2001;81(3):453–62.

182. Cooper MS, Walker EA, Bland R, Fraser WD, Hewison M, Stewart PM. Expression and functional consequences of 11beta-hydroxysteroid dehydrogenase activity in human bone. Bone. 2000;27(3):375–81.

183. Eijken M, Hewison M, Cooper MS, de Jong FH, Chiba H, Stewart PM, et al. 11beta-Hydroxysteroid dehydrogenase expression and glucocorticoid synthesis are directed by a molecular switch during osteoblast differentiation. Mol Endocrinol. 2005;19(3):621–31.

184. Cooper MS, Bujalska I, Rabbitt E, Walker EA, Bland R, Sheppard MC, et al. Modulation of 11beta-hydroxysteroid dehydrogenase isozymes by proinflammatory cytokines in osteoblasts: an autocrine switch from glucocorticoid inactivation to activation. J Bone Miner Res. 2001;16(6):1037–44.

185. Kaur K, Hardy R, Ahasan MM, Eijken M, van Leeuwen JP, Filer A, et al. Synergistic induction of local glucocorticoid generation by inflammatory cytokines and glucocorticoids: implications for inflammation associated bone loss. Ann Rheum Dis. 2010;69(6):1185–90.

186. Justesen J, Mosekilde L, Holmes M, Stenderup K, Gasser J, Mullins JJ, et al. Mice deficient in 11beta-hydroxysteroid dehydrogenase type 1 lack bone marrow adipocytes, but maintain normal bone formation. Endocrinology. 2004;145(4):1916–25.

187. Sher LB, Harrison JR, Adams DJ, Kream BE. Impaired cortical bone acquisition and osteoblast differentiation in mice with osteoblast-targeted disruption of glucocorticoid signaling. Calcif Tissue Int. 2006;79(2):118–25.

188. Cooper MS, Blumsohn A, Goddard PE, Bartlett WA, Shackleton CH, Eastell R, et al. 11beta-hydroxysteroid dehydrogenase type 1 activity predicts the effects of glucocorticoids on bone. J Clin Endocrinol Metab. 2003;88(8):3874–7.
189. Windahl SH, Andersson N, Borjesson AE, Swanson C, Svensson J, Moverare-Skrtic S, et al. Reduced bone mass and muscle strength in male 5alpha-reductase type 1 inactivated mice. PLoS One. 2011;6(6), e21402.
190. Tiganescu A, Walker EA, Hardy RS, Mayes AE, Stewart PM. Localization, age- and site-dependent expression, and regulation of 11beta-hydroxysteroid dehydrogenase type 1 in skin. J Invest Dermatol. 2011;131(1):30–6.
191. Vukelic S, Stojadinovic O, Pastar I, Rabach M, Krzyzanowska A, Lebrun E, et al. Cortisol synthesis in epidermis is induced by IL-1 and tissue injury. J Biol Chem. 2011;286(12):10265–75.
192. Slominski A, Zbytek B, Nikolakis G, Manna PR, Skobowiat C, Zmijewski M, et al. Steroidogenesis in the skin: implications for local immune functions. J Steroid Biochem Mol Biol. 2013;137:107–23.
193. Terao M, Murota H, Kimura A, Kato A, Ishikawa A, Igawa K, et al. 11beta-Hydroxysteroid dehydrogenase-1 is a novel regulator of skin homeostasis and a candidate target for promoting tissue repair. PLoS One. 2011;6(9), e25039.
194. Cirillo N, Prime SS. Keratinocytes synthesize and activate cortisol. J Cell Biochem. 2011;112(6):1499–505.
195. Napolitano A, Voice MW, Edwards CR, Seckl JR, Chapman KE. 11Beta-hydroxysteroid dehydrogenase 1 in adipocytes: expression is differentiation-dependent and hormonally regulated. J Steroid Biochem Mol Biol. 1998;64(5-6):251–60.
196. Tiganescu A, Tahrani AA, Morgan SA, Otranto M, Desmouliere A, Abrahams L, et al. 11beta-Hydroxysteroid dehydrogenase blockade prevents age-induced skin structure and function defects. J Clin Invest. 2013;123(7):3051–60.
197. Tiganescu A, Hupe M, Uchida Y, Mauro T, Elias PM, Holleran WM. Increased glucocorticoid activation during mouse skin wound healing. J Endocrinol. 2014;221(1):51–61.
198. Smith RE, Maguire JA, Stein-Oakley AN, Sasano H, Takahashi K, Fukushima K, et al. Localization of 11 beta-hydroxysteroid dehydrogenase type II in human epithelial tissues. J Clin Endocrinol Metab. 1996;81(9):3244–8.
199. Shimojo M, Ricketts ML, Petrelli MD, Moradi P, Johnson GD, Bradwell AR, et al. Immunodetection of 11β-hydroxysteroid dehydrogenase type 2 in human mineralocorticoid target tissues: Evidence for nuclear localization. Endocrinology. 1997;138:1305–11.
200. Perogamvros I, Keevil BG, Ray DW, Trainer PJ. Salivary cortisone is a potential biomarker for serum free cortisol. J Clin Endocrinol Metab. 2010;95(11):4951–8.
201. Bocchi B, Kenouch S, Lamarre-Cliche M, Muffat-Joly M, Capron MH, Fiet J, et al. Impaired 11-beta hydroxysteroid dehydrogenase type 2 activity in sweat gland ducts in human essential hypertension. Hypertension. 2004;43(4):803–8.
202. Leckie C, Chapman KE, Edwards CR, Seckl JR. LLC-PK1 cells model 11 beta-hydroxysteroid dehydrogenase type 2 regulation of glucocorticoid access to renal mineralocorticoid receptors. Endocrinology. 1995;136(12):5561–9.
203. Ferrari P. The role of 11beta-hydroxysteroid dehydrogenase type 2 in human hypertension. Biochim Biophys Acta. 2010;1802(12):1178–87.
204. Watson Jr B, Bergman SM, Myracle A, Callen DF, Acton RT, Warnock DG. Genetic association of 11 beta-hydroxysteroid dehydrogenase type 2 (HSD11B2) flanking microsatellites with essential hypertension in blacks. Hypertension. 1996;28(3):478–82.
205. Mongia A, Vecker R, George M, Pandey A, Tawadrous H, Schoeneman M, et al. Role of 11betaHSD type 2 enzyme activity in essential hypertension and children with chronic kidney disease (CKD). J Clin Endocrinol Metab. 2012;97(10):3622–9.
206. Quinkler M, Bumke-Vogt C, Meyer B, Bahr V, Oelkers W, Diederich S. The human kidney is a progesterone-metabolizing and androgen-producing organ. J Clin Endocrinol Metab. 2003;88(6):2803–9.

207. Whorwood CB, Ricketts ML, Stewart PM. Epithelial cell localization of type 2 11 beta-hydroxysteroid dehydrogenase in rat and human colon. Endocrinology. 1994;135(6):2533–41.
208. Zhang MZ, Xu J, Yao B, Yin H, Cai Q, Shrubsole MJ, et al. Inhibition of 11beta-hydroxysteroid dehydrogenase type II selectively blocks the tumor COX-2 pathway and suppresses colon carcinogenesis in mice and humans. J Clin Invest. 2009;119(4):876–85.
209. Walker BR, Connacher AA, Lindsay RM, Webb DJ, Edwards CR. Carbenoxolone increases hepatic insulin sensitivity in man: a novel role for 11-oxosteroid reductase in enhancing glucocorticoid receptor activation. J Clin Endocrinol Metab. 1995;80(11):3155–9.
210. Andrews RC, Rooyackers O, Walker BR. Effects of the 11beta-hydroxysteroid dehydrogenase inhibitor carbenoxolone on insulin sensitivity in men with type 2 diabetes. J Clin Endocrinol Metab. 2003;88(1):285–91.
211. Tomlinson JW, Sherlock M, Hughes B, Hughes SV, Kilvington F, Bartlett W, et al. Inhibition of 11β-HSD1 activity in vivo limits glucocorticoid exposure to human adipose tissue and decreases lipolysis. J Clin Endocrinol Metab. 2007;92(3):857–64.
212. Rosenstock J, Banarer S, Fonseca VA, Inzucchi SE, Sun W, Yao W, et al. The 11-beta-hydroxysteroid dehydrogenase type 1 inhibitor INCB13739 improves hyperglycemia in patients with type 2 diabetes inadequately controlled by metformin monotherapy. Diabetes Care. 2010;33(7):1516–22.
213. Feig PU, Shah S, Hermanowski-Vosatka A, Plotkin D, Springer MS, Donahue S, et al. Effects of an 11beta-hydroxysteroid dehydrogenase type 1 inhibitor, MK-0916, in patients with type 2 diabetes mellitus and metabolic syndrome. Diabetes Obes Metab. 2011;13(6):498–504.
214. Shah S, Hermanowski-Vosatka A, Gibson K, Ruck RA, Jia G, Zhang J, et al. Efficacy and safety of the selective 11beta-HSD-1 inhibitors MK-0736 and MK-0916 in overweight and obese patients with hypertension. J Am Soc Hypertens. 2011;5(3):166–76.
215. Courtney R, Stewart PM, Toh M, Ndongo MN, Calle RA, Hirshberg B. Modulation of 11β-hydroxysteroid dehydrogenase (11β-HSD) activity biomarkers and pharmacokinetics of PF-00915275, a selective 11β-HSD1 inhibitor. J Clin Endocrinol Metab. 2008;93(2):550–6.
216. Stefan N, Ramsauer M, Jordan P, Nowotny B, Kantartzis K, Machann J, et al. Inhibition of 11beta-HSD1 with RO5093151 for non-alcoholic fatty liver disease: a multicentre, randomised, double-blind, placebo-controlled trial. Lancet Diabetes Endocrinol. 2014;2(5):406–16.
217. Morgan SA, McCabe EL, Gathercole LL, Hassan-Smith ZK, Larner DP, Bujalska IJ, et al. 11beta-HSD1 is the major regulator of the tissue-specific effects of circulating glucocorticoid excess. Proc Natl Acad Sci U S A. 2014;111(24):E2482–91.
218. Wang Y, Yan C, Liu L, Wang W, Du H, Fan W, et al. 11Beta-hydroxysteroid dehydrogenase type 1 shRNA ameliorates glucocorticoid-induced insulin resistance and lipolysis in mouse abdominal adipose tissue. Am J Physiol Endocrinol Metab. 2015;308(1):E84–95.

Obesity and Metabolic Syndrome: A Phenotype of Mild Long-Term Hypercortisolism?

Vincent L. Wester and Elisabeth F.C. van Rossum

Abstract In clinical practice, a considerable overlap can be observed between the sequelae of obesity and an excess of glucocorticoids (i.e., Cushing's syndrome). In Cushing's, all aspects of the metabolic syndrome are frequently seen: abdominal obesity, insulin resistance, dyslipidemia, and hypertension. Furthermore, common variants in the glucocorticoid receptor which affect sensitivity to cortisol also affect adiposity and related metabolic characteristics. Overall, published research investigating the associations between adiposity and cortisol in blood, saliva, and urine have not provided consistent evidence that cortisol levels are associated with obesity in the general population. This lack of consistent associations may be because cortisol levels are highly variable due to acute stress, the diurnal rhythm, and day-to-day variations. This variability is reflected in cortisol levels measured in human fluid matrices. Over the past decade, the analysis of cortisol in scalp hair has emerged as a way to estimate cumulative cortisol exposure over prolonged periods of time. Hair cortisol levels have been found to be increased in obese individuals and are positively associated with body mass index and abdominal fat mass. Furthermore, increased hair cortisol has been associated with metabolic syndrome and cardiovascular disease in population-based studies. Although it is theoretically likely that a subtle chronic hypercortisolism contributes to the genesis of obesity and related cardiometabolic disturbances, causality has not been established yet. Future studies investigating hair cortisol levels, in particular those involving longitudinal designs and interventions, may greatly expand knowledge about the relationship between cortisol exposure and cardiometabolic health in the general population.

Keywords Obesity • Metabolic syndrome • Cardiovascular disease • Hypothalamic–pituitary–adrenal axis • Glucocorticoids • Cortisol • Cortisone • Hair analysis • hair cortisol • stress

V.L. Wester, M.D. (✉) • E.F.C. van Rossum, M.D., Ph.D.
Department of Internal Medicine, Division of Endocrinology, Erasmus MC, University Medical Center Rotterdam, Rotterdam, The Netherlands

Obesity Center CGG, Erasmus MC, University Medical Center Rotterdam, Rotterdam, The Netherlands
e-mail: v.wester@erasmusmc.nl; e.vanrossum@erasmusmc.nl

© Springer International Publishing Switzerland 2017
E.B. Geer (ed.), *The Hypothalamic-Pituitary-Adrenal Axis in Health and Disease*, DOI 10.1007/978-3-319-45950-9_15

Introduction: Obesity

Obesity is one of the biggest challenges in individual health care and public health policy of the twenty-first century. Obesity is associated with an increased risk of cardiovascular disease (CVD), diabetes mellitus, osteoarthritis, and certain cancers [1]. An individual is considered obese when his or her body mass index (BMI) exceeds 30 kg per square meter, and by this definition more than 640 million people worldwide are obese [2]. This definition does not take into account body composition (i.e., the ratio between lean and fat mass), the distribution of fat tissue across the body (e.g., centripetal versus peripheral fat), and the clinical consequences of increased weight and adiposity. Consequently, there have been attempts to create a definition of *clinically relevant obesity*.

One commonly used definition of clinically relevant obesity is the metabolic syndrome (MetS), which is focused on the cardiometabolic sequelae of central adiposity. MetS is a complex of five obesity-related risk factors that are associated with CVD: increased waist circumference, elevated blood pressure, elevated triglycerides, decreased high density lipoprotein (HDL) cholesterol, and elevated fasting glucose. Although definitions and cutoff values vary slightly, an individual is considered to have MetS if he or she meets three out of five criteria. MetS has an estimated point prevalence of 34 % in adult US individuals [3]. A large scale meta-analysis of prospective studies showed that MetS is associated with a 2.35-fold increased risk of CVD and a 1.58-fold increased risk of all-cause mortality [4].

Combating obesity is challenging, for obese individuals as well as for the health care professionals taking care of them. Recently, a large cohort study in the UK showed that after exclusion of bariatric surgery, the probability that obese individuals attain normal weight is extremely low. Morbidly obese persons (BMI>40) were even less likely to have clinically meaningful and sustained weight loss than obese persons with a BMI below 40 [5]. In most countries, access to behavioral interventions for obesity is limited. Bariatric surgery is by far the most effective intervention in obesity in terms of weight loss and glycemic control, but is associated with long-term sequelae such as dumping syndrome and nutritional deficiencies, and, although the risk is low, a chance of potentially life-threatening postoperative complications [6–8].

The etiology of obesity is manifold and complicated. It is generally assumed that a strong genetic component underlies obesity, as exemplified by twin concordance studies which show an estimated heritability of approximately 40–70 % [9]. However, this cannot explain the strong increase in obesity prevalence in the developed and undeveloped world over the past decades. Presumably, a so-called *obesogenic* environment promotes obesity in genetically prone individuals. Well-recognized environmental influences on obesity include calorie-rich food consumption, physical activity, societal influences and psychological factors [10]. Interestingly, several of these factors are known to increase cortisol. In particular, consumption of carbohydrate-rich food, sleep deprivation, and stress have been found to increase cortisol levels [11–13].

In this chapter, we review the available evidence that the activation of the HPA axis, possibly due to physical or psychological stressors, may promote obesity and its metabolic sequelae. We will focus on recent advances, in particular the introduction of hair cortisol measurements, and their contribution to the understanding of the relationship between long-term cortisol exposure and obesity.

Chronic Stress and HPA Axis Activity in Obesity and MetS

One of the psychological factors that have most often been associated with obesity and an adverse cardiometabolic risk profile is increased psychosocial stress. Studies investigating these relationships are widely divergent in terms of the populations investigated and the way stress is measured. Unsurprisingly, reported results are not always consistent. However, in a recent meta-analysis of longitudinal studies, increased psychosocial stress was associated with a small overall increase in adiposity [14]. Furthermore, in a meta-analysis which aggregated evidence from over a hundred thousand individuals who were on average followed for over a decade, high perceived stress significantly increased the incidence of coronary heart disease with a risk ratio of 1.27 [15]. One of the mechanisms that is suggested to explain these associations is increased activity of the hypothalamus–pituitary–adrenal (HPA) axis associated with chronic stress, resulting in increased levels of cortisol.

Since many of the effects of the stress response are caused by increased cortisol levels, Cushing's syndrome can be considered a biological model of extreme stress [16]. All of the features of metabolic syndrome, including hypertension, abdominal obesity, dyslipidemia, and insulin resistance, frequently occur in Cushing's (see Fig. 1), either due to endogenous hypercortisolism or due to corticosteroid therapy. As an expected result of the cardiometabolic derangements, cardiovascular causes of death are common in Cushing's syndrome [17]. It is therefore theoretically likely that part of the association between stress and cardiometabolic risk may be effected

Fig. 1 Overlap between Cushing's syndrome and metabolic syndrome

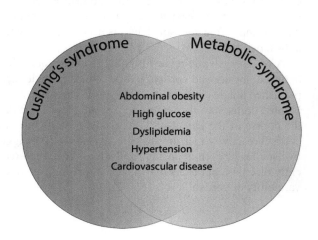

Cushing's syndrome   Metabolic syndrome

Abdominal obesity
High glucose
Dyslipidemia
Hypertension
Cardiovascular disease

through activation of the HPA axis and increased levels of cortisol. Obesity is a recognized cause of pseudo-Cushing's syndrome; however, most obese individuals do not have overt hypercortisolism [18].

Although the example of Cushing's syndrome makes a link between cortisol and obesity in the general population theoretically plausible, it represents an extreme example of chronically high cortisol exposure which does not occur in normal physiology. Further evidence that cortisol may have an adverse effect on the cardiometabolic phenotype and adiposity stems from studies investigating sensitivity to cortisol. Cortisol exerts its effects by binding to the glucocorticoid receptor (GR) and mineralocorticoid receptor (MR). Most of the metabolic effects of cortisol, including the effects on body composition leading to truncal obesity, are thought to arise from gene transactivation by the GR after ligand binding [19]. Over the past two decades, several polymorphisms in the GR have been described that influence the sensitivity to glucocorticoids. Approximately half of the general population carries either the N363S or BcII polymorphism, both of which have been associated with an increased sensitivity to glucocorticoids [20, 21]. Carriage of either of these variants has been associated with adiposity, supporting the concept that an increased activity of cortisol at the tissue levels promotes obesity [22–24]. In contrast, the ER22/23EK polymorphism, which is carried by about 8–9 % of the population, is associated with a relative resistance to glucocorticoids [25–27]. ER22/23EK carriers appear to be relatively protected against the deleterious cardiometabolic effects of cortisol, exemplified by increased lean body mass and insulin sensitivity, and lower cholesterol levels [26, 27].

There have been numerous attempts to unravel the association between obesity and exposure to systemic cortisol levels, using measurements in urine, saliva and blood. To interpret the results of these studies, it is important to take note of several situational and physiological factors that influence cortisol measurements. Cortisol follows a diurnal rhythm, characterized by a peak in the early morning (the cortisol awakening response, CAR) and generally declining levels during the day. Cortisol rises in response to physical or psychological factors, which causes cortisol levels to be variable within and across days [28]. Saliva and blood measurements can be used to obtain information about time-point cortisol levels, while urinary free cortisol (UFC) is used to estimate the total cortisol output over a 24-h period [18, 28].

A recent systemic review highlighted that studies investigating the associations between obesity and cortisol in body fluids provide inconsistent results [29]. Most published studies indicate that obesity is characterized by a diurnal rhythm with a blunted cortisol awakening response and a less sharp decline in cortisol levels over the course of the day. 24-h UFC tends to be higher in obese individuals, and the cortisol reactivity to acute stressors appears to be exaggerated. In most cases, negative studies or even opposing results have been reported as well [29]. These apparently inconsistent results may not be surprising, when we take into account the high variability of cortisol levels (Fig. 2).

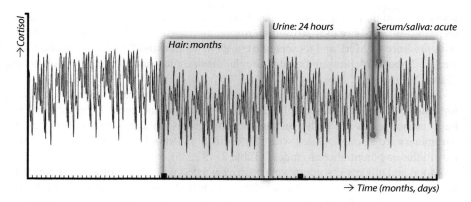

Fig. 2 Conceptual overview of the different matrices in which cortisol can be assessed: serum and saliva (time-point), urine (intermediate term output), and scalp hair (long-term cumulative levels). The line depicting circulating cortisol levels over a period of 3 months is fictional

Measuring Long-Term Exposure to Cortisol: Hair Analysis

A relatively novel way to account for the high variability in cortisol levels is scalp hair analysis. Scalp hair grows at a relatively stable rate of about 1 cm per month. During hair growth, substances are incorporated into the hair. This makes hair a suitable matrix to retrospectively assess long-term exposure to substances, depending on the length of the hair, up to several months back in time. Over the past decades, hair analysis has become an established method to retrospectively examine exposure to drugs of abuse and environmental toxins [30].

The first published report of endogenous glucocorticoids measured in human hair dates back to 2004 [31]. Scalp hair steroid analysis has since been performed in a number of labs and has greatly expanded the time frame of cortisol exposure that can be examined in a single measurement, as shown in Fig. 2. It is assumed that circulating free steroid hormones diffuse from the bloodstream into the hair shaft, although there may be minor contributions from sebum and sweat as well. Although, at first, in most studies immunoassays were used to measure cortisol, more recent studies report both hair cortisol (F) and cortisone (E) analyzed using liquid chromatography–tandem mass spectrometry (LC-MS/MS). Besides offering information about multiple simultaneously measured steroids, LC-MS/MS has higher sensitivity and is not hindered by antibody cross reactivity [32].

In the past decade, hair analysis has been used to measure long-term cortisol (and sometimes cortisone), most often in hair segments of 3 cm length, corresponding to cumulative levels over a period of 3 months.

In both obese adults and children, we found that hair cortisol concentrations (HCC) were increased compared to nonobese controls [33, 34]. Furthermore, in the largest population-based studies, HCC were positively associated with BMI and waist circumference, indicating that long-term cortisol exposure is on average increased in adiposity [35, 36]. Furthermore, increased HCC have been associated with presence of MetS and its separate components in a middle-aged population, as well as diabetes mellitus and cardiovascular disease presence in elderly populations (Table 1) [36–38].

Besides cardiometabolic parameters, a range of other clinical and situational factors have been investigated in relation to long-term cortisol levels. In larger studies, hair cortisol levels are higher in men and increase with age. Various hair-related parameters are also associated with hair cortisol and should be considered in clinical studies as potential confounders (Table 2) [32, 35].

Mood and anxiety disorders have been associated with alterations in hair cortisol levels (Table 1) [39]. Psychosocial stress, measured using standardized questionnaires such as the Perceived Stress Scale, has to date not been consistently associated with hair cortisol concentrations [32]. However, exposure to several physical and mental stressors has been associated with increases, including chronic pain, intensive aerobic exercise, and major life events (Table 1) [40–43]. This suggests that it may be the stressor itself, more than the subjectively experienced stress level that is associated with an increase in HPA axis activity.

Table 1 Published associations between health and situational factors and hair cortisol levels (adapted with permission from: Wester and van Rossum, Eur J Endocrinol 2015 [32])

	Increased hair cortisol	Decreased hair cortisol
Somatic health factors	Cushing's syndrome	Childhood asthma with inhalation glucocorticoids
	Hydrocortisone use	
	Obesity	
	Metabolic syndrome	
	Diabetes mellitus	
	Cardiovascular disease	
	Heart failure severity	
	Recent myocardial infarction	
Chronic and acute stressors	Intensive aerobic exercise	Traumatic experience
	Trauma	
	Life events	
	Unemployment	
	Shift work	
	Severe chronic pain	
Psychopathology	Posttraumatic stress disorder[a]	Posttraumatic stress disorder[a]
	Major depressive disorder	Generalized anxiety disorder
	Bipolar disorder, late onset	Panic disorder

[a]Posttraumatic stress disorder has been associated with both increased and decreased hair cortisol concentrations (depending on the type of traumatic event, characteristics of the patient sample examined, and the time span between the trauma and assessment), when compared to controls

Table 2 Overview of demographic and confounding factors that (potentially) affect hair cortisol concentrations

Factor	Significance for hair cortisol levels
Age	Increase with age
Sex	Higher in males
Season	Spring and summer may increase levels
Hair treatment	Inconsistent results
Hair washing frequency	Slightly lower with higher hair washing frequency
Sweating on the scalp	Experimental evidence is mixed
Use of corticosteroids	Both lower and higher hair cortisol levels have been reported; dependent on the corticosteroid and used method

Future Directions and Unresolved Issues

The studies involving scalp hair cortisol support the concept that obesity and its adverse cardiometabolic risk profile are associated with an increase in long-term systemic cortisol exposure. Whether this subtle hypercortisolism contributes to the development of obesity, insulin resistance, dyslipidemia, and cardiovascular disease is unknown, but it is likely from a pathophysiological perspective. Longitudinal studies may shed further light on this issue and determine whether hair glucocorticoid measurements deserve a place in cardiovascular risk stratification.

Obesity is known to be associated with increased psychological distress, social stigma, and psychopathology [44, 45]. This may explain part of the relationship between long-term cortisol and obesity. However, the evidence to date indicates that the subjective perception of stress has little impact on long-term cortisol levels [32]. Perhaps this association is modulated by individual factors, and only prone individuals suffer from increased long-term cortisol exposure. Furthermore, cortisol metabolism may be altered in obese individuals. Cortisol is primarily metabolized in the liver, and nonalcoholic fatty liver disease associated with obesity may influence cortisol metabolism [46]. Additionally, obesity is associated with low-grade inflammation, which may increase cortisol levels [47]. However, even if increased cortisol follows, rather than precedes cardiometabolic derangements, it is likely to at least contribute to the maintenance of an unfavorable risk profile.

The fact that hair collection is easily applicable in a clinical practice or research setting, with minimal burden to the participant, makes this method ideal to study the effects of behavioral, medical, or surgical interventions on long-term glucocorticoid exposure. Well-designed intervention studies involving hair analysis may greatly improve our understanding of the role of subtle variations in chronic HPA axis activity in health and disease, possibly paving the way to a more tailored treatment of obesity and cardiometabolic risk.

Several mechanistic questions regarding hair cortisol remain unresolved. It is assumed that steroids can incorporate into the hair through passive diffusion from the circulation, but there may also be contributions from sweating and sebum. The relative contribution of these three mechanisms is currently not known. Although sweating challenges do not seem to acutely influence hair cortisol, it is conceivable that repeated sweating over prolonged periods of time may affect hair levels measured [48]. Furthermore, the influence of conversion from cortisol to cortisone and vice versa by 11 beta hydroxysteroid dehydrogenases (11β-HSD) on hair cortisol and cortisone is not fully understood. Both local (e.g., skin or hair follicle) and overall systemic 11β-HSD could theoretically impact the ratio between cortisol and cortisone in hair [49]. At present, methods are available that measure both hair cortisol and cortisone using LC-MS/MS, yielding the potential to explore the ratio between these two as a marker for systemic 11β-HSD activity [50, 51]. We expect that both experimental and epidemiological studies may help understand how glucocorticoids are incorporated into the hair, as well as unravel the contribution of peaks in circulating hormone levels and local regulation to hair glucocorticoids.

Conclusion

Recent studies provide evidence for a firm link between high long-term HPA axis output and an adverse cardiometabolic risk profile. Novel developments in scalp hair analysis offer the opportunity to investigate long-term activity of the HPA axis. In addition to widely available short-term measurements such as cortisol in blood, urine, or saliva, hair cortisol analysis provides researchers and clinicians with retrospective information about glucocorticoids over months of time, with a single hair sample collection and analysis. We expect that future studies involving hair cortisol measurements, especially when used in intervention studies and longitudinal designs, will help unravel the role of long-term cortisol exposure in obesity and its implications for health.

Conflict of Interest Statement The authors declare no competing interests.

References

1. World Health Organization. Obesity and overweight. Fact sheet N 311. Updated January 2015. Accessed April 11, 2016.
2. NCD Risk Factor Collaboration. Trends in adult body-mass index in 200 countries from 1975 to 2014: a pooled analysis of 1698 population-based measurement studies with 19.2 million participants. Lancet. 2016;387(10026):1377–96.
3. Ervin RB. Prevalence of metabolic syndrome among adults 20 years of age and over, by sex, age, race and ethnicity, and body mass index: United States, 2003-2006. Natl Health Stat Report. 2009;13:1–7.
4. Mottillo S, Filion KB, Genest J, Joseph L, Pilote L, Poirier P, et al. The metabolic syndrome and cardiovascular risk: a systematic review and meta-analysis. J Am Coll Cardiol. 2010;56 (14):1113–32.

5. Fildes A, Charlton J, Rudisill C, Littlejohns P, Prevost AT, Gulliford MC. Probability of an obese person attaining normal body weight: cohort study using electronic health records. Am J Public Health. 2015;105(9):e54–9.

6. Colquitt JL, Pickett K, Loveman E, Frampton GK. Surgery for weight loss in adults. Cochrane Database Syst Rev. 2014;8, CD003641.

7. Ukleja A. Dumping syndrome: pathophysiology and treatment. Nutr Clin Pract. 2005;20(5):517–25.

8. Shankar P, Boylan M, Sriram K. Micronutrient deficiencies after bariatric surgery. Nutrition. 2010;26(11-12):1031–7.

9. Loos RJF. Genetic determinants of common obesity and their value in prediction. Best Pract Res Clin Endocrinol Metab. 2012;26(2):211–26.

10. Butland B, Jebb S, Kopelman P, McPherson K, Thomas S, Mardell J, et al. Tackling obesities: future choices: Project report. London: Government Office for Science; 2007.

11. Martens MJ, Rutters F, Lemmens SG, Born JM, Westerterp-Plantenga MS. Effects of single macronutrients on serum cortisol concentrations in normal weight men. Physiol Behav. 2010;101(5):563–7.

12. Minkel J, Moreta M, Muto J, Htaik O, Jones C, Basner M, et al. Sleep deprivation potentiates HPA axis stress reactivity in healthy adults. Health Psychol. 2014;33(11):1430–4.

13. Belda X, Fuentes S, Daviu N, Nadal R, Armario A. Stress-induced sensitization: the hypothalamic-pituitary-adrenal axis and beyond. Stress. 2015;18(3):269–79.

14. Wardle J, Chida Y, Gibson EL, Whitaker KL, Steptoe A. Stress and adiposity: a meta-analysis of longitudinal studies. Obesity (Silver Spring). 2011;19(4):771–8.

15. Richardson S, Shaffer JA, Falzon L, Krupka D, Davidson KW, Edmondson D. Meta-analysis of perceived stress and its association with incident coronary heart disease. Am J Cardiol. 2012;110(12):1711–6.

16. Charmandari E, Tsigos C, Chrousos G. Endocrinology of the stress response 1. Annu Rev Physiol. 2005;67:259–84.

17. Lacroix A, Feelders RA, Stratakis CA, Nieman LK. Cushing's syndrome. Lancet. 2015;386 (9996):913–27.

18. Nieman LK, Biller BM, Findling JW, Newell-Price J, Savage MO, Stewart PM, et al. The diagnosis of Cushing's syndrome: an Endocrine Society Clinical Practice Guideline. J Clin Endocrinol Metab. 2008;93(5):1526–40.

19. Newton R, Holden NS. Separating transrepression and transactivation: a distressing divorce for the glucocorticoid receptor? Mol Pharmacol. 2007;72(4):799–809.

20. Huizenga NATM, Koper JW, de Lange P, Pols HAP, Stolk RP, Burger H, et al. A polymorphism in the glucocorticoid receptor gene may be associated with an increased sensitivity to glucocorticoids in vivo. J Clin Endocrinol Metab. 1998;83(1):144–51.

21. Van Rossum EFC, Koper JW, Van Den Beld AW, Uitterlinden AG, Arp P, Ester W, et al. Identification of the BclII polymorphism in the glucocorticoid receptor gene: association with sensitivity to glucocorticoids in vivo and body mass index. Clin Endocrinol (Oxf). 2003;59(5):585–92.

22. Geelen CC, van Greevenbroek MM, van Rossum EF, Schaper NC, Nijpels G, 't Hart LM, et al. BclII glucocorticoid receptor polymorphism is associated with greater body fatness: the Hoorn and CODAM studies. J Clin Endocrinol Metab. 2013;98(3):E595–9.

23. Ukkola O, Perusse L, Chagnon YC, Despres JP, Bouchard C. Interactions among the glucocorticoid receptor, lipoprotein lipase and adrenergic receptor genes and abdominal fat in the Quebec Family Study. Int J Obes Relat Metab Disord. 2001;25(9):1332–9.

24. Marti A, Ochoa MC, Sánchez-Villegas A, Martínez JA, Martínez-González MA, Hebebrand J, et al. Meta-analysis on the effect of the N363S polymorphism of the glucocorticoid receptor gene (GRL) on human obesity. BMC Med Genet. 2006;7(1):1–11.

25. Russcher H, Smit P, van den Akker ELT, van Rossum EFC, Brinkmann AO, de Jong FH, et al. Two polymorphisms in the glucocorticoid receptor gene directly affect glucocorticoid-regulated gene expression. J Clin Endocrinol Metab. 2005;90(10):5804–10.

26. van Rossum EFC, Voorhoeve PG, te Velde SJ, Koper JW, Delemarre-van de Waal HA, Kemper HCG, et al. The ER22/23EK polymorphism in the glucocorticoid receptor gene is associated with a beneficial body composition and muscle strength in young adults. J Clin Endocrinol Metab. 2004;89(8):4004–9.
27. van Rossum EFC, Koper JW, Huizenga NATM, Uitterlinden AG, Janssen JA, Brinkmann AO, et al. A polymorphism in the glucocorticoid receptor gene, which decreases sensitivity to glucocorticoids in vivo, is associated with low insulin and cholesterol levels. Diabetes. 2002;51(10):3128–34.
28. Tsigos C, Chrousos GP. Hypothalamic–pituitary–adrenal axis, neuroendocrine factors and stress. J Psychosom Res. 2002;53(4):865–71.
29. Incollingo Rodriguez AC, Epel ES, White ML, Standen EC, Seckl JR, Tomiyama AJ. Hypothalamic-pituitary-adrenal axis dysregulation and cortisol activity in obesity: a systematic review. Psychoneuroendocrinology. 2015;62:301–18.
30. Cooper GAA, Kronstrand R, Kintz P. Society of Hair Testing guidelines for drug testing in hair. Forensic Sci Int. 2012;218(1):20–4.
31. Raul JS, Cirimele V, Ludes B, Kintz P. Detection of physiological concentrations of cortisol and cortisone in human hair. Clin Biochem. 2004;37(12):1105–11.
32. Wester VL, van Rossum EF. Clinical applications of cortisol measurements in hair. Eur J Endocrinol. 2015;173(4):M1–10.
33. Wester VL, Staufenbiel SM, Veldhorst MA, Visser JA, Manenschijn L, Koper JW, et al. Long-term cortisol levels measured in scalp hair of obese patients. Obesity (Silver Spring). 2014;22(9):1956–8.
34. Veldhorst MA, Noppe G, Jongejan MH, Kok CB, Mekic S, Koper JW, et al. Increased scalp hair cortisol concentrations in obese children. J Clin Endocrinol Metab. 2014;99(1):285–90.
35. Staufenbiel SM, Penninx BWJH, de Rijke YB, van den Akker ELT, van Rossum EFC. Determinants of hair cortisol and hair cortisone concentrations in adults. Psychoneuroendocrinology. 2015;60:182–94.
36. Stalder T, Kirschbaum C, Alexander N, Bornstein SR, Gao W, Miller R, et al. Cortisol in hair and the metabolic syndrome. J Clin Endocrinol Metab. 2013;98(6):2573–80.
37. Manenschijn L, Schaap L, van Schoor NM, van der Pas S, Peeters GM, Lips P, et al. High long-term cortisol levels, measured in scalp hair, are associated with a history of cardiovascular disease. J Clin Endocrinol Metab. 2013;98(5):2078–83.
38. Feller S, Vigl M, Bergmann MM, Boeing H, Kirschbaum C, Stalder T. Predictors of hair cortisol concentrations in older adults. Psychoneuroendocrinology. 2014;39:132–40.
39. Staufenbiel SM, Penninx BW, Spijker AT, Elzinga BM, van Rossum EF. Hair cortisol, stress exposure, and mental health in humans: a systematic review. Psychoneuroendocrinology. 2013;38(8):1220–35.
40. Van Uum SH, Sauve B, Fraser LA, Morley-Forster P, Paul TL, Koren G. Elevated content of cortisol in hair of patients with severe chronic pain: a novel biomarker for stress. Stress. 2008;11(6):483–8.
41. Skoluda N, Dettenborn L, Stalder T, Kirschbaum C. Elevated hair cortisol concentrations in endurance athletes. Psychoneuroendocrinology. 2012;37(5):611–7.
42. Grassi-Oliveira R, Pezzi JC, Daruy-Filho L, Viola TW, Francke IDA, Leite CE, et al. Hair cortisol and stressful life events retrospective assessment in crack cocaine users. Am J Drug Alcohol Abuse. 2012;38(6):535–8.
43. Staufenbiel SM, Koenders MA, Giltay EJ, Elzinga BM, Manenschijn L, Hoencamp E, et al. Recent negative life events increase hair cortisol concentrations in patients with bipolar disorder. Stress. 2014;17(6):451–9.
44. Puhl RM, Heuer CA. The stigma of obesity: a review and update. Obesity (Silver Spring). 2009;17(5):941–64.
45. de Wit L, Luppino F, van Straten A, Penninx B, Zitman F, Cuijpers P. Depression and obesity: a meta-analysis of community-based studies. Psychiatry Res. 2010;178(2):230–5.

46. Westerbacka J, Yki-Järvinen H, Vehkavaara S, Häkkinen A-M, Andrew R, Wake DJ, et al. Body fat distribution and cortisol metabolism in healthy men: enhanced 5β-reductase and lower cortisol/cortisone metabolite ratios in men with fatty liver. J Clin Endocrinol Metab. 2003;88(10):4924–31.
47. Gregor MF, Hotamisligil GS. Inflammatory mechanisms in obesity. Annu Rev Immunol. 2011;29:415–45.
48. Grass J, Kirschbaum C, Miller R, Gao W, Steudte-Schmiedgen S, Stalder T. Sweat-inducing physiological challenges do not result in acute changes in hair cortisol concentrations. Psychoneuroendocrinology. 2015;53:108–16.
49. Smith RE, Maguire JA, Stein-Oakley AN, Sasano H, Takahashi K, Fukushima K, et al. Localization of 11 beta-hydroxysteroid dehydrogenase type II in human epithelial tissues. J Clin Endocrinol Metab. 1996;81(9):3244–8.
50. Gao W, Stalder T, Foley P, Rauh M, Deng H, Kirschbaum C. Quantitative analysis of steroid hormones in human hair using a column-switching LC-APCI-MS/MS assay. J Chromatogr B Analyt Technol Biomed Life Sci. 2013;928:1–8.
51. Noppe G, de Rijke YB, Dorst K, van den Akker EL, van Rossum EF. LC-MS/MS-based method for long-term steroid profiling in human scalp hair. Clin Endocrinol (Oxf). 2015;83(2):162–6.

Index

A

Abiraterone inhibitor, 170
Abnormal steroidogenesis, 70–78
    armadillo proteins (*see* Armadillo proteins)
    cAMP/PKA signaling pathway
        (*see* cAMP/PKA signaling pathway)
    Carney complex phenotype, 78
    gene mutations, 78
ACRD. *See* Apparent cortisone reductase
    deficiency (ACRD)
Activator protein-1 (AP-1), 44, 46, 259
Adaptive immunity
    acquired immunity, 51
    autoimmune diseases, 52
    dendritic cells (DCs), 51
    immunological memory, 51
    peripheral tolerance failure, 52
    in pituitary gland, 52
    surface markers, 52
    synthetic GC treatment, 51
    thymocyte maturation, 52
    tolerogenic DCs inhibition, 52
5′ Adenosine monophosphate-activated
    protein kinase (AMPK) activity, 8
Adiponectin, CS patients, 203
Adipose tissue, 278–279
    AMPK activity, 8
    augmented insulin-mediated activation, 8
    cardiovascular disease, 4
    chronic GC exposure, 8
    corticosterone, 7
    cortisol concentrations, patients, 6
    description, 5
    dexamethasone pre-treatment, 8
    GC effects, 6
    GC-mediated adiposity, 6–7
    healthy controls, 6
    hormone/neuronal signals, 8
    hydrocortisone administration, 8
    increased lipolysis, 7
    LPL activity, 8
    metabolism, GCs, 7
    obesity, prevalence of, 4
    phenotypical changes, 4
    stromal vascular cells, 5
    subcutaneous, 5
    tissue-specific cortisol metabolism, 278–279
    WAT and BAT, 6
Adrenal adenoma
    and early cortisol autonomy, 188
    nonfunctioning, 189
    nonsecretory, 184
    scintigraphic uptake, 185
Adrenal incidentalomas (AI). *See* Mild adrenal
    cortisol excess (MACE)
Adrenocorticotropic hormone (ACTH), 243–244
Aggressive corticotroph pituitary tumors, 95,
    96, 101
    ACTH-secreting tumors, 101
    adrenal steroidogenesis inhibitors, 102
    behavior, 99
    benign tumors, 93
    biomarker candidates, 100
    carcinoma, 93
    chemotherapy, 103–104
    classification, 94
    clinical, radiological and histopathological
        criteria, 94
    dysfunctional hormonal and growth factor
        signaling pathways, 99

Printed by Books on Demand, Germany